T0192438

Innovation and the rise of the tunnelling industry

Frontispiece – Thames Tunnel and shield (1826 or 1836)

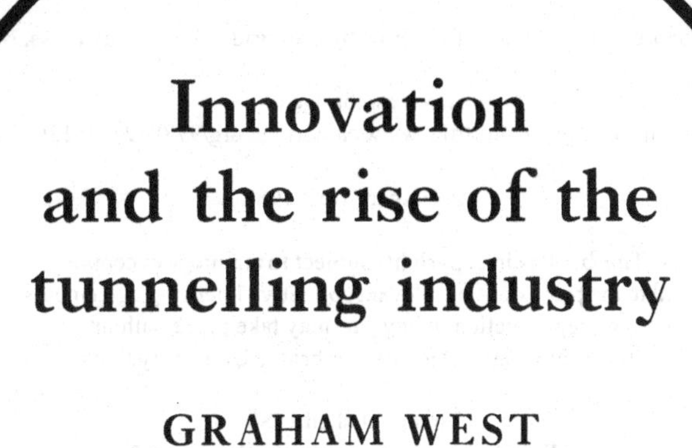

Innovation and the rise of the tunnelling industry

GRAHAM WEST

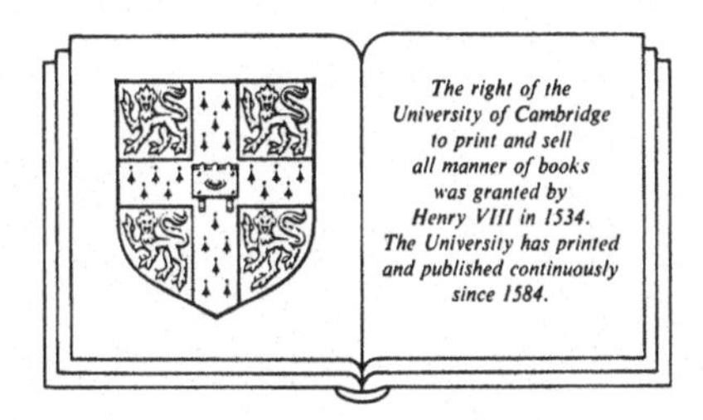

CAMBRIDGE UNIVERSITY PRESS

Cambridge

New York New Rochelle Melbourne Sydney

CAMBRIDGE UNIVERSITY PRESS
Cambridge, New York, Melbourne, Madrid, Cape Town, Singapore, São Paulo

Cambridge University Press
The Edinburgh Building, Cambridge CB2 2RU, UK

Published in the United States of America by Cambridge University Press, New York

www.cambridge.org
Information on this title: www.cambridge.org/9780521335126

First published 1988
This digitally printed first paperback version 2005

A catalogue record for this publication is available from the British Library

Library of Congress Cataloguing in Publication data
West, Graham.
Innovation and the rise of the tunnelling industry.
Includes bibliographies and index.
1. Tunnelling – History. 2. Tunnels – History.
I. Title.
TA803.W47 1987 624.1'93'09 87-8089

ISBN-13 978-0-521-33512-6 hardback
ISBN-10 0-521-33512-4 hardback

ISBN-13 978-0-521-67335-8 paperback
ISBN-10 0-521-67335-6 paperback

For Sheila, Caroline and Jonathan

Now among all the benefits that could be conferred upon mankind, I found none so great as the discovery of new arts, endowments, and commodities for the bettering of man's life. For I saw that among the rude people in the primitive times the authors of rude inventions and discoveries were consecrated and numbered among the Gods. And it was plain that the good effects wrought by founders of cities, law-givers, fathers of the people, extirpers of tyrants, and heroes of that class, extend but over narrow spaces and last but for short times; whereas the work of the Inventor, though a thing of less pomp and show, is felt everywhere and lasts for ever.

Francis Bacon (1603)
From the preface to *On the Interpretation of Nature*

Contents

Figures

Tables

Preface

It is a truism that everyone is interested in holes in the ground, but for me an original interest in holes in the ground has developed into a further interest, namely an interest in how holes in the ground are made. For a decade I was fortunate in being associated with the tunnelling industry, albeit as a research worker, and this gave me the opportunity to find out about tunnelling technology and its fascinating history of development. This book is the outcome of that endeavour.

The book has been written with the interests of three kinds of reader in mind. Firstly, civil and mechanical engineers, engineering geologists and others who are involved in present-day tunnelling, but who wish to know something of the history of the technical development of the tunnelling industry. Secondly, students of the history of technology who will find in the book a case history with important lessons that can be seen to have a relevance beyond the confines of the tunnelling industry. And thirdly, a more general and growing readership consisting of those who see in the history of technology a subject which is not only interesting and worthy of study in its own right, but one which can be properly regarded as part of the culture of our civilisation.

I would like to express my gratitude to two particular colleagues and friends: Dr Noel G. Coley of the Open University, for guidance and advice on research in the history of technology and on historical writing, and Dr Myles P. O'Reilly of the Transport and Road Research Laboratory, for support and encouragement and for reading through the whole of the book in draft form and commenting on the technical matters. However, I hasten to add, any shortcomings and errors in the work are entirely my own.

Colleagues in the tunnelling industry are thanked for their help with information and for their forbearance in setting aside present concerns to answer historical questions. The following libraries are thanked for access to their collections: Institution of Civil Engineers, Science Museum, Lyon Playfair (Imperial College), Institution of Mechanical Engineers, Transport and Road Research Laboratory, and Reading University.

My employer, the Transport and Road Research Laboratory, is thanked for assistance towards the higher degree studies that eventually led to this book; however, any views expressed in the book are entirely my own and not those of the Laboratory or of the Department of Transport, of which the Laboratory is a part.

Much of the book is based on a thesis submitted to the Open University with whom I am proud to have been a part-time postgraduate student.

Dr Richard L. Ziemacki and the editorial staff of the Cambridge University Press are thanked for their help in preparing the book for publication.

October 1986 Graham West

Introduction

A problem without a solution may interest the student, but can hardly fail to annoy the casual reader.

The rise of the modern tunnelling industry has been made possible by the introduction of a succession of remarkable technical innovations including new machinery and novel methods of working. This book investigates the most significant of such innovations introduced between about 1825 and 1985. It aims to examine the innovations themselves, the problems which made them necessary and their success in overcoming the practical difficulties encountered in driving particular tunnels. In addition, their contributions to growing technical expertise in the tunnelling industry and its ability to meet the demands for tunnels of all kinds during the period covered is considered. Thus, whilst each of the technical innovations treated in this work has been chosen for its contribution to the solution of a particular problem, the choice has been carefully made to illustrate the development of tunnelling techniques and the tunnelling industry in general. In some cases the practical development of the innovation itself caused problems that had to be solved before the device or process could be made to work satisfactorily. Approached from the viewpoint of solving problems, this study throws light on the origins, reasons for and processes of technical innovation both in the tunnelling industry and more generally. Moreover, it exposes many aspects of the rich field that the tunnelling industry offers for study within the history of technology and it also reveals how past innovations may still have relevance to current problems. Indeed, the potential contributions of the history of technology to the solution of current and future technological problems is a recurring theme throughout this work. Although this book is not intended to be a history of tunnelling,[1] a broadly chronological sequence of important technical developments in the industry has been followed and this serves to emphasise the crucial nature of each of the chosen innovations. In Chapters 2–12 the origins and technical details of these innovations are explored, together with the problems they were intended to solve and the difficulties encountered in putting them into practice. Some of the world's most important tunnels feature in this account because important innovations in tunnelling were introduced during their construction, but other equally important tunnels are not mentioned because they involved no important technical innovations of significance for this study.

It has been remarked that although the objective of the historian of technology is to define the nature of the engineering problem as it was faced in the past and to analyse the ideas and techniques that were used to achieve a solution, the method he must use is actually to study the way that machines, structures and processes were designed, fabricated and developed.[2] To this we should add that it is also necessary to study the way they worked and it is for this reason that throughout this book detailed technical descriptions of the machines and methods used have been given. These accounts are as accurate as it has been possible to make them given the lack of clarity of detail in some of the available contemporary (and later) drawings and the sometimes sketchy contemporary descriptions of nineteenth-century machines and processes. It is emphasised that these technical details are regarded as an important aspect of this work. They have been included because a proper understanding of the design and operation of each innovation is essential if it is to be assessed correctly as a solution to a particular problem. A clear understanding of the technical detail is also necessary if the innovations discussed are to be accorded their due significance in the history of tunnelling technology. Moreover, the ways in which the engineers solved their problems need to be judged by contemporary rather than present standards.[3] Thus, for example, when considering how the engineers of the 1850s strove to produce compressed air rock drilling machines (Chapter 2), we must remember they were operating with the understanding of mechanical principles and practices of the 1850s and not as we know them today. Later, in the 1970s, when the hydraulic rock drilling machine (Chapter 5) was developed, mechanical engineering had advanced so considerably that an altogether more sophisticated machine was possible. Each chapter in this book is devoted to a particular type of innovation. During the course of studying these innovations it was observed that they fell into certain well-defined categories depending on their origins (see Chapter 14). Although derived exclusively from tunnelling, it may be that such a system of classification could be applied to technical innovation more generally and it may therefore prove to have a wider value in helping to identify the potential origins of technical innovation in other branches of engineering and industry.

Several books dealing wholly or in part with the history of tunnelling and touching upon some of the technical innovations discussed here have recently appeared. *Tunnelling history and my own involvement*[4] is by the late Sir Harold Harding (1900–1986), formerly the doyen of British tunnelling. The first part of the book is a précis of the history of tunnelling, whilst the second part is an account of Harding's personal involvement in tunnelling, both as a contractor's engineer and as a consultant, extending over the period from 1922 to 1977. The particular interest of the book lies in some of the comments that Harding was able to make from personal experience on particular tunnels and machines with which he was involved and which do not appear elsewhere in the published literature. The two-volume *Tunnels: planning, design, construction*[5] by T.M. Megaw and J.V. Bartlett is a modern textbook of tunnelling, although the first chapter gives an historical introduction and there are other historical notes at the beginnings of some of the other chapters. One interest of the book is that Bartlett was the inventor of one of the innovations discussed here in Chapter 9. Also, Volume 2 contains an extensive bibliography of books, reports and papers on tunnelling which is a valuable reference source. Barbara Stack's comprehensive 700-page *Handbook of mining and tunnelling machinery*[6] is especially noteworthy. The *Handbook* is clearly a monumental work: the collection, listing and orderly presentation of all the information it contains has produced an encyclopaedia which will be an essential reference work for any worker in the field of mining and tunnelling for many years to come. Some of the research for this book was made easier by the appearance of the *Handbook* – for instance Stack's bringing to light the details of Maus' tunnelling machine which is discussed in Chapter 11 here. Lastly, the *Tunnel engineering handbook*[7] edited by J.O. Bickel and T.R. Kuesel is a modern tunnelling textbook with contributions on the constituent subjects by specialists. There are occasional brief historical notes in the introductions to some of the chapters, but the main interest of the book is that it provides a comprehensive outline of present-day practice in tunnel design and construction.

1.1 Innovations in tunnelling technology

For the purposes of this book an important innovation has been taken to be one which enabled something to be done that could not be done before, or one which allowed something to be done so much better than before that it constituted a great technical advance. For example, compressed air tunnelling (Chapter 7) was an innovation that enabled subaqueous tunnels to be constructed, an operation that was not previously possible, whilst the compressed air rock drilling machine (Chapter 2) was an innovation that allowed hard rock tunnelling to be done very much better than the previous method of hand drilling. So both of these are considered to be major innovations in tunnelling in this book. The majority of tunnel engineers would probably agree that most of the technical innovations selected satisfy the given criteria, although some might wish for some other developments to be included. For example, nothing is said about tunnel surveying because it followed the well-established surveying methods used in civil engineering in general and produced no major innovations in its own right. This is not to say that tunnel surveying is not an important subject and that great ingenuity has not been shown in adapting surface surveying techniques to underground application, but there were no significant innovations in this area which meet the criteria outlined above. One exception to this statement is the laser, which has been such a boon to tunnel engineers, and which is discussed in Chapter 13. Similarly, nothing has been said about ground treatment and similar geotechnical processes because these were innovations in civil engineering more generally and not specific to tunnelling. Again, this is not to say that these processes are not important, indeed some of them have probably become more important in tunnelling than in other branches of civil engineering. Nevertheless, they are basically civil engineering innovations and not tunnelling ones. It so happens that the history of ground treatment in civil engineering has been dealt with by R. Glossop.[8] Lastly, except in passing, cut-and-cover tunnelling is not considered here, this method falling within the purview of general civil engineering rather than of tunnelling.

Whilst this work is a synthesis in that it brings together a number of diverse innovations from different parts of the tunnelling industry and from different periods of time, it is also analytical in that each innovation is examined in an effort to identify its origins and the problems associated with it. The work concerns itself with innovations in the tunnelling industry partly because the author's association with this industry[9] allows a well-informed view to be taken on the innovations in it, but also because of the need to confine the subject to a manageable size. Taking a wider view, the classification of technical innovation that has emerged from the work might be applied to other industries and possible uses for this are discussed in Chapter 14. Also, from time to time throughout the book, but particularly in Chapters 2, 6 and 13, the rise of the tunnelling industry has been related to the social and economic factors which gave rise to the need for tunnels and therefore provided the spur for innovation in the tunnelling industry.

Chapter 13 looks at some other developments, but only the perspective of time will show whether these are to be ranked as major innovations in tunnelling. Also included in Chapter 13 is a brief description of the emergence of a vigorous and innovative tunnelling industry in Japan; as we will see, it would seem that the centre of gravity of innovative tunnelling technology has shifted to Japan.

This work does not deal with the economics of innovation, nor except in passing, with the economic factors associated with the rise of the tunnelling industry. Nor are the costs of tunnels, machinery or construction materials discussed. This is not to say that these matters are not important but rather that they are not of central relevance to the themes here being studied. Nevertheless, it seems likely that an examination of the economics of the tunnelling industry would be a fruitful field of study. Similarly, except for the instances referred to earlier, the work does not concern itself with the social causes of the technical innovations discussed, nor does it deal with the social changes that the construction of tunnels made possible by these innovations may have brought about. Again, this is not to say these matters are not important and, as for economics, it would seem that the social causes and effects of the rise of the tunnelling industry would also be worthy of investigation.

Technical innovation is not only the concern of the historian of technology, but is of vital importance to the well-being of present-day industry. Consequently the state of contemporary technical innovation is frequently the subject of enquiry. For example, the economist Kerry Schott has reviewed the process of innovation in present-day industry in the United Kingdom, the United States and Canada with the objective of making recommendations for action by government and business to stimulate more of it.[10] In 1981 in Great Britain the National Research Development Corporation and the magazine *New Civil Engineer* jointly sponsored a Civil Engineering Innovation Competition, with a first prize of £10 000, the aim of which was to stimulate innovation in civil engineering[11] (see Chapter 13). These approaches take it for granted that technical innovation is a process that can be artificially promoted, fostered, stimulated or directed, but it is not at all certain that this is so, and the point will be taken up again in Chapter 14.

1.2 Source materials

R.L. Hills, Curator of the Manchester Museum of Science and Industry, has stated that there are basically three kinds of source material for the history of technology.[12] These are (i) written records, (ii) pictorial sources and (iii) the machines themselves. To this list we might add, for the recent past, a fourth category – personal experience, and in this book use has been made of all four kinds of source material. In the main, written records and pictorial sources in the form of drawings, prints and photographs form the principal sources used here and the nature of these is discussed below. Except in very rare instances such as the Brandt hydraulic drill (see Section 5.1), old tunnelling machinery has not survived, so that nearly all the old machines themselves are no longer available for inspection. However, for the more recent technical innovations considered here, every opportunity has been taken to examine the actual hardware. This includes most of the British tunnelling machinery described here from 1974 to the present, together with many of the tunnels described and the tunnel linings and support systems. Some foreign tunnelling machinery dating from the same period has also been examined at first hand. The virtually complete disappearance of old tunnelling machinery is in striking contrast to the survival of nearly all the old tunnels themselves, most of which have remained in service to the present, and many of which are therefore well maintained. The Victorian

Technology Survey,[13] carried out in Great Britain in 1968–70, commented on this latter fact, but it is noteworthy that the Survey recorded no examples of the machinery with which the tunnels were built.

In the references cited at the end of each Chapter, wherever possible, published material has been cited. However, during the research for this work numerous unpublished sources have been used such as manufacturers' leaflets, brochures, technical specifications, drawings and photographs, but since these are ephemeral they have not been specifically cited unless no published information is available, when they are referred to in the notes. Correspondence and discussion with people in the tunnelling industry, likewise, have not been specifically cited, although important instances where a crucial point is at issue, are referred to in the notes. Contemporary accounts in engineering journals have been the most useful published sources; some details of the most important of these are listed in Table 1.1.

Patent specifications[14] have been a most useful primary source for the work, but they need to be used with circumspection. This is because many patented ideas or machines were never realised in practice, and because sometimes the actual working machine differed from the patent specification. Most of the patents discussed in this book are either of machines that were made and used or they embody ideas that were subsequently used. Some machines were the subject of multiple patents, specifications being filed both in their country of origin and abroad. Since January 1980, the introduction of a single European patent[15] has obviated much of this duplication. Specific patents used are cited in the notes and references.

Among the primary sources consulted special mention should be made of the seventeen long-quarto volumes comprising Marc Isambard Brunel's diaries for the years he was engaged on the construction of the Thames Tunnel which are now in the Library of the Institution of Civil Engineers, London. The diaries for the years 1828–35 inclusive were examined in order to establish Brunel's early ideas on compressed air tunnelling as discussed in Chapter 7. The quotations from the diaries given there are the author's translations of Brunel's entries in French, although most of the other diary entries are in English. Brunel's diaries have been used by several of his biographers and other writers on the Thames Tunnel (see Chapter 6). Other primary sources include Richard

Table 1.1 *Sources of tunnelling information*

Sources	Period	Volumes
Institution of Civil Engineers		
Minutes of Proceedings	1837–1935	1–240
Journal	1935–1951	1–36
Proceedings	1952–1985	1–79
American Society of Civil Engineers		
Transactions	1872–1981	1–146
Proceedings	1894–1955	20–81
Journals	1956–1985	82–111
Others		
Engineering	1866–1985	1–225
The Engineer	1856–1985	1–261
The Illustrated London News	1842–1985	1–287
Tunnels and Tunnelling	1969–1985	1–17
Tunnels et Ouvrages Souterrains	1974–1983	1–72[a]
Underground Space	1977–1985	1–9
Advances in Tunnelling Technology and Subsurface Use	1981–1985	1–5

[a]Numbers

Trevithick's original drawing of the Thames driftway (see Chapter 6), the contemporary model of Brunel's shield and Thames Tunnel (see Frontispiece), Hawkins' report of an immersed tube trial (see Chapter 10), Harding's unpublished report on one of the Oahe Dam tunnelling machines (see Chapter 11), and John Price's manuscript description of his proposed tunnelling machine for the Central London Railway (see Chapter 12).

Engineering drawings are an important kind of source material for the history of technology and extensive use of them has been made in this work. They are particularly valuable when, as is the case with tunnelling technology, few of the early machines have survived. Original old engineering drawings are rare but the author was fortunate in acquiring from the firm Markham and Co Ltd dyeline prints of some relating to two historically important tunnelling machines: the Beaumont–English tunnelling machine (Figure 11.4) and the Whitaker tunnelling machine (Figure 11.6). Both these are examples of the form of engineering drawing known as a general assembly, that is they show the whole machine rather than the parts. These are the most useful form of drawing for the historian of technology since they contain internal evidence of how the machine was intended to work and can resolve points that may be ambiguous or missing in a written description. A considerable amount of this kind of historical research was necessary during the compiling of the detailed descriptions given in Chapters 2–12. On one occasion a careful study of the drawing provided corroborative evidence of the origin of the machine (see Section 2.7). The old engineering drawings used in this study have mainly been of historical technical interest, but old engineering drawings can sometimes assume crucial present-day importance for another reason. For example, the safety of old earth embankment dams has become a subject of great concern, both to the civil engineering profession and to the public at large because of the potentially disastrous consequences of failure. Both an historical knowledge of the way these old dams were made and contemporary construction drawings of them and their associated works are vital to an analysis of their current safety,[16] and these records, where they still exist, show the value of preserving historical technical documents and demonstrate that the history of technology can sometimes have a striking relevance to the present. In certain circumstances, such as when tunnels are to be repaired or enlarged, contemporary construction drawings of old tunnels may have the same importance as those of old dams. For the later tunnelling machines, photographs have also been a useful source, and a selection of these illustrate this book. Photographs are valuable because machines were not always manufactured exactly as detailed in the engineering drawings, but photographs, especially if they are taken of the machine actually on the job, show the machine as it was built.

Good use has been made of secondary sources as well as primary ones, and where secondary sources have been used this is made clear in the notes and references. An example of an excellent secondary source is Glossop's paper on the early use of compressed air in civil engineering works (see Chapter 7); this is a scholarly paper showing the history of technology at its best, and with its comprehensive notes and references is an essential starting point for its subject. Another fine secondary source is the monograph on tunnel engineering by R.M. Vogel of the Smithsonian Institution (see Chapter 2); it has only limited references, but does contain good descriptions of early hard rock and soft ground tunnelling methods which were researched by Vogel for the purpose of producing museum reconstructions. It also contains several reproductions of contemporary drawings and prints of the tunnelling machinery and tunnels with which it deals, together with photographs of the museum reconstructions. It is likely that the discipline of making such reconstructions leads to a very complete understanding of the working of the devices being reconstructed, which lends authority to the descriptions given by Vogel. Nineteenth- and early twentieth-century textbooks[17] on tunnelling have also proved to be a useful source; they often contain a wealth of information on certain technical developments of the time that can now be seen as important innovations. They sometimes contain details of machines or of tunnel construction that are not now available anywhere else. Some of the old tunnelling textbooks are well illustrated. Good illustrations of early tunnels and tunnelling machinery, mainly European, are also given in the collection compiled by Louis Figuier.[18] The sources of some of the illustrations used as Figures here are listed at the end of the book. It should be noted that the dates given in the captions to the Figures are the dates of the items depicted and not the dates of the sources of the illustrations which, in some cases, are later.

Throughout this book many personalities are mentioned: scientists, tunnel engineers, inventors, consultants, contractors etc, and for some of these, brief biographies have been given.[19] This has been done partly as a change from what seems to be a dichotomy in the history of technology when some writers give accounts either of the engineers or of the engines but not of both. An exception to this which must be mentioned is the admirable series of biographies of engineers by L.T.C. Rolt.

1.3 Units of dimensions and technical terminology

In this book the dimensions given in descriptions of tunnel construction and machinery dating from before January 1969 are given in Imperial units, whilst those of tunnels constructed after this date are in metric units. This is because in January 1969 the British construction industry adopted the metric system of units in place of the Imperial system. The following approximate conversions are given; they are accurate enough for general mental comparison purposes.[20]

Length	*Force*
1 mile = 1.6 km	1 ton f = 10 kN
1 ft = 0.3 m	*Pressure*
1 in = 2.5 cm	1 atm = 15 lb f/in^2 = 100 kN/m^2
Volume	*Power*
1 gal = 4.5 l	1 hp = 750 W
1 pt = 0.6 l	*Temperature*
Mass	n°F = $\frac{5}{9}$ (n − 32) °C
1 ton = 1000 kg (1 tonne)	
1 lb = 0.5 kg	
1 cwt = 50 kg	

Because this book deals with tunnelling, the use of some of the technical terms used in tunnelling has been necessary. However, these have been kept to a minimum and the meaning of most of them has been defined in the text where they first appear or will be obvious from the context. In other cases, reference can be made to the Glossary or to one of the dictionaries of civil engineering.[21] Also, in a book dealing with tunnelling through the ground, the use of some geological terms has been necessary. In this case concise definitions of the terms used are given in the Glossary or fuller definitions can be found by reference to a geological dictionary.[22]

1.4 The modern tunnelling industry

Tunnelling as a civil engineering activity can be divided into two branches: hard rock tunnelling and soft ground tunnelling. In hard rock tunnelling the main objective is to produce an opening – the tunnel – in the rock mass by breaking out and transporting away fragments of rock. This is often a laborious business. The opening is usually, but not always, self-supporting, but it is customary nowadays to provide the tunnel with a lining whether this is the case or not, although many early tunnels were left unlined. As will be discussed in Section 1.6, before about 1850, the tunnels in hard rock that had been constructed were driven by methods that were essentially those of the metalliferous mining industry and which date back to the Middle Ages; this close association is reflected in the fact that the face-workers in tunnels right down to the present day are referred to as 'miners'.[23] Traditionally in hard rock tunnelling, the main problem facing the engineer was the task of breaking out the rock to form the excavation, but once the excavation had been formed it could usually support itself if the rock mass was reasonably free from joints. Even if the rock mass was broken up by an intersecting network of joints, the excavation could still be made safe by erecting some temporary timber support. If in competent rock, the tunnel was left unlined after any loose blocks had been 'scaled down' by the miners using iron bars. A permanent brick or masonry lining was installed where the rock was judged to be unsafe or where it was not competent. The portals of tunnels were often provided with a brick or masonry portico, often graced with architectural embellishments.[24] For example, Isambard Kingdom Brunel provided the west portal of the Box Tunnel with a fine Bath Stone portico flanked by elegantly curved retaining walls. This portal was prominently visible from the Old Bath Road which the Great Western Railway[25] had been built to replace as the major route from London to Bristol, and was intended to impress travellers misguided enough to be still using the Bath Road with the grandeur of the new railway.

In soft ground tunnelling the excavation is usually easily accomplished compared to hard rock tunnelling but the problem is often to prevent the ground from collapsing into the tunnel. Erection of a strong durable lining as soon as possible after excavation is essential. Traditionally, soft ground tunnelling was carried out using hand excavation by spade or pick and shovel followed by very elaborate timbering to support the walls, roof and sometimes the face as well. These old methods have now almost vanished but examples can be seen in old tunnelling text book illustrations and small timbered headings are occasionally driven still. Close behind the miners came a team of bricklayers or masons who erected a massive structural brick or masonry lining to support the ground permanently. Soft ground tunnelling by the traditional method could not be carried out beneath the water table in permeable strata and this severely limited its use and prevented it from being used to provide tunnels for river crossings. In both hard rock and soft ground tunnelling the debris from the excavation, called muck, was usually removed in rail-mounted wagons pulled by ponies or, more often, mules, but sometimes pushed by men.[26]

Although the practices of hard rock tunnelling emerged from those of mining, there were notable developments in the mid-nineteenth century which established hard rock tunnelling as a modern industry in its own right, and these can be conveniently dated from the commencement of construction of the Hoosac Tunnel in the United States in 1851. In the case of soft ground tunnelling, we can confidently date the birth of that branch of the modern industry as 1825, the date of the commencement of Marc Isambard Brunel's Thames Tunnel in London. During the following 150 years or so, the modern tunnelling industry has risen to the position it holds today and indeed, this short period encompasses the whole of its history. The social and economic changes that gave rise to the need for tunnels, and hence called the industry into being, were different for hard rock tunnelling than for soft ground tunnelling, but in both cases they were of the general kind that produced the improvements in public utilities that were a feature of the Victorian age. The circumstances that gave rise to hard rock tunnelling are discussed in the introduction to Chapter 2 and those that gave rise to soft ground tunnelling are discussed in the introduction to Chapter 6.

The division of tunnelling into these two branches of hard rock tunnelling and soft ground tunnelling works reasonably well in practice, except that, inevitably, there are instances of soft rock or hard ground where there is overlap and where techniques from either branch of tunnelling would be equally applicable.[27] The technical innovations described in Chapters 2–5 are from hard rock tunnelling whilst those described in Chapters 6–9 are from soft ground tunnelling. Tunnelling machines, described in Chapters 11 and 12, have found application in both hard rock and soft ground tunnelling.

Although we can clearly identify the emergence of a distinct tunnelling industry by the mid-nineteenth century, it was, of course, part of the civil engineering industry which was in turn part of the larger construction industry. Consultants, contractors and individual engineers concerned with tunnelling were usually involved in other civil engineering works as well as tunnels and this has continued to be true right up to the present day. Because of this it was not until fairly recently that the special concerns of tunnel engineering were formally recognised by the formation of societies and associations specially devoted to tunnelling. In 1971 the British Tunnelling Society was formed[28] with the aim of improving the practice of British tunnelling. To start with, the Society was formed as a branch of the Institution of Civil Engineers, but in 1979 the Society became independent, although continuing to use the premises and facilities of the Institution. The journal *Tunnels and Tunnelling* is used as the official journal of the Society. In France the Association Française des Travaux en Souterrain was formed much earlier – in 1901; their official journal is *Tunnels et Ouvrages Souterrains*. Other countries too have national tunnelling associations. To foster international co-operation and exchange of information in tunnelling, the International Tunnelling Association was formed in 1974.[29] The aims of the Association are similar to those of the national tunnelling societies but its activities tend to concentrate on its international role and on promoting the wider use of tunnels in general. The official journal of this Association is *Advances in Tunnelling Technology and Subsurface Use*.

The main participants in a tunnelling scheme in Great Britain are the Promoter (sometimes called the Client) for whom the tunnel is to be built, the Consultant (or Consulting Engineer) which is usually a firm having experience in the design of tunnels and the supervision of their construction, and the Contractor, again usually a firm, who undertakes to construct the tunnel including the provision of all plant, labour and materials to do so. One person, usually from the firm of Consultants, is designated the Engineer and he is responsible to the Promoter for the effective, safe and economic design and construction of the tunnel. In addition to these principals, there are others who are also involved, important among them being the tunnelling machine designers and manufacturers and the suppliers of tunnelling equipment of all kinds. Sometimes a tunnelling machine manufacturer may be partly or wholly the subsidiary of a Contractor. Also, there are sub-contractors who undertake specialised work for which the main Contractor is not equipped. Sometimes, where the Promoter is a large public authority having its own professional engineering staff, the Promoter acts as Consultant as well, providing its own engineers to design the tunnel and supervise construction; in this case one of the Promoter's senior engineers will be designated as the Engineer. When construction of a tunnel actually commences, the senior representative of the Consultant on site is designated the Resident Engineer and the senior representative of the Contractor on site is designated the Agent. Latterly there has been a trend for Consortia to be set up for large tunnel schemes: these are groups of Consultants or Contractors who by coming together can offer a range of expertise or capability that any one alone could not. In the United States the three main participants are termed Owner, Designer and Contractor.

Elsewhere abroad it is common for one firm to design the tunnel as well as build it, these being called 'design and build' contracts. In the nineteenth century these distinctions with their division of responsibility were not so clear cut and it is not always possible to identify who was who in modern terms. For example, on the Thames Tunnel, which is described in Section 6.4, the Thames Tunnel Company seem to have been Promoter, Consultant and Contractor all at the same time, although it is fairly clear that Marc Isambard Brunel was the Engineer. In the pages that follow we will see that the inventors of tunnelling innovations came from engineers on the staffs of promoters, consultants, contractors and machine manufacturers as well as from persons outside the tunnelling industry including scientists, notably chemists, and army and navy officers.

1.5 Some general remarks on technical innovation

The emergence of industries or technologies that are based on scientific discoveries is comparatively recent, dating from the second half of the nineteenth century and can be exemplified by the rise of the electricity, chemical and communications industries.[30] If one examines earlier periods, examples can be found of the development of industries or technologies that were entirely based on improvements made from within the particular area of technology and which owed little or nothing to the application of science. For example, the whole history of the sailing ship is one of continuous improvement based on observation of the performance of ships, a process which led eventually to the perfection of the sailing ship in the form of the clippers of the mid-nineteenth century. This development was entirely as a result of the activities of shipbuilders, shipwrights, riggers and mariners and not due to the application of science. Scientific research into the shape of ships' hulls, for example, did not commence until the 1830s when it was almost too late to affect the sailing ship, the new knowledge being applied to the improvement of steamship hulls.[31]

Civil engineering is an example of an industry occupying a position mid-way between these two extremes. In the beginning it was essentially craft-based, stemming from the work of largely self-taught engineers like Thomas Telford (1757–1834) and John Loudon Macadam (1756–1836). However, as the nineteenth century unfolded, the application of science to civil engineering steadily gained ground, and scientific principles began to take equal place with the intuitive genius of later engineers such as Isambard Kingdom Brunel (1806–59).[32] We have noted that tunnelling is a branch of civil engineering and, therefore, following from what has just been said, it might be thought that the technical innovations in tunnelling that are the subject of this book would be found to have their origin in either the older craft-based empiricism or in the newer science-based knowledge.

This simple division of technical innovation into two categories is very close to the view of the seventeenth-century philosopher Francis Bacon. Bacon considered that there were two kinds of invention: those that depend on a knowledge of things and those that do not.[33] Examples of the first that Bacon gave were the mariner's compass, gunpowder and the production of silk from silkworms; an example of the second was the printing press. These two kinds of invention correspond to science-based and technology-based innovation respectively. Much later, Lyon Playfair (1818–98) held a view that was the same as Bacon's, considering that technological advances were either scientific or empirical.[34]

Therefore, *a priori*, we might expect to see what can be termed technology-based innovations and science-based innovations. As the work described in Chapters 2–13 progressed it was found that these two categories of innovation could indeed be recognised, but that by themselves they could not account for the origin of all the innovations, and some other origins of innovation could be seen. Each of Chapters 2–12 concludes with a Summary in which the perceived origin of the particular innovation which is the subject of the Chapter is identified. The results of these observations are given and their significance is discussed in Chapter 14, but it is stressed that this classification of technical innovation into different categories based on origin must be regarded as speculative for the present.

In this book I have taken the word science to mean the systematic study of the properties and behaviour of the natural world (for example, chemistry), whilst the word technology is used to mean the practice and application of the technical arts (for example, mechanical engineering). This is somewhat different from current popular usage in which science is taken to include technology,[35] and should be borne in mind, particularly when some aspects of the interaction of the two are considered in the final Chapter.

C.J. Jackson has made a study of the history of drill bits and drilling techniques in which he has considered the demands for holes arising from social and economic developments, the problems of working various kinds of materials (wood, metal, rock) and the skills of craftsmen, both in devising tool shapes and in changing them to suit perceived needs.[36] Jackson's approach and that of the author have the same underlying aim, namely to make a break with the traditional approaches to the history of technology – which has either been concerned with descriptions of machines or the lives of engineers, or has been in the hands of economic historians – and to try to explore basic themes and fundamental issues.[37] It is, of course, clear that if the whole field of technology is considered there are reasons other than the solving of problems for the introduction of technical innovations. Nevertheless, it is argued here that the concept of technical innovations as solutions to problems can be particularly fruitful when applied to the tunnelling industry, especially when the problem to be solved arose from a compelling need as a number did. As the argument is developed, however, it becomes clear that the concept is at once more subtle and more complex than may appear at first sight. The importance of the major technical innovations discussed in Chapters 2–12 cannot be overestimated. Indeed they gave rise to the modern tunnelling industry, with all the benefits this has conferred upon society.[38]

Although the book has been concerned with innovations in the tunnelling industry, it has by no means confined itself to a narrow view. Attempts have been made, throughout the work, to show the links with other subjects where appropriate and relevant because, of course, tunnelling exists not in isolation but as part of man's activities more generally. Many of the subjects impinge in numerous places but others are of more limited relevance. Thus, for example, geology and mechanical engineering crop up throughout the book whereas chemistry and hydrostatics are confined to specific Chapters. The geographical range of the innovations considered is world-wide. In this way, it is hoped that although the work has been concerned with tunnelling history it has avoided the dangers of the blinkered approach to historical writing which has been dubbed 'tunnel history'.[39]

Finally, an important caveat: in this book the expression 'the first' and 'the second', 'the next', etc, when discussing a particular innovation does not necessarily mean the first example ever, but rather the first example of the innovation discussed in this book. This is because the author has sometimes had to exercise judgement as to which particular item can be considered the first practical example of an innovation, or the first design of one worth seriously considering. In some cases there is no conflict, for instance Brunel's Thames Tunnel shield (Chapter 6) can probably be called the first tunnelling shield in all senses without fear of contradiction. By contrast, to call the Detroit River Tunnel (Chapter 10) the first immersed tube tunnel is to disregard some earlier proposals and some small earlier projects.

1.6 Ancient tunnelling

Although this book deals with the rise of the modern tunnelling industry, as remarked above a period from about 1825 onwards, it is helpful to an understanding of the modern industry and its technical innovations to review briefly the earlier periods of tunnelling and look at some aspects of their technology. In its earliest days, tunnelling was closely related to mining so that the one cannot properly be considered without the other and some general remarks will first be made about this relationship. Mining and tunnelling are the two main types of underground excavation that man has engaged in, but it is interesting to contrast the aims of these two activities. In mining the objective is to recover a useful or desirable mineral (metal ores, coal, precious stones, etc) and the shafts and galleries excavated to obtain the mineral are of no use to the miner save as access routes to more of the mineral. In tunnelling, however, the tunnel itself is the purpose of the excavation, which is made in order to provide a transport route for people, materials, water, etc, and the spoil arising from the tunnel construction is usually of no use and has to be disposed of as an unwanted by-product. In spite of these completely different aims, mining and tunnelling have much in common, particularly in methods of excavation, but also to a lesser extent in the methods of support. The main difference in this latter aspect lies in the fact that mine roadways have only to last the life of the mine whereas tunnels are usually constructed to have a very long life as they are intended to be permanent features of some transport system or public utility.

The history of mining is almost as old as the history of man himself and in Great Britain there is a remarkable group of Neolithic flint mines at Grime's Graves, Norfolk which provide evidence of this.[40] The purpose of the mines was to obtain flint from which to make tools and weapons, and the period during which the 700–800 pits in the whole mined area were being worked, as indicated by radio-carbon dating, extended from about 2330 BC to 1740 BC. A typical pit at Grime's Graves consisted of a vertical shaft some 14 ft in diameter extending to a depth of 30 ft below ground surface. At the bottom, galleries about 3–4 ft high were driven off in radial directions. Some of the galleries led from the bottom of one pit to the bottoms of others nearby; these main galleries were up to 7 ft wide by 5 ft high. Fifteen pits in all have been investigated. The objective of the miners was to obtain flint from the Chalk that occurs on the site beneath about 10 ft of silt, sand and clay. The miners were seeking a particularly desirable layer of tabular flint known as 'floorstone' which was of the best quality for making flint implements. The large pieces of tabular flint were brought to the surface where they were used in the manufacture of flint tools and weapons, this activity being carried out on flint-knapping 'floors' set out in the space around the pitheads. In sinking their mines, once they had dug through the soft overlying deposits the miners were faced with the problem of excavating a shaft some 20 ft down through fairly hard chalk, and then to drive the galleries out through the same material, and finally to prise out the flint. Although there is some evidence that they occasionally used stone axes, from the great abundance of them found on the site it is clear that the principal tool used to excavate both chalk and flint was a pick made from the antler of the red deer (Figure 1.1). The shape of the section of antler with brow tine is well suited to use as a pick, and the substance from which it is made being both tough and reasonably hard is suitable for the work of excavation provided the rock is not too hard. This implement, therefore, can be considered to be the first technical innovation in the history of tunnelling; there is no doubt that it made possible the flint-mining industry of Grime's Graves.

Tunnelling, as distinct from mining, can be traced back to antiquity and perhaps the first instance of tunnelling that we know of was the construction of *sinnors* in the ancient fortified cities of Syria and Palestine. These date from about 1200 BC and consisted of a flight of stairs leading down a shaft to a gallery which was driven to intercept an underground spring beneath the city. Sometimes the spring was deepened into a well or underground cistern. The *qanat*, which seems to be a natural development of the *sinnor*, is a tunnel or adit of low gradient driven into a hillside to intercept the water table and thereby provide a source of water (see Figure 1.2). *Qanats* were constructed by excavating the tunnel from the bottoms of a series of shafts sunk along the line of the proposed *qanat* at regular intervals.[41] The top of each shaft is surrounded by a spoil heap which makes the course of a *qanat* across country easy to trace, especially on air photographs of which there is a striking example in Volume 1 of Megaw & Bartlett's book.[5] Where the *qanat* emerged from the base of the hill the water was conveyed to the place it was required, usually a city some way out on the plain, by an aqueduct which could be up to 35 miles long. Some typical dimensions of a *qanat* which have been quoted are: a 4 ft high by 2 ft wide adit driven on a rising slope of 1–3 per cent, with 3½ ft diameter shafts spaced about 120 ft apart and up to 350 ft deep. *Qanat* construction seems to have originated in Armenia where it is known to have been established well before 700 BC and from here the technique spread throughout the Near and Middle East where conditions were right for its use; they are reported to be in frequent use in India by about 300 BC. *Qanats* are still built in Iran where the builders, known as *muquanni*,

Figure 1.1 Antler pick

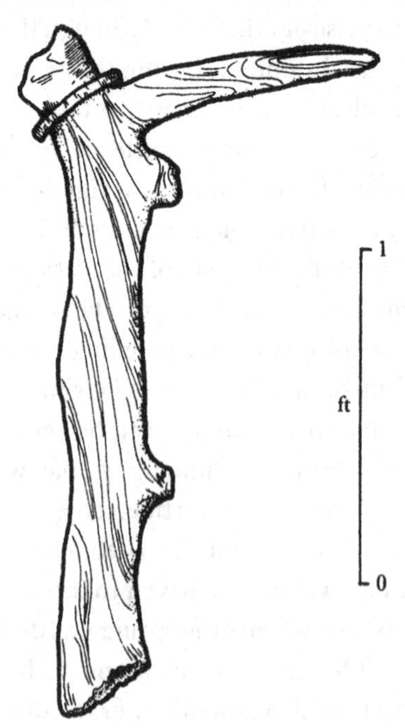

usually a two-man team, are held in high esteem by the local communities. A fascinating account of modern *qanat* construction has been given by Donald Hill[42] who gives the following technical details: a 5 ft high by 3 ft wide tunnel dug on a grade of from 1:1000 to 1:500, with vertical shafts about 3 ft in diameter spaced some 90–150 ft apart. These dimensions are very similar to the ancient ones. Hill describes the ground conditions as being 'firm soil' which explains why *qanat* construction is and was possible by simple hand excavation methods. At the top of each shaft a winch is set up for spoil haulage, the material then being simply dumped around the mouth of the shaft to give the characteristic tell-tales of the *qanat* line that we have already noted. Hill describes the prospecting and surveying operations of modern *qanat* construction which also seem little changed from the past.

The Greeks also constructed tunnels for water supply purposes, amongst which can be mentioned that for the water supply of the town of Samos on the island of the same name. The Samos aqueduct tunnel, probably constructed in the period 535–522 BC, was 3300 ft long and was driven through limestone from both ends of a hill simultaneously. The cross section of the tunnel was 5½ ft high by 5½ ft wide. When tunnel driving was finished it was found that the gradient was too shallow for a swift flow of water and therefore another conduit was constructed beneath the tunnel floor with a steeper gradient. We actually know who the Engineer was for the Samos aqueduct tunnel: Eupalinus, son of Naustrophus of Megara. The Romans also built tunnels for water supply and drainage, but are notable for being the first people who regularly constructed road tunnels, for in constructing their famous roads the Romans often found it necessary to excavate rock cuttings and tunnels; an example is the Roman road that runs along the Danube and through the pass known as the Iron Gate, completed in AD 103 by Trajan. An outstanding and well-known Roman tunnel is the Pausilippo road tunnel

Figure 1.2 The principle of the *qanat*

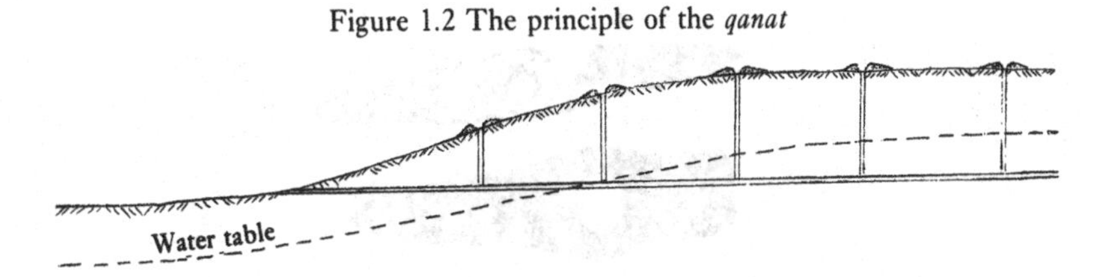

between Naples and Pozzuoli, 25 ft wide by 30 ft high and almost one mile long, built by the engineer Lucius Cocceius Auctus in 36 BC. The methods the Romans used for tunnelling were the same as they used for mining. The principal tools were the hammer and the wedge, which by Roman times were made of iron (Figure 1.3). The hammer had a long handle like the present-day sledge hammer, and some of the types of wedge seem also to have been fitted with a similar handle, probably so that they could be held safely in position against the rock by one miner whilst being struck by a second miner. Using hammers and wedges the rock was broken out from the tunnel face and split into blocks of about 150 lb weight which were then carried or dragged out of the tunnel in baskets, sacks or chests. The workings were illuminated by amphorae lamps suspended from the roof or placed in niches. Progress rates were very slow, as little as 25–30 ft per year when in hard rock has been estimated, and considering their other technical achievements in civil engineering, the Roman mining and tunnelling excavation methods seem very primitive. Indeed, Neuburger[43] remarks that (with the exception of mine drainage) the art of mining appeared to have made almost no technical progress from the date of the earliest traces recorded to the fall of the Roman Empire. In spite of this rather severe stricture we can note that the metal hammer and wedge

Figure 1.3 Roman iron mining tools

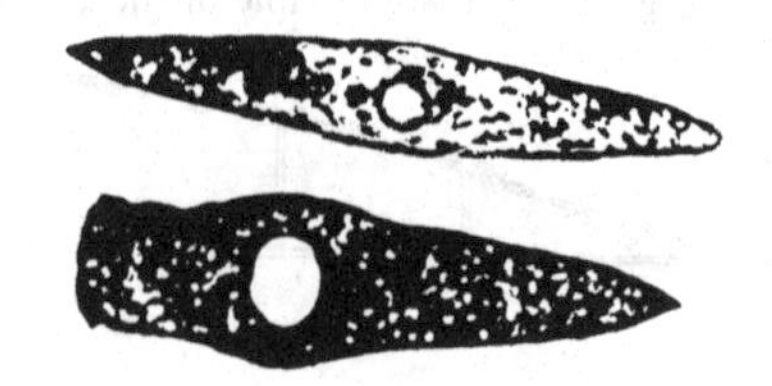

had appeared, and can be considered the second technical innovation in the history of tunnelling.

When we come to the Middle Ages (600–1500) the primitive picture of mining and tunnelling that has just been described changes in a dramatic fashion. For in Europe, particularly in a belt extending from southern Germany through Austria and Czechoslovakia to Hungary, there arose a metal-mining industry that developed many technical advances in the methods of mining. Not only that, but the miners engaged in this activity were a highly skilled, well-respected and well-paid workforce in stark contrast to the previous Egyptian, Greek and Roman miners who were for the most part slaves, captives, prisoners of war or criminals. The principal reason why we know so much about this flowering of European mining technology in the Middle Ages is because of the writings of the famous scholar of mining Georg Bauer, better known under the Latin version of his name Georgius Agricola. Agricola was born on 24 March 1494 at Glauchau in Saxony. After studying at the Universities of Leipzig, Bologna and Padua he took a degree in medicine at the University of Ferrara. He went to practise as a physician in the mining town of Joachimsthal in Bohemia from 1527–33, and in Chemnitz from 1534 till the end of his life. Living in the heart of this great mining region, Agricola wrote six great books on geology, mineralogy and mining. The one to concern us here was *De re metallica*,[44] published posthumously in 1556, which deals with the technologies of mining and smelting metal ores. Agricola died on 21 November 1555. From the pen of Agricola we read of a competent, effective and well-organised mining industry, but above all for the first time, one that was technically proficient. The details of the technical advances that Agricola describes, most of which are illustrated by attractive woodcuts (often reproduced in modern books on the history of mining), fall outside the scope of this book. However, we can note that rock excavation was still mainly carried out during the Middle Ages using hammer and wedge, although 'fire-setting', a technique whereby the rock was weakened by alternately heating it and then rapidly cooling it with water, was used when particularly hard rock was encountered. What it is important to realise is that any tunnels driven during the Middle Ages and for a long time afterwards were excavated using essentially these mining techniques. Certain of the tunnelling techniques that will be touched upon later in the book, notably hand drilling of rock (Chapter 2), gunpowder (Chapter 3) and timber tunnel supports (Chapters 6 and 8) form a transition between the mining methods of the Middle Ages and those of the beginnings of the modern tunnelling industry.

Notes and references for Chapter 1

1 By far the best book on the history of tunnelling from a technical standpoint is:
Sandström, G.E. (1963) *The history of tunnelling*. London (Barrie and Rockliff).
A good summary of tunnelling history is given in the following paper:
Halcrow, W.T. (1942) A century of tunnelling.
Proceedings of the Institution of Mechanical Engineers, **146**, (3), pp. 100–16.

2 Hall, A.R. & N.A.F. Smith (1980) Preface. *History of Technology*, **5**, unnumbered page.

3 Smith, N. (1971) *A history of dams*. London (Peter Davies) p. 112.

4 Harding, Sir Harold (1981) *Tunnelling history and my own involvement*. Toronto (Golder Associates).

5 Megaw, T.M. & J.V. Bartlett (1981) *Tunnels: planning, design, construction*. Volume 1. Chichester (Ellis Horwood Ltd).
Megaw, T.M. & J.V. Bartlett (1982) *Tunnels: planning, design, construction*. Volume 2. Chichester (Ellis Horwood Ltd).

6 Stack, B. (1981) *Handbook of mining and tunnelling machinery*. Chichester (John Wiley and Sons).

7 Bickel, J.O. & T.R. Kuesel, Editors (1982) *Tunnel engineering handbook*. London (Van Nostrand Reinhold Co).

8 Glossop, R. (1960) The invention and development of injection processes. Part 1: 1802–1850. *Géotechnique*, **10**, (3), pp. 91–100.
Glossop, R. (1961) The invention and development of injection processes. Part 2: 1850–1960. *Géotechnique*, **11**, (4), pp. 255–79.

9 The author is on the staff of the Transport and Road Research Laboratory, and during the period 1974–85 carried out research in tunnelling; his interest in innovations in the tunnelling industry, their origins and the problems associated with them stems from this involvement.

10 Schott, K. (1981) *Industrial innovation in the United Kindom, Canada and the United States*. London (British–North American Committee).
This report contains numerous references to works which deal with the subject of present-day innovation in industry.

11 Details of the competition together with the rules were set out in a leaflet entitled *Civil Engineering Innovation Competition* published and distributed by the sponsors.

12 Hills, R. (1977) Museums, history and working machines. *History of Technology*, **2**, pp. 157–67.

13 Smith, N.A.F. (1970) *Victorian technology and its preservation in modern Britain*. Leicester (Leicester University Press) pp. 22–3.

14 Haddan, R. (1937) *Patents for inventors*. London (Sir Isaac Pitman and Sons Ltd).
Grace, H.W. (1971) *A handbook on patents*. London (Charles Knight and Co Ltd)

15 European patent office (1985) *Protecting inventions in Europe*. Munich (European Patent Office) 5th edition.

16 Boden, J.B. & J.A. Charles (1984) The safety of old embankment dams in the United Kingdom – some geotechnical aspects. *Municipal Engineer*, **111**, (2) pp. 46–50.

17 One textbook deserves special mention as a source of information on the history of tunnelling and its technology up to 1893; this is the monumental 1500-page work:
Drinker, H.S. (1878, 1882, 1893) *Tunnelling, explosive compounds and rock drills*. New York (John Wiley and Sons) 1st, 2nd and 3rd editions.
Other classical textbooks which are useful are:
Copperthwaite, W.C. (1906) *Tunnel shields and the use of compressed air in subaqueous works*. London (Archibald Constable and Co Ltd).
Hewett, B.H.M. & S. Johannesson (1922) *Shield and compressed air tunnelling*. New York (McGraw-Hill Book Co Inc).
Simms, F.W. (1896) *Practical tunnelling*. London (Crosby Lockwood and Son) 4th edition, revised and greatly extended by D.K. Clark.
Gripper, C.F. (1879) *Railway tunnelling in heavy ground*. London (E. and F.N. Spon).

18 Figuier, L. (1890) *Les nouvelles conquêtes de la science: grands tunnels et railways métropolitans*. Paris (La Librairie Illustrée).

19 Together with others cited in the notes and references, the following biographical works have been found generally useful:
Gillispie, C.C. Editor (1970–6) *Dictionary of scientific biography*. 24 Volumes, New York (Charles Scribner's Sons).
Committee on History and Heritage of American Civil Engineering (1972) A biographical dictionary of American civil engineers. *ASCE Historical Publication* No. 2. New York (American Society of Civil Engineers).
Marshall, J. (1978) *A biographical dictionary of railway engineers*. Newton Abbot (David and Charles).
The card-index biographical references compiled by Engineer-Captain Edgar Charles Smith (1872–1955), now in the Science Museum Library, London, have also proved useful.

20 The exact equivalents of corresponding units in the Imperial and metric systems are given in:
Anderton, P. & P.H. Bigg (1969) *Changing to the metric system*. London (H M Stationery Office) 3rd edition.

21 For example:
Scott, J.S. (1980) *The Penguin dictionary of civil engineering*. Hardmondsworth (Penguin Books) 3rd edition.

22 For example:
Whitten, D.G.A. & J.R.V. Brooks (1972) *The Penguin dictionary of geology*. Harmondsworth (Penguin Books).

23 A consequence of this tradition, often remarked upon, is that the patron saint of miners, St Barbara, is also the patron saint of tunnellers.

24 Tunnel portals and views of tunnels under construction were often the subjects of Victorian artists; some examples are given in:
Klingender, F.D. (1972) *Art and the industrial revolution*. London (Paladin) Figures 86–90.

25 An extremely fine lithographically illustrated description of the Great Western Railway is given by:

Bourne, J.C. (1846) *The history and description of the Great Western Railway.* London (David Bogue).

26 There is a vivid account of traditional tunnelling on the early railways given in:
Coleman, T. (1968) *The railway navvies.* Harmondsworth (Penguin Books) pp. 49–52 and 115–38.

27 Examples of these in Great Britain are the Chalk and Keuper Marl.

28 Harding, Sir H., A.M. Muir Wood & D.J. Lyons (1971) British Tunnelling Society. *Tunnels and Tunnelling,* **3**, (3), pp. 181–90.

29 Gosselin, C. (1974) The International Tunnelling Association is born. *Tunnels and Tunnelling*, **6**, (3), p. 47.

30 The point that these industries were crucially dependent on chemistry and physics is well made in:
Bernal, J.D. (1969) *Science in history.* Volume 3. Harmondsworth (Penguin Books) pp. 705, 773–83 and 823–4.

31 Well described by:
Goodman, D.C. & C.A. Russell (1973) *Unit 15: Science and engineering.* Milton Keynes (The Open University Press) pp. 69–80.

32 A good account of the history of early civil engineering in Great Britain is given by:
Gregory, M.S. (1971) *History and development of engineering.* London (Longman) pp. 34–45.
Illustrations are given in:
Upton, N. (1975) *An illustrated history of civil engineering.* London (Heinemann) *passim.*

33 Francis Bacon (1561–1626) is considered to be the first person who thought seriously about industrial science. He believed that the proper purpose of knowledge, or science, was to improve the material conditions of civilisation and that technical innovations were of more use to mankind than other more traditionally hallowed benefits (see page v). His works and beliefs are discussed in:
Farrington, B. (1951) *Francis Bacon: philosopher of industrial science.* London (Lawrence and Wishart Ltd).
Bacon's two kinds of invention are described by:
Cardwell, D.S.L. (1972) *Technology, science and history.* London (Heinemann Educational Books Ltd) Chapter 2.

34 Playfair, Sir Lyon (1852) The study of abstract science essential to the progress of industry. *Records School of Mines*, **1**, (1), pp. 23–48.
Playfair wrote 'Newton, by his exposition of the law of gravitation, produced more real practical benefits to industry than all the preceding ages of empiricism'. He was trying to win support for the newly opened Government School of Mines (now the Royal School of Mines of the Imperial College of Science and Technology) and clearly had high hopes for the application of science to mining, but it can be noted that at the time he made this statement mining owed far more to empiricism than to science.

35 Consider, for example, the use of the word science in 'The Science Museum', a museum in which most of the exhibits are technological rather than scientific. Also, the title of the following well-known work, in which the role of technology as much as that of science is described:
Bernal, J.D. (1969) *Science in history.* 4 Volumes. Harmondsworth (Penguin Books).

36 Jackson, C.J. (1984) A history of the development of drill-bits and drilling techniques in 19th century Britain. *PhD thesis*, The Open University.
37 Smith, N.A.F. (1977) The origins of the water turbine and the invention of its name. *History of Technology*, 2, p. 216.
38 Tunnelling in all its forms has made an important contribution to modern civilisation. This point is well made for civil engineering more generally by: Sprague de Camp, L. (1977) *Ancient engineers*. London (Tandem Publishing Ltd) p. 26.
39 Marwick, A. (1970) *The nature of history*. London (Macmillan) p. 178.
40 Rainbird Clark, R. (1970) *Grime's Graves, Norfolk*. London (H M Stationery Office).
41 Forbes, R.J. (1955) *Studies in ancient technology*. Volume 1. Leiden (E.J. Brill) pp. 152–6.
42 Hill, D. (1984) *A history of engineering in classical and medieval times*. London (Croom Helm) pp. 33–6.
43 Ancient Greek and Roman tunnels are described in:
Neuburger, A. (1930) *The technical arts and sciences of the ancients*. London (Methuen and Co Ltd) pp. 409–36 and pp. 461–2.
44 Agricola, G. (1556) *De re metallica*. Translated by H.C. Hoover and L.H. Hoover (1912) London (The Mining Magazine).

Compressed air rock drilling machines

Of all inventions, the alphabet and the printing press excepted, those which abridge distance have done most for the civilisation of our species.

The compressed air rock drilling machine, here considered to be the first major innovation in hard rock tunnelling, is the subject of this Chapter, but before discussing it, and by way of introduction, some remarks will be made on the need for hard rock tunnels that occurred in the nineteenth century and which gave rise to this branch of the modern tunnelling industry.

2.1 The need for hard rock tunnels

The demand for hard rock tunnels in the nineteenth century arose mainly as a consequence of the coming of the railways and of the extremely rapid and widespread expansion of railway networks, first throughout the United Kingdom and then over most of the civilised world. If the reason for the rise of a technology can be defined as the coincidence of a need and a tool – then, as we now see it, in the case of the railways the need was that of carrying raw materials, manufactured products and people more rapidly than canals and more reliably than roads were capable of doing in the nineteenth century. The 'tool' was the combination of iron rail and steam locomotive. It is generally agreed that the first railway was the Stockton and Darlington, opened in September 1825, although there had been colliery lines, cableways and tramways before this. The first trunk railways, the 110 mile long London and Birmingham, built by Robert Stephenson (1803–59) and completed in 1838, and the 118 mile long Great Western from London to Bristol, built by Isambard Kingdom Brunel and completed in 1841, marked the beginning of the construction of a vast network of railway lines that was to cover Great Britain and extend to its farthermost limits; the whole being virtually complete by

Table 2.1. *Route mileage of railway in England and Wales in the nineteenth century*

Year	Route mileage
1840	1 300
1850	5 000
1860	7 000
1870	10 200
1880	13 400
1890	14 000
1900	15 000

1900. To give some idea of the scale of this remarkable development, Table 2.1 shows the approximate cumulative route mileage of railway constructed in England and Wales during the nineteenth century.[1] The last British trunk railway, the Grand Central, promoted by Edward William Watkin (1819–1901) and built from Manchester to London but intended eventually to run on to Paris through a proposed Channel Tunnel (Watkin was also a Channel Tunnel enthusiast) was completed in 1899.

This remarkable expansion of a national network of railways was paralleled in every civilised country of the world so that by the end of the century it could be said that the interiors of each of these countries were interconnected by railway – sometimes via a multitude of ports – and no longer was commerce between nations restricted to cities and towns near the borders or the coast. For example, the startling growth of railways in the United States is shown in Table 2.2, which gives the approximate cumulative route mileage of railway constructed in the United States during the nineteenth century.[2]

The railways brought with them greater demands on the skill of the civil engineer than had the previous roads and canals. Whereas the coach roads had simply taken the easiest gradient up hill and down dale and the canals had wound around the contours, the railways had more stringent requirements for line and level requiring gradual curves and, more particularly, very shallow gradients. This meant that the engineers who designed the routes, when faced with hills or mountains, were increasingly called upon to build tunnels when even the expedient of deep cuttings

Table 2.2. *Route mileage of railway in the United States in the nineteenth century*

Year	Route mileage
1840	2 800
1850	9 000
1860	30 600
1870	52 900
1880	93 700
1890	159 300
1900	192 900

would not suffice. It is true that the canal engineers in the eighteenth century had built tunnels, notably some very long ones like the Standedge Tunnel on the Huddersfield Canal, over 3 miles long, which was started in 1794 and completed in 1811.[3] However, canal tunnels were constructed entirely by the traditional methods which had developed from mining practice, and because the innovations in tunnelling techniques considered here did not appear until the railway tunnels were constructed in the nineteenth century we will not be discussing canal tunnels any further; a full account of British canal tunnels has been given by P.K. Roberts.[4]

The first trunk railway lines were built with gradients limited to about 1 in 300 because of the caution of the early railway engineers on the question of the development of steam locomotive power.[5] The result of this was that in order to keep to shallow gradients when constructing a railway through hilly or mountainous terrain, cuttings and tunnels were frequently necessary. In particular, when the railways began extending towards the Alps in Europe and towards the Appalachians in America, the need for very long tunnels through hard rock became pressing and it was in response to this need that innovations in hard rock tunnelling were introduced. Two tunnels especially, the Hoosac Tunnel in the United States and the Mont Cenis Tunnel in Europe on the frontier between Italy and France – the first Alpine tunnel, are landmarks in the history of tunnelling, because they were sites where innovations were made that are of particular importance and we shall be returning to each of these tunnels several times in the course of this work.

2.2 The traditional method of hard rock tunnelling

Before the 1850s the method of driving tunnels in rock for roads, canals and the emerging railways was essentially the same method that had been used in mining and which had persisted virtually unchanged for the previous 200 years. The basic features of the method are illustrated in Figure 2.1. The heading was advanced by first drilling a number of holes (about 8–14) into the face to a depth of about 3 ft. A typical pattern of shotholes is shown in Figure 2.2. The drilling was done by one man holding the drill steel which was then struck by two other men each wielding an 8 lb sledge hammer called a 'double jack'; this method of drilling was therefore termed 'double jacking'. Sometimes holes were drilled by 'single jacking', in which the miner swung a 4 lb sledge hammer (a 'single jack') in one hand and held the drill steel in the other. From time to time the drill steel was withdrawn and the borings were raked out of the hole using a long tool called a 'spoon' or a 'scraper'. A drill steel would only drill about 1 ft before becoming blunt and a procession of men, or boys, carried sharp drill steels into the face and the blunt steels out to the blacksmith to be resharpened and tempered. Drilling rates for a single hole were about 1 ft per hour, and it was this slow progress that limited the rate at which the whole tunnel could be advanced. When the required number of holes had been drilled, all except sometimes a centre hole which was left empty, were charged with gunpowder (also called 'black

Figure 2.1 Tunnel driving in rock (*c* 1850)

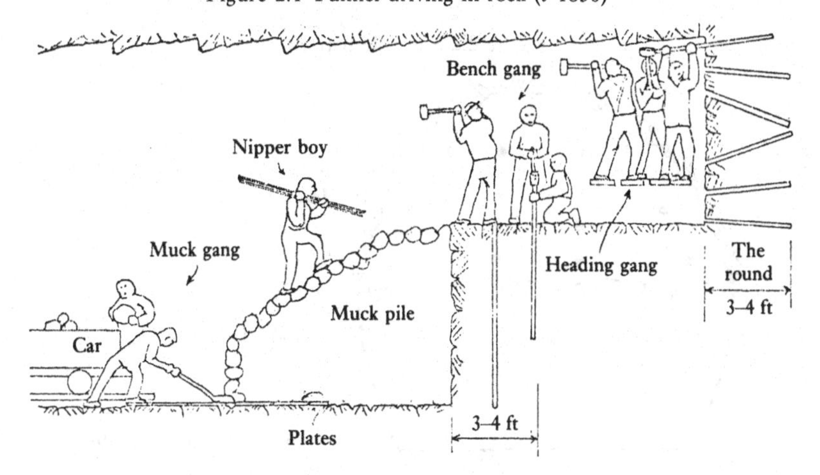

powder'). Fuses were inserted and the holes were then stemmed with clay or fine rock debris. The round was fired from a safe distance, and when the smoke had cleared away the miners returned to the face and loaded the broken rock into rail-mounted wagons or mine cars which when full were hauled by mules or horses to the tunnel portal or shaft bottom. If the completed tunnel was to be larger than the heading, a benching team followed behind, carrying out the same operation as the heading team, except that they drilled vertical holes in the floor of the already excavated heading (see Figure 2.1). Using this simple method, many road, canal and railway tunnels were driven through rock in Europe and North America but advance rates were very low, typically 40 ft per month, and the method was too slow for it to be a practicable proposition to drive long rock tunnels. Good descriptions of the early methods of tunnel driving are given by R.S. Mayo for American tunnels[6] and by G.E. Sandström for both European and American tunnels.[7]

Although, because it was so slow, hand drilling was bound eventually to be replaced by machine, it did reach a peak of refinement with the advent of specialised miners known as 'steel-driving men'. Nearly all books on the history of tunnelling tell of the most famous of these, the legendary American Negro John Henry, and of his remarkable exploit (14 ft in 35 minutes) in the driving of the Big Bend Tunnel for the Chesapeake and Ohio Railroad in 1870. Later on, towards the end of the nineteenth century rock drilling by hand developed into a sort of sport rather like wood chopping contests in present-day Australia, there being events for

Figure 2.2 Shothole pattern

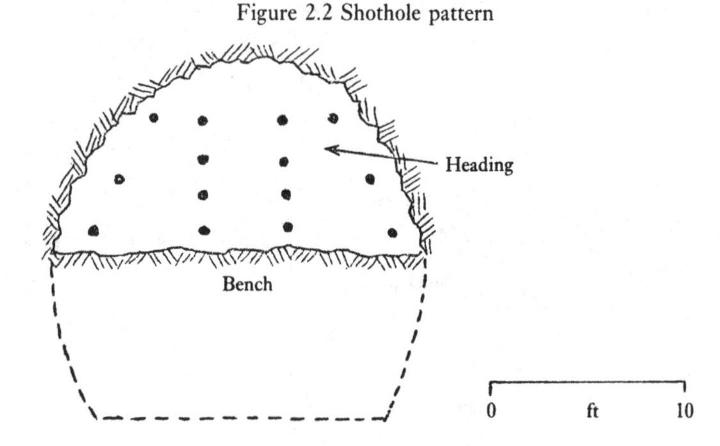

both double jacking and single jacking. It is said that no record was recognised unless the rock used was granite from Gunnison in Colorado which therefore enjoyed a vogue for this particular use, being shipped to the venues of important competitions.

From the 1850s onwards, tunnel driving in rock was revolutionised by a number of extremely important innovations. These were (i) the compressed air rock drilling machine, *c* 1860 (this Chapter), (ii) the use of nitroglycerine explosive, *c* 1860 (Chapter 3), (iii) the tungsten carbide drill bit, *c* 1950 (Chapter 4), and (iv) the hydraulic rock drilling machine, *c* 1980 (Chapter 5). The introduction of each of these measures had a great effect on the rates at which rock tunnels could be advanced. (The present-day rate is over 1000 ft per month.)

Looking at the method of driving rock tunnels as it was practised up to the 1850s, it is clear that the problem that was limiting the rate of advance was the operation of drilling the shotholes in the face of the heading. If this could be speeded up, then the whole rate of advance of the tunnel could be improved. Although the steam engine was available as a source of power for operating a rock drill, for, as will be described later, steam operated rock drills were invented for use out-of-doors in cuttings and quarries, a steam boiler could not be operated in the confined space of a tunnel under construction, nor could steam be piped for long distances into the tunnel. Furthermore, even if these difficulties had been overcome, the exhaust steam from the drills would soon have made the atmosphere at the working face intolerable. A solution to the problem was at hand in the suggestion made first by C. Brunton in 1844 that compressed air could be used for driving machine drills in tunnels. Brunton proposed his idea to alleviate the working conditions of the miners drilling shotholes in driving the deep levels of Cornish mines. His proposed machine was really a compressed air hammer consisting of a small cylinder of 3 or 4 in diameter whose piston would be a hammer of about 12 lb weight. Air was to be supplied in sufficient quantity to move the piston at 200 blows per minute, and the piston was to be applied to the end of an ordinary drill rod which was to be held and turned by hand in the traditional manner. Brunton intended mounting the machine on a piece of timber which would be braced across the heading in such a position that the hammer was aligned with the drill rod. Regarding the manufacture of the machine, Brunton says:

> Any mechanic well acquainted with the construction of the steam engine and slide valve, would be at no loss how to construct the cylinder and piston hammer, – the slide valve of which would, I think, be best operated upon by a boy, who would regulate the velocity, start and stop, under the direction of a miner.[8]

This extract clearly shows that Brunton's idea for a simple compressed air drilling machine was based directly on the steam engine cylinder and piston, and that the steam engine slide valve was the proposed method of regulating it. We have then, in this earliest conception of a primitive rock drilling machine, evidence that its origin stemmed from the steam engine. Returning to the idea that compressed air could be used to drive rock drills, it can be seen that this idea had two great advantages: the first was that because compressed air could be piped for long distances without much loss of power the compressors could be sited outside the tunnel well clear of the works, and the second was that the exhaust air from the drills would actually assist in ventilating the heading. However, it remained to produce rock drilling machines that would work with compressed air and the origin of these will now be described. Before doing so, a brief history and description will be given of the two tunnels, for the driving of which, the compressed air rock drill was invented.

2.3 The Hoosac Tunnel

The $4\frac{3}{4}$ mile long Hoosac Tunnel,[9] constructed through Hoosac Mountain in northwest Massachusetts, USA, between 1855 and 1876 is of great importance because of two crucial innovations: the compressed air rock drilling machine and the use of nitroglycerine explosive. The Hoosac Tunnel was required to complete a rail link between Boston in Massachusetts and Troy on the Hudson River in New York State (Figure 2.3). Construction commenced in 1855 and the tunnel was finished in 1876 after an extremely ill-fated construction period during which at least six contractors as well as the Commonwealth of Massachusetts itself were involved in the works and 200 men lost their lives. Construction was from both portals and from two shafts sunk along the tunnel line, one at the centre and the other near the west portal. The rock encountered for most of the tunnel drive was micaceous gneiss or schist but a harder and more coarsely textured granitoid gneiss was found in the west-central region.

Hard seams of quartz varying from a few inches to several feet in thickness occurred throughout.

Initially rock excavation on the Hoosac Tunnel was as follows: holes were drilled by the technique of single or double jacking as previously described, the hole positions being selected by the foreman based on the shape of the rock at the face and the positions of any joints in the rock. After drilling about ten holes the men were withdrawn behind timber screens, the holes were charged with gunpowder and the fuses lighted. The fuses used were probably Bickford's Safety Fuse, developed by William Bickford in 1831, consisting of a powder thread spun in jute yarn and water-proofed with coal tar.[10] During the blast the foreman would count the explosions to be sure that a hole was not 'hanging fire'. When the all-clear signal was given the men returned to the face and loaded the broken rock into mule cars. Drilling more blastholes then commenced and the cycle was repeated. In this way a heading 14 ft wide by 6 ft high was advanced which was later enlarged to the full single-track dimensions of 14 ft by 18 ft. (In 1862 it was decided to enlarge the tunnel to double-track.) The advance rate using this method was 50 ft per month, and based on this figure and assuming four drives, i.e. from both portals and a central shaft, it can be seen that excavation of the tunnel would have taken over ten years – clearly a more rapid rate of advance was desirable. In 1866 the first attempt at compressed air rock drilling at the Hoosac Tunnel was tried at the east end, and in 1866 the first experiments of blasting with

Figure 2.3 Location of the Hoosac Tunnel

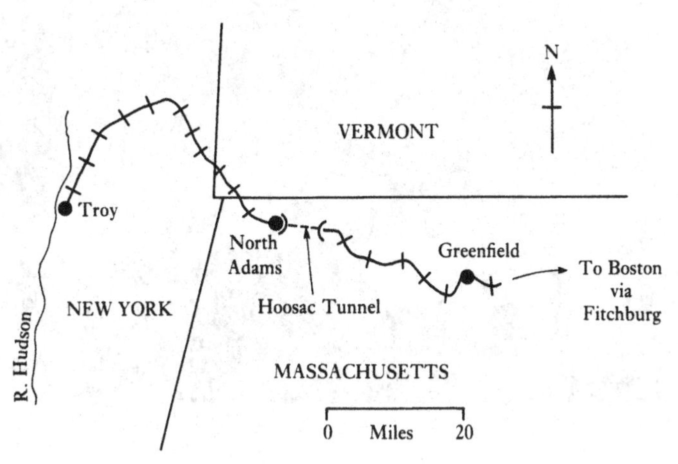

Figure 2.4 Excavation of the Hoosac Tunnel (1872)

nitroglycerine as the explosive took place at the east heading of the west shaft. When teething troubles had been overcome, these two innovations greatly increased the excavation rate and advances of up to 148 ft per month were achieved in 1872. In 1873 advance rates varied from 131 to 184 ft per month, and had this progress been available from the start the tunnel could have been excavated in just over three years. The fact that, in the event, the tunnel took 21 years to build was due to a series of accidents including numerous changes of contractor, fire, influx of water, alterations in tunnel design and extremely poor ground conditions at the west portal area. Section 2.5 describes in detail the invention of the compressed air rock drilling machine at the Hoosac Tunnel, and Chapter 3 will deal with the introduction of nitroglycerine blasting at the Hoosac Tunnel, both accounts concentrating attention on the origins of these innovations and the problems associated with them.

Figure 2.4 is a contemporary engraving of work going on in one of the headings of the Hoosac Tunnel. In the centre, rock drilling is taking place using the new compressed air rock drilling machines, four of which are shown mounted on the frame of a drill carriage for support. On either side of the heading, hand drilling is also taking place using the method of double jacking. Miners are also shown loading the spoil into a rail-mounted skip. The compressed air lines to the drills can be seen leading back out of the heading.

The excavation of the Hoosac Tunnel was completed by the end of 1874, but enlargement and lining of the tunnel took a further eighteen months, and the tunnel was not opened for rail traffic until July 1876. The Hoosac Tunnel is still in use today.

2.4 The Mont Cenis Tunnel

The $7\frac{1}{2}$ mile long Mont Cenis Tunnel,[11] constructed through the Alps on the border between France and Italy, between 1857 and 1871 is of comparable importance to the Hoosac Tunnel because of the independent invention there of the compressed air rock drilling machine. The Mont Cenis Tunnel was built to provide a rail connection between Chambéry in France and Turin in Italy (Figure 2.5). Construction commenced in 1857 at both portals and the tunnel was constructed entirely from both ends, there being no intermediate shafts to enable other headings to be driven as at Hoosac. The rock over the drive consisted of quartzite, limestone, dolomite and schist. Construction was completed in 1871 after a construction period relatively free from the troubles which beset the Hoosac Tunnel, for example in this case there were 28 fatalities compared with the 200 deaths associated with the driving of the American tunnel.

The Mont Cenis Tunnel construction commenced with the driving of a pilot heading 11 ft wide by 8 ft high. This was done at first by manual drilling. Holes were hand drilled to a depth ranging from $1\frac{1}{2}$ to 3 ft depending on the nature of the rock by single jacking. The holes were charged with gunpowder and ignited with a fuse. The pilot heading was then broken out to finished size. The advance rate obtained by this method during the early years was 23 ft per month. This was painfully slow – at this rate the tunnel would have taken over 71 years to complete. While this slow progress was being made, the Engineer for the tunnel and his assistants applied their minds to the problem of mechanising the rock drilling operation.

In 1857 when work on the Mont Cenis Tunnel started, the whole of the tunnel was situated in Piedmont and Savoy and these were then provinces of the Kingdom of Sardinia. In 1860, at the end of the Italian War of Liberation, Savoy was ceded to France and the French agreed to pay half

Figure 2.5 Location of the Mont Cenis Tunnel

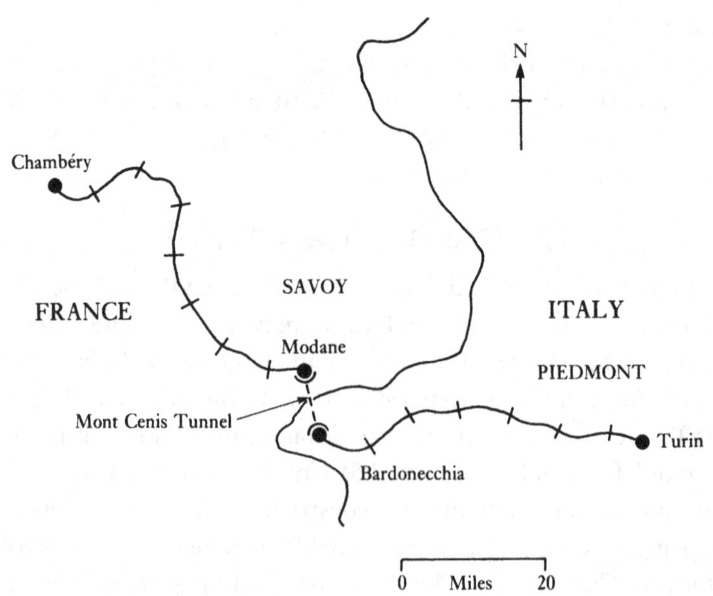

the cost of the tunnel. However, the Italians remained responsible for the construction.

No account of the Mont Cenis Tunnel would be complete without mention of its Engineer, Germano Sommeiller,[12] born on 15 February 1815 at Saint-Jeoire-en-Faucigny, a small parish of the Bonneville district in the Département of Haute-Savoie. Left an orphan at the age of 15, he was looked after by his sister who made sacrifices so that he could be educated at the college at Annecy which he entered in 1834. In October 1836 he went on to Turin University, where, several years later, he obtained his diploma of engineering. At the age of 29, not having found a position, he supported himself by giving mathematics lessons to young men preparing for a scientific education. He then joined the Army as a sub-lieutenant. At last, in 1845, he had an opportunity to prove himself when the Chief of Service of the Ministry of the Interior, who was charged with organising a Corps of Civil Engineers, offered him a place. Sommeiller's ability was noticed by his supervisors and in due course he was appointed to a post in the Transport Department. In 1846 the Sardinian Government decided to send several engineers to Belgium to study railway engineering, and Sommeiller was one of those selected. On his return to Piedmont in 1850, he was assigned to assist the engineer Henri-Joseph Maus working on the Turin–Chambéry railway line. In 1857 Sommeiller was appointed Engineer for the Mont Cenis Tunnel, the final plans for which were prepared by him and his second-in-command, Severino Grattoni. Sommeiller was responsible for the design of the rock drilling machines that were introduced into the tunnel in 1861. He died on 11 July 1871, just two months before the tunnel was ceremonially opened in September 1871.

By the middle 1860s when mechanised rock drilling in the Mont Cenis Tunnel had settled down to a routine operation the rate of advance averaged 231 ft per month. In contrast to the Hoosac Tunnel, gunpowder was used as the explosive for the whole drive even though dynamite was available towards the end of the period. Section 2.6 describes in detail the invention and development of the compressed air rock drilling machine used on the Mont Cenis Tunnel and discusses its origin and the problems of its development. The Mont Cenis Tunnel is still in use today.[13]

The first great transmontane railway tunnels like the Mont Cenis and the Hoosac dramatically improved communications, and can in that sense be said to have had the effect of shortening distance. For example, the second Alpine tunnel, the St Gotthard, reduced the travelling time between Germany and Italy by 36 hours. And after the railway systems of France and Italy had been linked by the Mont Cenis Tunnel the route to India via Suez was much shortened, the mail from England reaching Brindisi by rail in two days instead of the week taken by steamship.

2.5 Rock drilling machines at the Hoosac Tunnel

Before examining specific rock drilling machines, it is worth considering the three basic operations a successful rock drilling machine had to perform, since these gave rise to the design problems the inventors had to overcome. Firstly, the machine had to impart a succession of blows to the drill rod in much the same manner as the hammer blows in manual drilling. Secondly, the machine had to turn the drill rod slightly on its axis between each blow so as to prevent the drill rod from becoming jammed in the hole – this was done by the miner turning the rod by hand in the manual drilling method. Thirdly, the machine had to be provided with some means of advancing the drill rod progressively as the hole was drilled deeper – this was usually referred to as the 'feed' mechanism. In the accounts of particular rock drilling machines that follow, we shall see how the designers solved or attempted to solve these problems.

The first mechanical rock drilling machine in the United States was invented by J.J. Couch of Philadelphia in 1848. This was a steam driven machine that hurled a 'lance' at the rock and was not a practicable proposition, although Joseph W. Fowle, Couch's assistant, built a working model which drilled some holes in granite. It was patented in March 1849. Couch's drill consisted of a steam engine above which was mounted a large wooden frame. The engine drove a flywheel which, via a crank, propelled a gripper box backwards and forwards in a reciprocating motion. The drill rod was gripped on the backward stroke and hurled freely against the rock on the forward stroke. The drill rod was turned by means of a ratchet wheel having a hole in its centre through which the rod passed. The wheel had a stud which engaged in a groove cut in the drill rod, and this caused the rod to rotate with the wheel. Because the drill rod was freed from the gripper box at the end of the forward stroke, there was no need to provide any feed mechanism, the drill rod being free to follow the hole as it was drilled deeper. One serious disadvantage of this drilling machine was that the force of the blow that the drill rod could impart to the rock was, in practice, limited by the weight of the drill rod itself.

In 1849 Couch and Fowle separated and in May 1849 Fowle filed a caveat for an improved rock drilling machine having the following features (i) a drill rod directly attached to the crosshead of the engine, (ii) a simpler rotating mechanism for the drill rod, (iii) an automatic feed, (iv) an arrangement to prevent the piston from being driven through the head of the cylinder and (v) an S-shaped edge of the drill bit. Fowle's invention, finally patented in March 1851, was the precursor of all later American rock drills, and he testified as to the original invention of his drill as follows:

> My first idea of ever driving a rock drill by direct action came about in this way: I was sitting in my office one day, after my business had failed, and happening to take up an old steam cylinder, I unconsciously put it in my mouth and blew the rod in and out, using it to drive in some tacks with which some circulars were fastened to the wall. That was my first idea on the subject.[14]

It is important to note the difference between Couch's drill and Fowle's new idea. In Couch's drill the steam engine was merely a prime mover, providing power for a mechanical drill, whereas in Fowle's machine the piston of the steam engine was to drive the rock drill directly. In Fowle's drill the drill rod was connected directly to the crosshead of the steam engine and was, therefore, driven directly by the entire force of the steam engine's piston rod; a ratchet and pawl mechanism was used for rotating the dill rod. The feed system in Fowle's drill was provided by allowing the whole engine with its frame to move forward, impelled by the momentum of the crosshead. The feed motion was restrained by a chain running over sprocket wheels in front and behind. In a direct action drilling machine such as Fowle's, there was a danger of the piston hitting, or even being driven through, the end of the cylinder (see item (iv) above): Fowle anticipated this and obviated it in the mechanical design of the engine. The S-shaped cutting edge of the drill bit (item (v)) was a small improvement to cut well-rounded drill holes. When an Italian and French Commission came to the United States to see what ideas they could pick up for driving the Mont Cenis Tunnel they examined the models filed in the US Patent office, including the Fowle drill. Fowle's drill, like that of

Couch, was driven by steam, but in 1850 or 1851 Fowle used compressed air for driving his drill. Fowle, however, lacked the means to bring his drill into practical use and the ideas he had conceived lay in abeyance until taken up by Charles Burleigh (1824–83).

During the period that the rock drilling machine was being developed in the United States, an engineer, Charles Storer Storrow (1809–1904), made a trip to Europe for the Commonwealth of Massachusetts to visit the Mont Cenis Tunnel and investigate the compressed air rock drill which had been introduced there in 1861. He saw the Sommeiller drill, the drill carriage and noted the efficiency of compressed air as a power source. Storrow reported that although the Sommeiller drill had improved the rate of advance of tunnelling, the cost was still high because the drill required much maintenance. Storrow believed the Americans could develop a more efficient drill and the Burleigh drill was the result.

The first attempt at compressed air rock drilling at Hoosac was at the east end of the tunnel in the summer of 1866. The drill used, known as the Brooks, Gates and Burleigh drill, proved a failure because it lacked mechanical strength and was always breaking down. Burleigh made extensive observations on this machine and proceeded to invent another compressed air drill which incorporated all the best ideas of Fowle's drill, Burleigh having in 1865 purchased Fowle's patent for this purpose. Burleigh's improvements were (i) to dispense with the crosshead and attach the drill rod directly to the piston rod, and (ii) to rotate the drill rod by rotating the piston. Burleigh's drill is shown in Figure 2.6 in which the drill is depicted mounted on a small carriage frame for benching work. (For heading work it was mounted as shown in Figure 2.4.) Burleigh's first improvement was to get rid of the crosshead of the engine and attach the drill rod directly to the piston rod, a fairly obvious step once the principle of direct action had been accepted. It made for a simpler and more compact engine. The same can be said of his second improvement – with it a separate external mechanism for rotating the drill rod was no longer required. Burleigh's drill, like the Brooks, Gates and Burleigh drill before it, was a compressed air drill, the steam driven drill having been finally abandoned.

Two versions of the Burleigh drill were manufactured.[15] In the standard model automatic feed was provided, in which by rotation of a nut on the fixed screw at the rear of the drill, the cylinder was moved slowly forward. In the other model the automatic feed gear was omitted from the construction and hand feed was substituted. The latter version of the drill can be readily identified in contemporary drawings by the presence of a hand crank on the end of the screw (see Figure 2.6), this crank being

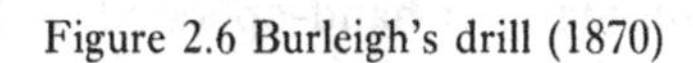

Figure 2.6 Burleigh's drill (1870)

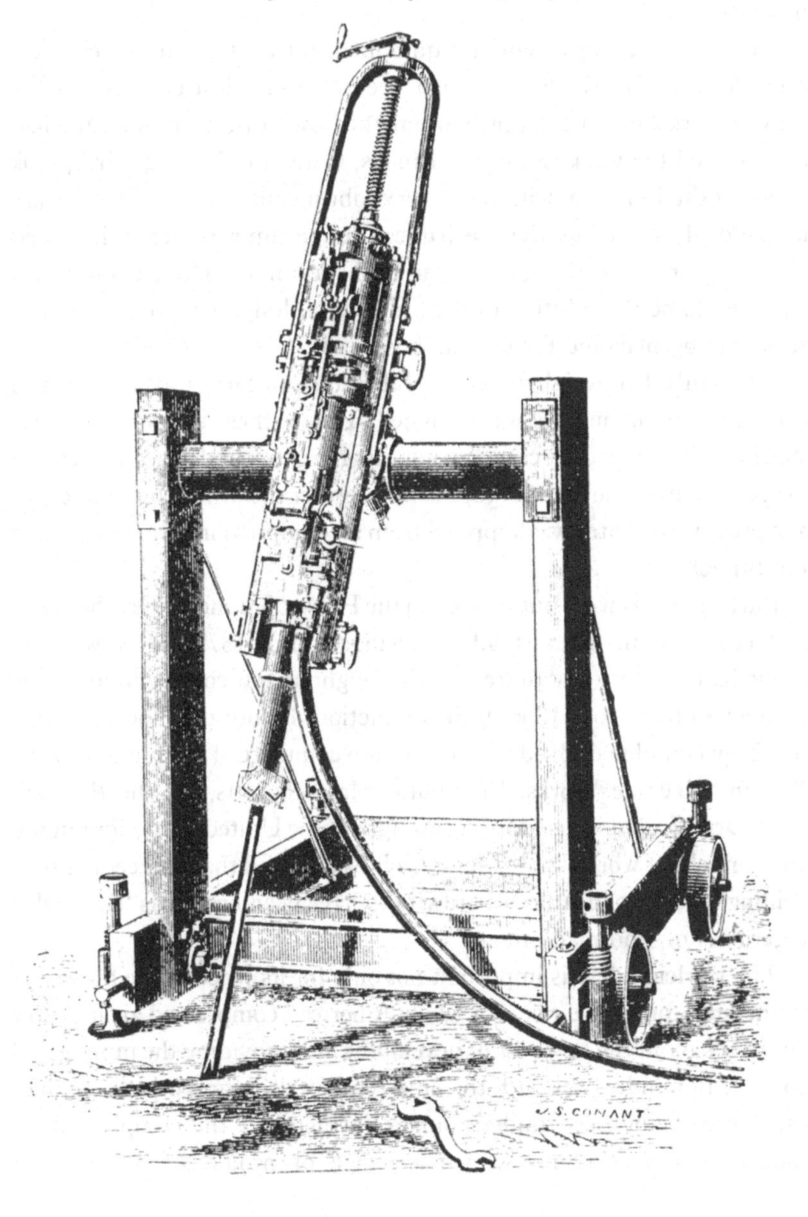

absent on the version with automatic feed (see Figure 2.4). Burleigh had found solutions to each of the three problems of drill design: in his drill compressed air acted on a piston attached directly to the drill rod, he rotated the drill rod by rotating the piston, and he provided screw-feed mechanisms (either automatic or hand-operated) in both versions of his machine.

Durability was an essential requirement for a successful mechanical rock drill, and Burleigh, who had been a mechanical engineer in the railway workshops at Fitchburg where the Fowle drill was made, and had experienced the weakness of the Brooks, Gates and Burleigh drill, took care that the Burleigh drill was of very robust construction. This feature of the drill, as well as the mechanical design improvements discussed above, ensured that when it was put to the test in the Hoosac Tunnel, it was seen to be the solution to the problem of designing a fully practical rock drilling machine for use underground.

Burleigh's drill weighed over 500 lb and so to carry the drills, a drill carriage running on rails was designed. Two of these operated side-by-side in the heading, each carriage mounting two to four drills. One of these drill carriages is shown in Figure 2.4. To power the drills, compressed air at a pressure of 4 atm was supplied from air compressors situated outside the tunnel.

Burleigh drills were put to work in the Hoosac Tunnel in October 1866 and remained in use virtually unchanged until 1872. They were an immediate success and were used throughout the construction of the Hoosac Tunnel from 1866 until completion. Before the Hoosac tunnel had been completed, the drill was put into commercial production by the Putnam Machine Works, Fitchburg, Massachusetts, for the Burleigh Rock Drill Co and its use spread throughout the United States for mining and tunnelling work. It became so well-established that succeeding rock drilling machines in America are said to have been known as 'Burleighs' regardless of who made them.[16]

The Burleigh drill is important not only for its place in the history of tunnelling, for it gave rise to a whole American compressed air industry which set about the business of developing and providing the mining and construction industries with the only types of drills and other machines capable of being used underground. Furthermore the compressed air industry soon developed a whole range of products for civil and mechanical engineering quite outside of tunnelling – an example of what today would be called 'spin-off'.

2.6 Rock drilling machines at the Mont Cenis Tunnel

The story of rock drilling machines at Mont Cenis begins with an Englishman, Thomas Bartlett. Bartlett was an engineer who at this time was the resident agent in Savoy of the large English civil engineering contracting firm, Thomas Brassey and Co.[17] In 1854 he had designed a small portable steam driven drilling machine for boring holes in rock. After trials at Brighton the machines were taken to Savoy and used in 1855 for excavating cuttings for the railway being constructed between Chambéry and St Jean de Maurienne. The drill was exhibited at work to a group which included the three engineers who were to build the Mont Cenis Tunnel, Someiller, Grattoni and Sebastian Grandis. For this demonstration steam was employed as the motive power. The Commissioners appointed by the Piedmontese Government reported after inspecting the drill that they had seen a machine at work, which satisfied them that the Mont Cenis tunnel could be made in a comparatively short time, and at moderate expense. In 1856 two Bartlett machines were sent to St Pierre d'Avena near Genoa for trials before an Italian Commission, compressed air instead of steam being used to run them. The Bartlett drill was judged to be somewhat inconvenient to handle, Colonel Menabrea, a member of the Commission summing up as in the following abridged account:

> The mechanism of Bartlett's drill worked equally well with compressed air as with steam. But being designed to work with steam the machine is more complicated than it needs to work with compressed air. The machine has to be regulated by hand which was a defect. M Sommeiller proposed a simpler drill which would be self-acting.[18]

Before examining Bartlett's drill a brief biographical note will be given of its inventor Thomas Bartlett[19] who was born on 7 July 1818. He was engaged by Thomas Brassey and Co to work on the construction of the South Eastern Railway and the Great Northern Railway, on completion of which he was entrusted with the works in Savoy already referred to, and for which he invented and developed his steam driven drilling machine. He then went to Spain and was working on the Bilbao Railway when he

Figure 2.7 Bartlett's drill (1855)

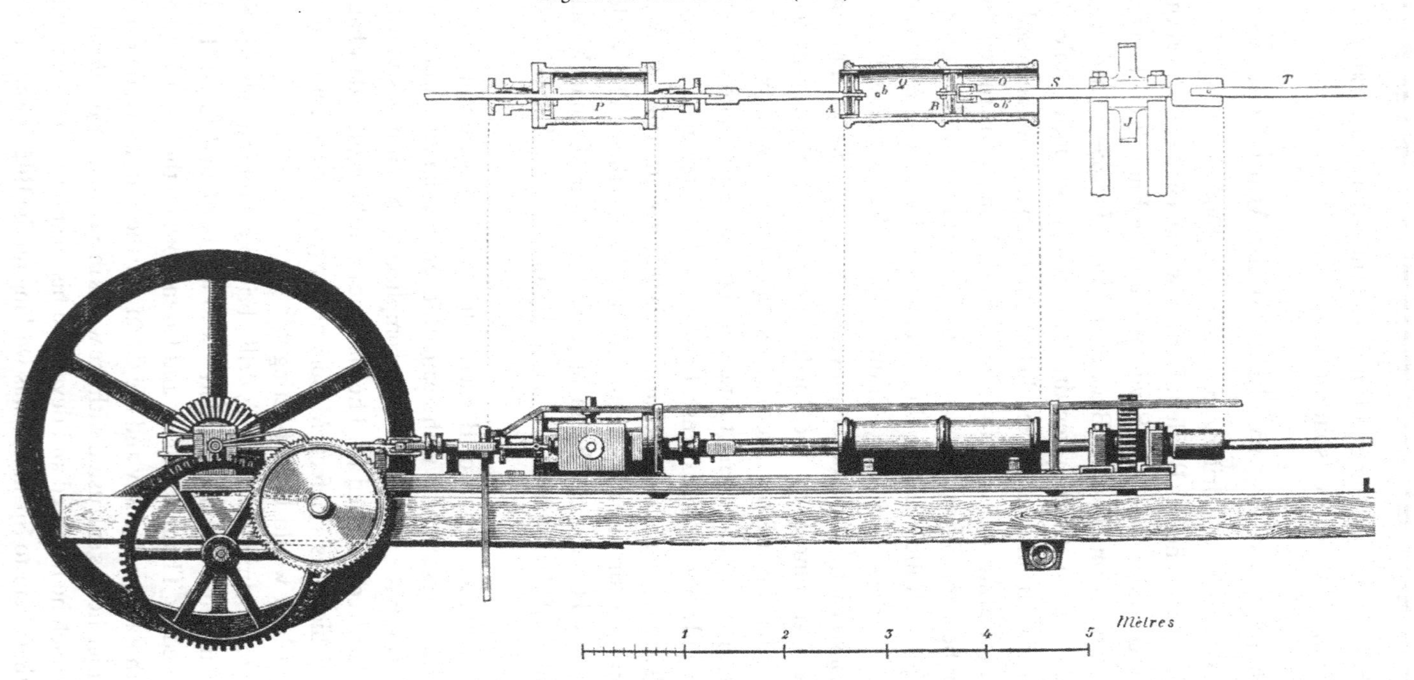

suddenly died, aged 46, on 23 July 1864, leaving seven children unprovided for, his wife having died the year before.[20]

Bartlett's steam driven rock drilling machine is shown in Figure 2.7. Bartlett's drill consisted essentially of two cylinders, mounted one behind the other on the same frame. The first cylinder *P* was part of a steam engine powered by steam supplied from a boiler located nearby. The second cylinder *QO* was a chamber having two pistons *A* and *B*. When the steam engine was in operation, it caused piston *A* to be moved forwards and backwards in a reciprocating motion. On the forward stroke, piston *A* compressed air in cylinder *QO* which drove piston *B* forward. Attached to piston *B* was a rod *S*, to which the chisel-ended drill rod *T* was fixed. Rotation of the drill rod was accomplished by the toothed wheel *J* which was driven by the engine via a train of gears. As has been noted, if the cylinder of the steam engine was supplied with compressed air instead of steam, the drill would still operate successfully; however, operated in this way the mechanism was more complicated than it needed to be, the second cylinder appearing to be unnecessary. It can be seen that Bartlett had solved two of the problems facing a rock drill designer, those of arranging for a method of applying blows to the drill rod and of making the drill rod rotate between blows. It has not been possible, from the drawing or from contemporary descriptions, to establish whether or not Bartlett provided a feed mechanism for his drill but from the remarks made by Menabrea quoted earlier, it would seem that the drill had a hand-regulated feed and not an automatic one.

Having observed Bartlett's drill and noted the fact that when it was running on compressed air instead of steam the second cylinder seemed to be superfluous, Sommeiller set himself the task of designing a drill in which the compressed air would act directly on the boring rod. Sommeiller patented his drill design in December 1858 and the first drill was put to work at the beginning of 1861 at the south end of the Mont Cenis Tunnel. The Sommeiller drill possessed the following features: (i) it imparted a sudden hard blow to the drill rod and withdrew the rod after each stroke, (ii) it had an automatic advance as the hole increased in depth, (iii) it had a means of getting rid of the borings, and (iv) it turned the drill rod slightly on its axis at each stroke. Sommeiller's drill is shown in Figure 2.8. The main mechanism of the drill consisted of a large cylinder in which the piston was directly coupled to the end of the drill rod. Compressed air

Figure 2.8 Sommeiller's drill (1863)

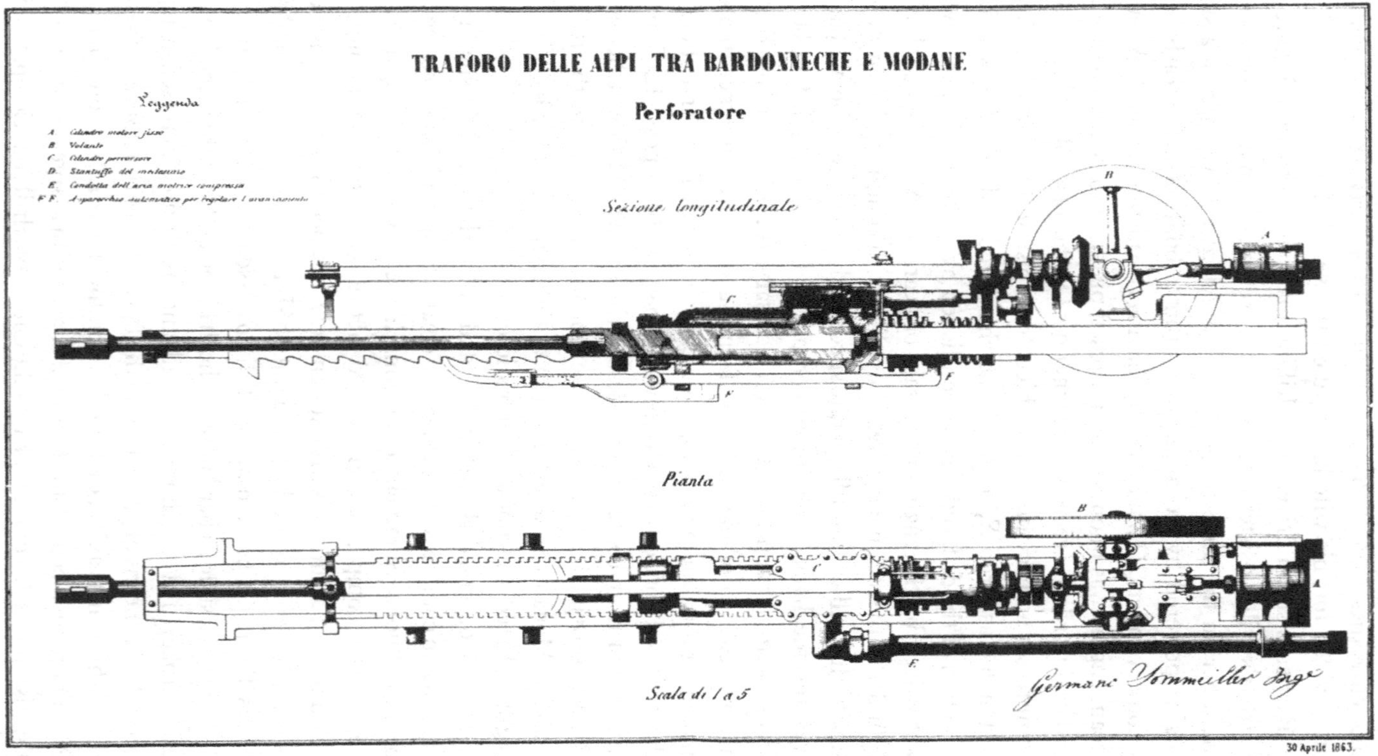

was supplied directly to this cylinder. A small fixed rotary engine, also driven by compressed air, operated the valve mechanism of the large cylinder, advanced and withdrew the large cylinder on the frame of the drill, and gave the rotary motion to the drill rod. In Figure 2.8 the large cylinder is labelled *C* and the small rotary engine *A*. The flywheel *B* could be used as a handwheel for manual control of the valve mechanism instead of using the rotary engine if required. Pipe *E* was the main compressed air supply to the large cylinder of the drill. Sommeiller had solved all three problems facing the rock drilling machine designer. His was a direct action drill in which compressed air acted on a piston (*D*) directly coupled to the drill rod. The small engine gave the desired slight rotary motion to the drill rod, and there was an automatic feed system that operated on the frame of the drill, being regulated by the ratchet and pawl *FF*. Sommeiller's drill had one further innovatory feature, namely a method of removing the borings from the drill hole (see item (iii) above) by means of a jet of water, the supply for which came from wrought-iron water tanks mounted in a truck behind the drill carriage (see below). It can be noted that the facility for powered withdrawal of the large cylinder on the frame of the machine allowed for rapid changing of the drill rod when it had become blunt – a most desirable feature in those days when the edge of drill steels became rapidly lost. (This point will be taken up again in Chapter 4.)

Each drill weighed 670 lb, so to support them and to provide the necessary reaction for drilling, a rail-mounted iron drill carriage was provided. The carriage weighed 15 tons and moved on rails which were laid up to the face of the heading. Up to eleven drills could be mounted on the drill carriage which also contained the compressed air connections for the drills and a large air reservoir. Compressed air was supplied individually to each drill by means of 2 in diameter flexible tubes. An engraving of the drill carriage in action is shown in Figure 2.9; in contemporary accounts it was often referred to as the 'gun carriage'. Behind the carriage was towed a truck containing water supply tanks for the water jets used to cool the drill bits and flush out the borings: this arrangement is shown in Figure 2.10 and is a detail not depicted in other contemporary drawings of the Sommeiller carriage. Compressed air for the drills was supplied at a pressure of 5 atm from compressors outside the tunnel and conveyed down the tunnel in $7\frac{5}{8}$ in diameter pipes to the

Figure 2.9 Carriage for Sommeiller drills (1863)

Figure 2.10 Water supply for Sommeiller drills (1863)

reservoir on the drill carriage; the loss in power in transmitting compressed air from the compressors to the drills was said to be only one-sixtieth.[21] After 8–10 shifts, say after drilling 60–80 holes, a drilling machine needed overhauling, four machines being kept in reserve for each one working.

The Sommeiller drill underwent considerable development during the construction period of the Mont Cenis Tunnel, there being at least two further surviving sets of engineering drawings of the Sommeiller 'perforator' as it was called. In the last of these, the characteristic handwheel has disappeared.

In the same way that the Burleigh drill was of interest beyond Hoosac, so Sommeiller's drill attracted attention from outside the Mont Cenis project. For example, as early as 1863 the Belgians investigated the drill with a view to using it in their coal-mining industry.[22]

2.7 Summary

We have seen that the compressed air rock drill was invented in Europe and the United States at almost the same time. In each case the need was the same – to find a solution to the problem of very slow manual drilling, and in each case the application was the same – for a projected long railway tunnel through hard rock. The same need, although in different continents, produced the same innovative solution. However, the inventions cannot be regarded as completely independent, because we have noted that the Italian and French Commission had examined the patent for Fowle's drill and that the Commonwealth of Massachusetts had investigated Sommeiller's drill. Nevertheless, there seems to be a clear line of innovation and development on each side of the Atlantic and these visits probably did no more than give the engineers confidence that they were working along the right lines.

In both Europe and the United States the compressed air rock drill had the same precursor – the steam driven rock drill. At the Hoosac Tunnel the precursors were Couch's and Fowle's drills and at the Mont Cenis Tunnel the precursor was Bartlett's drill. Furthermore we have Fowle's own words about the old steam cylinder giving him the idea for a direct action rock drill. In addition, internal evidence from an examination of the surviving engineering drawings of Sommeiller's drill shows that the valve mechanism for operating the main cylinder is very similar to the slide valve of a steam engine. It is therefore clear that nearly all the ideas for the compressed air rock drill had their origin in the technology of the steam engine. And we know this is also true of Brunton's simple idea for an air hammer which predated these drills.

Another point of similarity may be noted. Both the Sommeiller and the Burleigh drills were too heavy for a man to hold and operate so that some kind of support was necessary. At both tunnels the response was similar – to design a rail-mounted drill carriage. The rails kept the drills square on to the face as well as providing support.

The Burleigh and Sommeiller drills were the first American and European compressed air rock drilling machines respectively that were used successfully to drive tunnels. In both continents they were the first in a long succession of rock drills[23] which developed as compressed air technology and mechanical engineering improved. We shall see in Chapter 5 that the compressed air rock drilling machine was developed to the limit of technical efficiency and was then superseded by a further innovation – the hydraulic rock drilling machine.

Notes and references for Chapter 2

1 Morgan, B. (1973) *Railways: civil engineering.* London (Arrow Books Ltd) p. 138.

2 Thompson, S. (1925) *A short history of American railways.* Chicago (Bureau of Railway News and Statistics) Addenda B.

3 Rolt, L.T.C. (1973) *Navigable waterways.* London (Arrow Books Ltd) pp. 61–2.

4 Roberts, P.K. (1977) British canal tunnels (a geographical study). 2 volumes, *PhD thesis,* University of Salford.

5 Morgan, B. (1973) *op cit,* pp. 53–4.

6 Mayo, R.S. (1977) Early American tunnels. *Transportation Research News,* (72), pp. 2–6.

7 Sandström, G.E. (1963) *The history of tunnelling.* London (Barrie and Rockliff) *passim.*

8 Brunton, C. (1844) Design of wind hammer for boring rocks. *Mechanics Magazine,* **41**, 21 September, pp. 203–4.

9 The sources mainly used for this account of the construction of the Hoosac Tunnel, including rock drilling, are:
Drinker, H.S. (1893) *Tunnelling, explosive compounds and rock drills.* New York (John Wiley and Sons) 3rd edition, pp. 315–37.
Brierley, G.S. (1976) Construction of the Hoosac Tunnel 1855 to 1876. *Journal of the Boston Society of Civil Engineers Section, American Society of Civil Engineers,* **63**, (3), pp. 175–209.
Sandström, G.E. (1963) *op cit,* pp. 152–60.
Walker, F.N. and G.C. Walker, Editors (1957) *Daylight through the mountain.*

Montreal (The Engineering Institute of Canada).

10 It has not been possible to establish this fact. Mayo (Reference 6) says that Bickford's Safety Fuse, an English invention, was not widely used in early American tunnels, the miners making their own fuses from reeds, straw, goose quills or small paper tubes filled with gunpowder.

11 The sources mainly used for this account of the construction of the Mont Cenis Tunnel, including rock drilling, are:
Sopwith, T. (1864) The actual state of the works on the Mont Cenis Tunnel, and description of the machinery employed. *Minutes of Proceedings of the Institution of Civil Engineers*, **23**, pp. 258–319.
Sopwith, T. (1873) The Mont Cenis Tunnel. *Minutes of Proceedings of the Institution of Civil Engineers*, **36**, pp. 1–34.
Drinker, H.S. (1893) *op cit*, pp. 352–59.
Sandström, G.E. (1963) *op cit*, pp. 132–52.
Duluc, A. (1952) *Le Mont Cenis sa route, son tunnel.* Paris (Hermann et Cie).
Anon (1867) The Mont Cenis Tunnel. *Engineering*, 3, 12 April, pp. 329–30.

12 Duluc, A. (1952) *op cit*, pp. 53–4.

13 The Mont Cenis Tunnel is so called because it took its name from the old Mont Cenis road across the Alps, 15 miles away, that it replaced. Nowadays it is known as the Fréjus Rail Tunnel after the much closer Col de Fréjus. There is also a Fréjus Road Tunnel, completed in 1979, alongside. In this book the original name is retained.

14 Drinker, H.S. (1893) *op cit*, pp. 206–7.

15 André, G.G. (1876) *A practical treatise on coal mining*. Volume 1. London (E. and F.N. Spon) p. 159.

16 Vogel, R.M. (1964) Tunnel engineering – a museum treatment. *Contributions from the Museum of History and Technology*, Paper 41. Washington DC (Smithsonian Institution) pp. 203–39.

17 Helps, A. (1872) *Life and labours of Mr Brassey*. London (Bell and Daldy) pp. 178–81.
Walker, C. (1969) *Thomas Brassey: railway builder*. London (Frederick Muller) pp. 151–2.
Thomas Brassey (1805–70) was a notable British nineteenth-century railway contractor who built many lines abroad as well as at home. He partly financed Bartlett's steam drill.

18 Sopwith, T. (1864) *op cit*, p. 288.

19 Anon (1864–5) Memoirs. Mr Thomas Bartlett. *Minutes of Proceedings of the Institution of Civil Engineers*, **24**, pp. 526–7.

20 This tragic circumstance led directly to the founding of the Institution of Civil Engineers Benevolent Fund in 1864.

21 Anon (1869) The new overland route and the railway tunnel of the Alps. *The Illustrated London News*, **54**, (1525) pp. 152–3 and 166.

22 Devillez, A. (1863) *Des travaux de percement du tunnel sous les Alpes et de l'emploi des machines dans l'interieur des mines.* Liége (F. Renard) pp. 26–41 and 247–59.

23 Some examples are given in:
Hiscox, G.D. (1905) *Compressed air: its production, uses and applications.* London (Archibald Constable and Co Ltd) 4th edition, Chapter 22.

Nitroglycerine explosive

It is easy! Give us dynamite and drills!

The second major innovation in hard rock tunnelling was the introduction of nitroglycerine as the explosive in place of gunpowder. A general description of drill-and-blast operations using gunpowder was given in Section 2.2, but before going on to the introduction of nitroglycerine, some further details of how gunpowder was used for blasting will be considered. The following description is taken from the account given by D.K. Clark[1] of the method used in the Mont Cenis Tunnel. The holes to be charged were about $1\frac{1}{2}$ in in diameter and usually 3 ft long. The hole was thoroughly cleared of borings and into it was pushed a long cartridge made from a cylinder of paper containing the gunpowder, fuse and tamping. The tamping consisted of small pieces of broken stone or clean dry sand. The charge of gunpowder varied from one-third to one-half the length of the hole and weighed from $\frac{3}{4}$ to $1\frac{1}{2}$ lb. Between the gunpowder and the tamping was placed a wooden cone with its base towards the gunpowder, the gunpowder end of the cartridge, of course, being at the bottom of the hole. When the gunpowder was fired, the explosion acted on the base of the cone which then forced the tamping hard against the side of the hole. The cone, thus wedged in, presented a fixed surface against which the explosive force of the gunpowder reacted. The gunpowder used for blasting rock in tunnels, mines and quarries was made of a finely powdered intimate mixture consisting of 72 per cent saltpetre (potassium nitrate), 11 per cent sulphur and 17 per cent charcoal (carbon). This recipe, although with differing proportions of the three ingredients, had been known for about 500 years. Gunpowder was exploded by ignition with a flame, applied by means of a fuse. Gunpowder was a reasonably effective explosive for tunnelling but it had two disadvantages. These were firstly that it produced a large quantity of smoke and fumes which had to be dispersed before the miners could return to the face of the heading, and secondly that because of its limited explosive power a large number of shotholes had to be drilled in the face. It was clear that it would be of great benefit if an explosive of higher power than gunpowder could be found, particularly if it were one which would also produce less smoke and fumes. It so happened that the explosive nitroglycerine was available to meet this need at just the time when it was required, and the origins of the use of nitroglycerine as a tunnelling explosive will be discussed in this Chapter together with the problems that had to be solved before it could be safely used.

3.1 Discovery of nitroglycerine

The discoverer of nitroglycerine, Ascanio Sobrero,[2] was born at Casale, Piedmont, Italy on 12 October 1812. He originally qualified as a physician and surgeon, but later studied chemistry in Turin, Paris and Giessen. He returned to Turin and in 1849 was appointed Professor of Applied Chemistry in the Technical Institute. In 1860 he was appointed Professor of Pure Chemistry as well. In 1882 Sobrero retired; he died in Turin on 26 May 1888.

In Paris Sobrero had been the pupil of the French chemist Théophile-Jules Pelouze (1807–67), who in 1836 began to study the effect of nitric acid on organic substances, particularly cellulose. Sobrero had assisted him in this work and when he returned to Turin, he acquired a laboratory where he was able to continue with research along the same lines. Sobrero commenced a study of the effect of nitric acid on sugar, mannite (an alcohol that can be obtained from manna) and glycerine. As an ominous portent, Sobrero found that the effect of nitric acid on mannite was to produce a substance that could be violently detonated by a hammer blow.[3] In 1846 Sobrero turned his attention to the effect of nitric acid on glycerine and thereby discovered one of the most powerful explosives known – nitroglycerine.

Sobrero's method of making nitroglycerine was as follows.[4] Glycerine, concentrated to a syrupy condition, was added, drop by drop, to a mixture of two volumes of concentrated sulphuric acid (1.84 specific gravity) and one volume of concentrated nitric acid (1.50 specific gravity), the whole being kept well-cooled. The glycerine, in these circumstances, dissolved in the mixture of acids without noticeable reaction, but on pouring the solution into distilled water, an oily substance separated out and collected at the bottom of the vessel – this was nitroglycerine. Sobrero's initial observations on its properties were physiological: a tiny quantity put on the tongue produced a severe headache and a small quantity administered to a dog killed it. But he also observed that on heating nitroglycerine it exploded. It was later observed that it could also be detonated by a slight shock. Sobrero at first called the substance he had discovered 'piroglycerina' because of its fiery nature, but later changed the name to nitroglycerine. In his report to the Turin Academy[5] he cautioned that the substance was a violent explosive, and in fact it was so unpredictably dangerous to handle that Sobrero ceased working on it.

Nitroglycerine is an odourless yellow oily liquid of specific gravity 1.6 with a sweet burning taste. It freezes at about 8–10 °C, a fact that was to be of great importance in its safe handling as we shall later see. The chemical formula of nitroglycerine is $C_3H_5O_3(NO_2)_3$. In spite of its early recognition as a powerful explosive, for many years nitroglycerine had no use other than as a heart medicine, for it had been found that it caused dilation of the arteries and so was used for the treatment of *angina pectoris*, usually in the form of a dilute solution in alcohol.[6]

The reason why nitroglycerine was not used as an explosive was because it was so sensitive to detonation by shock. As an example of this, H.S. Drinker[7] described the explosion of a can of nitroglycerine at Yonkers, New York, which was caused when it was struck by a stone thrown by a boy. This property of nitroglycerine effectively prevented its industrial manufacture or use until 1863.

3.2 Manufacture of nitroglycerine

The first person to try to manufacture nitroglycerine was Alfred Bernhard Nobel, later to become famous as inventor and philanthropist.[8] Nobel was born at Stockholm, Sweden on 21 October 1833, the third of four sons of Immanuel and Caroline Nobel. As a youth he was sent for training as a mechanical engineer to the United States for four years, returning to Europe in 1854. During the late 1850s he commenced his long study of explosives, especially nitroglycerine, which is the aspect of his life that concerns us here. He died on 10 December 1896 at San Remo, Italy. Nobel acquired a large fortune from the manufacture of explosives and from the exploitation of the Baku oil fields: on his death he left the bulk of this to endow the now famous Nobel prizes for peace, physics, chemistry, physiology/medicine and literature.

Nobel was greatly impressed by the explosive power of Sobrero's nitroglycerine and was quick to see the huge commercial potential if it could be manufactured and sold as a blasting agent. Nobel commenced the manufacture of nitroglycerine at Heleneborg, near Stockholm in Sweden, in 1863 in a laboratory he shared with his father. In spite of an explosion here in 1864 which killed his younger brother and four workmen, Nobel persisted with the manufacture of nitroglycerine and established a further nitroglycerine factory at Kümmel on the River Elbe near Hamburg in Germany in 1865. Between 1865 and 1866, however, there were a series of spectacular and devastating explosions both on board ship and on shore during the transport of nitroglycerine.[9] International reaction was swift, most countries introducing laws forbidding or restricting the transport of nitroglycerine. For example, the United States Congress passed a law making infringement of the anti-nitroglycerine restrictions on interstate commerce a capital offence.[10] There were two ways out of this dilemma, either to manufacture nitroglycerine on site close to the place at which it was to be used – this was the solution adopted at the Hoosac Tunnel, or to find a means of making nitroglycerine safe to transport and handle – this was the solution that Nobel set out to find.

Also, nitroglycerine could only be used as an industrial explosive if a reliable means of causing it to explode when required was available. We have seen that gunpowder can be exploded simply by igniting it with a flame, but this was not the case for nitroglycerine which had to be detonated by a shock. Nobel realised nitroglycerine could not be marketed until there was an effective means of detonating it, and he applied himself to providing this need. In 1864 he introduced his mercury fulminate detonator which made possible the industrial use of nitroglycerine.[11] The mercury fulminate detonator, itself set off with an electric current, produced the sudden shock which in turn set off the nitroglycerine. The detonator, then, can be regarded as a necessary adjunct to the innovation of nitroglycerine itself.

3.3 Nitroglycerine at the Hoosac Tunnel

As was mentioned in Section 2.3, blasting at the Hoosac Tunnel was initially carried out with gunpowder. When the headings had advanced some distance from the portals, the problem of exhausting the large volumes of noxious fumes produced by the gunpowder blasts became more and more pressing. This was the main stimulus that led to the search for a replacement explosive that would be smokeless, but the need for a more powerful explosive to keep pace with the increased speed of machine drilling was an added incentive. For these reasons, in 1866 Thomas Doane (1812–97), then Chief Engineer of the Hoosac Tunnel, invited Nobel's American representative, Colonel Taliaferro Preston Schaffner, to conduct the first blasting experiments with nitroglycerine in the east heading from the west shaft of the Hoosac Tunnel. These were successful and in 1868 George Mordey Mowbray (1814–91) came to North Adams, Massachusetts to set up a nitroglycerine manufacturing plant near the western portal of the tunnel. Production of nitroglycerine at Mowbray's factory commenced in 1868 and continued throughout construction of the tunnel; more than 500 000 lb of nitroglycerine was made but not all was used in the Hoosac tunnel.[12] Mowbray had previously gained experience with the use of nitroglycerine in the Titusville, Pennsylvania oil field and this had extended to its manufacture.

Mowbray's method of manufacturing nitroglycerine – essentially that of Sobrero – was by adding glycerine, drop by drop, to a mixture of nitric and sulphuric acids. During the process the solution was agitated by a stream of air bubbles and cooled by being bathed in ice. After manufacture, the nitroglycerine was cleansed of impurities which were skimmed from the surface during periods of storage, first at 60–70 °F, and then after freezing. This process was patented by Mowbray in 1868. A contemporary visitor has given an account of Mowbray's nitroglycerine factory.[13] The dominant feature was a 50 ft long trough filled with ice and salt and containing a large number of 1 gallon earthenware jars; these jars contained the mixture of nitric and sulphuric acids. Inverted cans, each holding about 2 pints of glycerine, were positioned 2 ft above the jars, allowing the glycerine to fall slowly into the jars, drop by drop. The acids were agitated by a stream of cold air and by men stirring the mixture with glass rods. When the reaction was complete the contents of the jars were emptied into a reservoir holding 40 gallons of water in order to allow the nitroglycerine to separate out and to wash away any remaining traces of acid. The nitroglycerine was then stored in glass or earthenware jars, it being remarked that on the occasion of the visit there were some 1000 lb of nitroglycerine in store, in jars containing 3–5 gallons each. It was noted that 42 lb of glycerine made 94 lb of nitroglycerine.

For use in blasting, the nitroglycerine was put into metal tubes which were 1–1½ in in diameter and 4–6 ft long. These were placed in the drill holes together with the detonators and wires for electrical ignition. Mowbray laid down stringent rules for handling nitroglycerine, but even slight carelessness in following these could cause a premature detonation resulting in injury or death to the miners. Nitroglycerine was able to fracture harder rock and more rock than gunpowder and produced far less smoke and fumes. To give some idea of the relative power of the two explosives, 1 g of nitroglycerine on explosion produces 2000 cm^3 of gas while 1 g of gunpowder produces 210 cm^3 of gas – on this basis nitroglycerine is almost ten times as powerful.[14] If the effects of the temperature and pressure of the released gas are taken into account together with the rapidity with which the gases are generated, then other estimates put the power of nitroglycerine as even higher than ten times that of gunpowder.

A minor innovation that was introduced at the Hoosac Tunnel was the production there by Charles A. Browne and Isaac S. Browne of the electrical blasting cap for detonating the new nitroglycerine explosive. The electrical blasting cap (also called an 'exploder') was a small wooden plug fitted with two wires between which was a small amount of copper fulminate. This plug was inserted into an outer casing containing mercury fulminate, the whole assembly comprising the blasting cap. In fact, the principle is the same as that of Nobel's detonator – passing an electrical current through the wires sets off the blasting cap. The Browne brothers set up a small factory near the western portal of the tunnel and as tunnel construction reached its peak, up to 30 000 blasting caps a month were produced there.

We have seen that the construction of the Hoosac Tunnel gave rise to at least three ancillary industrial services: the manufacture of compressed air rock drilling machines (Section 2.5), Mowbray's nitroglycerine factory and the small factory for making blasting caps. This is an example of an aspect of tunnel construction which needs to receive proper consideration by those who weigh up the pros and cons for the construction of major tunnels on economic grounds. For the tunnel is not only of lasting benefit to the community as a whole (the Hoosac and Mont Cenis tunnels have been in service for well over a century now), and gives useful employment to many engineers, miners and others during its construction, but also gives rise to industrial undertakings which are established to provide the tunnel contractor with plant and materials; these firms are often able to become independent and prosper after the tunnel has been completed.

The effects of replacing gunpowder with nitroglycerine at the Hoosac Tunnel were striking. Its greater power required fewer shotholes to be drilled over the same area of working face, and it was also capable of blasting out rock from a deeper hole so that some 40 per cent more length of heading was excavated per cycle of operations.[15]

However, no matter how good nitroglycerine was as a tunnel explosive something had to be done about the great danger of handling it and transporting it, and it is an extraordinary fact that the solution to this problem was found quite literally by accident. The discovery of how nitroglycerine could be safely handled occurred in the following way. The account which follows is based on that given by Drinker,[16] and is corroborated by the historian of the Hoosac Tunnel, G.S. Brierley.[17]

During the winter of 1867–8, when the manufacture of nitroglycerine at North Adams had just commenced, William P. Granger then acting superintending engineer of the Hoosac Tunnel, obtained ten cartridges of nitroglycerine from Mowbray in order to try their effect at the east end of the tunnel. Because it was believed that nitroglycerine was more easily detonated when frozen, the cartridges were warmed to 90 °F and then carefully packed in sawdust to prevent them from cooling. Granger set off by sleigh on his journey over the mountain to the east side with the carefully insulated nitroglycerine. During the trip the sleigh was upset and the cartridges spilled out of the box into the snow. By the time Granger got the mishap rectified and the cartridges gathered up it was found that the nitroglycerine had frozen. Despite his now thinking that the nitroglycerine was much more dangerous than before, Granger replaced the cartridges in the box and continued his journey to the east end of the tunnel. When he tried to use a charge of frozen nitroglycerine he found it would not explode and, in fact, could not be made to do so until it had thawed out. In this way it was discovered that frozen nitroglycerine was safe, and from then on all nitroglycerine was transported in a frozen state; after this the standard method of transporting nitroglycerine for its journey from the factory to the tunnel was to pack it in ice and carry it on a mule-drawn wagon or sleigh. In fact, so important was the need for a supply of ice for this purpose that Mowbray even overcame the local ice-merchant's objection to trading on the Sabbath![18]

It has not been possible to discover why it was originally thought that frozen nitroglycerine was more dangerous to handle than the liquid form when in fact the reverse is true. One possible explanation for this belief may be in the fact that many accidents have occurred when frozen nitroglycerine was jarred while being thawed, the material being particularly sensitive during transition from one phase to another. This may have led to a mistaken observation that it was the frozen form that was the more sensitive to detonation.

However, although potentially so important, the discovery that frozen nitroglycerine was safe to handle and transport was only to have a short lived use, for in 1867 Nobel invented dynamite – a mixture of nitroglycerine and kieselguhr – which replaced the use of liquid nitroglycerine as an explosive.

3.4 Dynamite

In 1866 in an attempt to make liquid nitroglycerine safe to handle, Schaffner had made mixtures of nitroglycerine and sand, but these experiments were not successful because the sand did not absorb enough nitroglycerine to give a sufficiently powerful explosive. In 1897 Mowbray made mixtures of nitroglycerine and mica powder and found that mica powder would absorb up to its own weight of nitroglycerine, although he referred to the mica powder as being a 'carrier' rather than an absorber of nitroglycerine. These mixtures were made primarily as a means of diluting the nitroglycerine in order to reduce its power for use in blasting roof and bench holes and not specifically as a means of safer handling.[19] In 1867, as we shall see, Nobel solved the problem by using kieselguhr in place of the sand or mica powder. Kieselguhr, which is a fine-grained diatomaceous earth formed from the skeletons of minute plants, is extremely porous and is able to absorb up to three times its own weight of nitroglycerine while still remaining a powder. Nobel called nitroglycerine absorbed in kieselguhr by the name 'dynamite', after the Greek word *dynamis* meaning power. Nobel patented dynamite in 1867 and also went on to perfect the technique of detonating it using his mercury fulminate detonator. Dynamite is six times as powerful an explosive as gunpowder, and a safe material to transport, handle and store. Although it was available towards the end of the job, dynamite was not used in the Mont Cenis tunnel, but five years later, in 1872, it was chosen as the explosive to be used for driving the St Gotthard Tunnel – the second tunnel through the Alps. Dynamite has been employed in tunnels the world over for the last one hundred years, and still continues to be used even though there are now other explosives available.

The circumstances of the discovery of kieselguhr as the ideal absorbant for nitroglycerine are worth examination. It was in 1866 at Hamburg that Nobel began the search for an absorbant that would make nitroglycerine safe to handle. He tried mixing it with powdered charcoal, sawdust, brick dust, cement, paper pulp, wood waste, coal, clay and gypsum. He even tried mixing it with gunpowder as, in fact, his father Immanuel Nobel had done in 1862. All these substances, like the sand that Schaffner had tried and the mica powder that Mowbray had tried, did not absorb enough nitroglycerine to make a sufficiently powerful explosive. Finally Nobel tried kieselguhr,[20] a material that was readily to hand because at the nitroglycerine factory at Kümmel the nitroglycerine was put into zinc cans which were crated in wooden boxes, the space between the cans being packed with kieselguhr, which was used because of its light weight, cheapness, stability and ready availability in north Germany.[21] The new explosive, dynamite, could be detonated in the same way as nitroglycerine, but unlike the latter was safe to handle. Before leaving Nobel, a curiosity of history is worth noting.[22]

In Great Britain, dynamite production commenced in 1873 at a factory at Ardeer on the Ayrshire coast of south-west Scotland, this being the eighth of Nobel's world-wide network of explosive plants.[23] The Ardeer plant came under the control of the Home Office in 1875 following the passing of the Explosives Act in that year. It still manufactures nitroglycerine-based explosives today.

Dynamite and other explosives based on, or derived from, nitroglycerine are still very important explosives for blasting rock today.[24] Another development has been the discovery in 1947 of an explosive that is made by mixing together ammonium nitrate and fuel oil (and therefore called 'ANFO'). This explosive is relatively cheap, the constituents are themselves not explosives, and the ANFO itself can be prepared quickly and easily on site (one 80 lb bag of fertiliser-grade ammonium nitrate to 1 gallon of fuel oil), the ammonium nitrate readily absorbing the right amount of fuel oil to provide the correct oxygen balance.[25] However, compared to nitroglycerine-type explosives, ANFO is much inferior as a blasting agent and, lacking water resistance, can only be used in dry rock unless it is prepacked in waterproof cartridges. Although fairly recent in origin, ANFO had a precursor because as long ago as 1867 two Swedish chemists, Ohlsson and Norbein, patented an explosive called 'ammoniiakkrut' consisting of ammonium nitrate and charcoal,[26] which was an ammonium nitrate combustible mixture working on the same principle as ANFO.

It can be noted that all these explosives from gunpowder on depend for their action on the intimate proximity of oxygen-producing and readily combustible substances: nitrates or the nitro-group to provide oxygen, and carbon, hydrogen and sulphur to burn. This principle has, therefore, been in engineering use for over four centuries.[27]

Figure 3.1 shows the layout of shotholes for present-day blasting in a

Figure 3.1 Shothole pattern

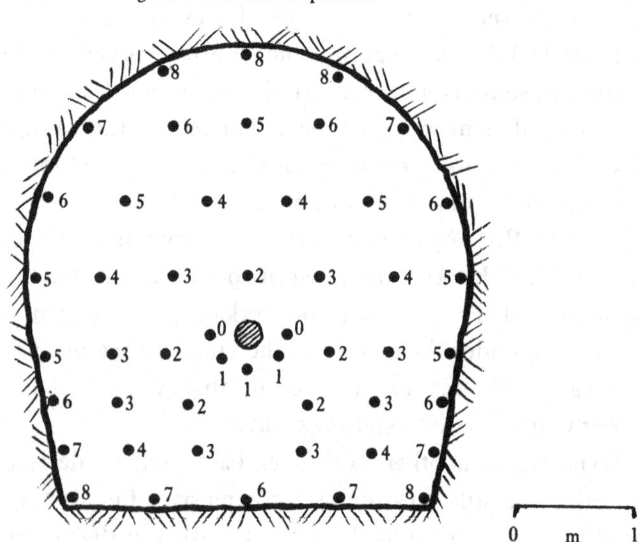

tunnel through granite; the shotholes are 2.8 m deep into the tunnel face and are 50 mm in diameter except for the centre hole which is 200 mm in diameter. Each, except for the centre hole which is left empty, is charged with a cartridge containing the explosive and detonator (see Figure 3.2). The numbers alongside the shotholes indicate the delay period, in half-

Figure 3.2 Inserting the cartridge in the shothole (1980)

second units (e.g. 6 denotes a delay of 3 seconds), the idea being that instead of all the explosions being initiated simultaneously, the charges in the centre of the face are set off first and those further out follow progressively towards the periphery whose charges are detonated last of all. In this way the explosion spreads from the centre of the face to the walls, allowing the rock from the outside to be blown into the loose zone created in the middle. Figure 3.1 can be compared with Figure 2.2; the tunnel faces are of roughly similar size. The shothole layout shown in Figure 3.1 is for what is called a 'burn cut', but other shothole patterns are used, in particular the 'wedge cut' in which holes are drilled symmetrically at converging angles to the face so that a wedge of rock is displaced in the direction of the tunnel axis when the charges are exploded.

3.5 Summary

The discovery of nitroglycerine was the result of purely scientific research. Sobrero was not investigating explosives, but was following a programme of chemical research aimed at systematically studying the effect of nitric acid on a series of organic substances. However, it should be recalled that Sobrero was Professor of Applied as well as Pure Chemistry and the possible application of his discovery would not have escaped him. In fact, he warned against using nitroglycerine because he considered it to be so dangerous. It was Nobel in Europe and Mowbray in the United States, both practical men, who made of nitroglycerine a usable explosive, but they could not have done this until Sobrero, a scientist, had discovered nitroglycerine and published the method of making it.

In spite of the dangers of transporting and handling nitroglycerine, it was actually used as a tunnel explosive, although the hazards remained a major problem and it is likely that safety legislation would have eventually prevented its widespread use. We have seen, however, that suddenly, two quite different ways of overcoming the problem were found. The first of these, the discovery that nitroglycerine could be safely handled and transported when frozen was, quite literally, accidental because it was the accident of Granger's sleigh spilling the nitroglycerine cartridges into the snow and their becoming frozen that led to this discovery. Furthermore, this instance is also an example of the accident being observed by someone with a prepared mind. The fact of the frozen nitroglycerine failing to explode would only have had significance to a person who was aware of the need for a method of making nitroglycerine safer to handle and transport. The second solution, the discovery of dynamite, is different. Here we have the fortuitous circumstance that the ideal absorbant for nitroglycerine was already being used at the nitroglycerine factory for another quite different purpose. When the kieselguhr was tried, it was found to absorb enough nitroglycerine to make a powerful explosive whilst being safe to handle. We must recognise, however, that Nobel was systematically seeking a substance to absorb nitroglycerine and it is probable that he would have hit on kieselguhr eventually even if it were not being used at his factory.

Notes and references for Chapter 3

1 Simms, F.W. (1896) *Practical tunnelling.* London (Crosby Lockwood and Son) 4th edition, revised and greatly extended by D.K. Clark, p. 251.

2 Williams, T.I. (1974) *A biographical dictionary of scientists.* London (Adam and Charles Black) 2nd edition, p. 481.

3 Sobrero, A. (1847) Sur la mannite nitrique. *Comptes rendus hebdomadaires des séances de l'académie des sciences (Paris)*, **25**, p. 121.

4 Anon (1848) Einwirkung der Mischung von Schwefelsäure und Saltpetersäure auf einige organische Substanzen. *Annalen der Chemie und Pharmacie (Liebig's Annalen)*, **64**, pp. 396–8.

5 An English translation of Sobrero's report to the Turin Academy is given by: Macdonald, G.W. (1912) *Historical papers on modern explosives.* London (Whittaker and Co) pp. 160–3.

6 The properties of nitroglycerine and details of how it is made are given by: Naoum, P. (1928) *Nitroglycerine and nitroglycerine explosives.* London (Baillière, Tindall and Cox) English translation by E.M. Symmes, pp. 1–10. Fieser, L.F. and M. Fieser (1956) *Organic chemistry.* New York (Reinhold Publishing Corporation) 3rd edition, p. 131. Partington, J.R. (1964) *A history of chemistry.* Volume 4. London (Macmillan and Co Ltd) p. 476. The author repeats the caution given by most writers on nitroglycerine that no attempt should be made to make it.

7 Drinker, H.S. (1893) *Tunnelling, explosive compounds and rock drills.* New York (John Wiley and Sons) 3rd edition, p. 66.

8 Williams, T.I. (1974) *op cit*, p. 394.

9 Chidsey, D.B. (1964) *Goodbye to gunpowder.* London (Alvin Redman) pp. 163–6.

10 For an example of United Kingdom legislation on nitroglycerine see: Anon (1869) Nitroglycerine. *Engineering*, **8**, 20 August, p. 127. The importation of nitroglycerine was prohibited, and the manufacture, sale and carriage of nitroglycerine was subject to new restrictions and regulations.

11 Cook, M.A. (1971) *The science of high explosives.* New York (Robert E. Krieger Publishing Co Inc) p. 8.

Fordham, S. (1980) *High explosives and propellants.* Oxford (Pergamon Press) 2nd edition, p. 13.

12 Brierley, G.S. (1976) Construction of the Hoosac Tunnel 1855 to 1876. *Journal of the Boston Society of Civil Engineers Section, American Society of Civil Engineers*, **63**, (3), pp. 175–209.

13 Parmelee, D.D. (1868) The manufacture of nitroglycerine. *Engineering*, **6**, 23 October, p. 376.

14 Drinker, H.S. (1893) *op cit*, pp. 58 and 67.

15 Vogel, R.M. (1964) Tunnel engineering – a museum treatment. *Contributions from the Museum of History and Technology*, Paper 41. Washington DC (Smithsonian Institution) pp. 203–39.

16 Drinker, H.S. (1893) *op cit*, p. 66.

17 Brierley, G.S. (1976) *op cit*, pp. 175–209.

18 Walker, F.N. & G.C. Walker, Editors (1957) *Daylight through the mountain.* Montreal (The Engineering Institute of Canada) p. 44.

19 Drinker, H.S. (1893) *op cit*, pp. 88–90.

20 De Mosenthal, H. (1899) The life-work of Alfred Nobel. *The Journal of the Society of Chemical Industry*, 31 May, pp. 443–51.

21 The following apocryphal story is told about Nobel's discovery of dynamite. It is said that one day a crate of nitroglycerine was damaged, and upon it being opened, Nobel saw that the kieselguhr packing had soaked up a whole canful of nitroglycerine from a leaking can whilst remaining perfectly granular. He took it to his laboratory and found it exploded perfectly when detonated with a percussion cap. This story is told by:
Pauli, H.E. (1942) *Alfred Nobel. Dynamite king – architect of peace.* New York (L.B. Fischer) p. 84.
Halasz, N. (1960) *Nobel: a biography.* London (Robert Hale Ltd) p. 59.
In later life Nobel refuted this account, saying that he had never seen an accidental leakage of this kind, and that kieselguhr just happened to be the last of a long list of potential absorbants he was trying. This denial is reported by de Mosenthal (Reference 20) and by:
Bergengren, E. (1962) *Alfred Nobel – the man and his work.* London (Thomas Nelson and Sons Ltd) English translation by A. Blair, p. 44.

22 Later in his life Nobel had nitroglycerine prescribed for his heart condition – he himself referred to this as 'an irony of fate'. This is described by:
Pauli, H.E. (1942) *op cit*, pp. 260–1.

23 Nobel's Explosives Company Limited (1971) *A century of explosives manufacture at Ardeer 1871–1971.* Ardeer (Nobel's Explosives Co Ltd).

24 Blasting rock for tunnel driving is described by:
Rankin, W.W. & R. Haslam (1957) Modern blasting practice in tunnelling operations. Glasgow (Imperial Chemical Industries Ltd, reprint from *Civil Engineering and Public Works Review*).

25 Greenland, B.J. & J.D. Knowles (1969) Rock breakage. *Mining Magazine*, **120**, (2), pp. 76–83.

26 Cook, M.A. (1971) *op cit*, p. 10.
Again, no attempt should be made to make either of these explosives.

27 The first reported use of gunpowder for blasting in civil engineering was in 1541 where it was said to have been used for blasting obstructions in the Yellow River in China. See:
Needham, J. (1971) *Science and civilisation in China.* Volume 4, Part 3. Cambridge (Cambridge University Press) p. 343.

Gunpowder was first used for blasting in mines in Europe in 1627. See:
Agricola, G. (1556) *De re metallica.* Translated by H.C. Hoover and L.H. Hoover (1912) London (The Mining Magazine) Footnote to p. 119.
Hollister-Short, G. (1983) The use of gunpowder in mining: a document of 1627. *History of Technology*, 8, pp. 111–15.
The first use of gunpowder for tunnel excavation is said to have been in 1679–81 at Malpas, France, for a 540 ft long canal tunnel on the Canal de Languedoc, a canal built by Pierre-Paul Riquet de Bonrepos (1604–80) to provide a link between the Atlantic and the Mediterranean. This canal is described by:
Kranzberg, M. & C.W. Pursell, Editors (1967) *Technology in western civilisation.* Volume 1. London (Oxford University Press) pp. 204–5.
Derry, T.K. & T.I. Williams (1973) *A short history of technology.* London (Oxford University Press) p. 188.

Tungsten carbide drill bits

I'm arm'd with more than complete steel.

We have seen how the invention of the compressed air rock drilling machine and the introduction of nitroglycerine explosive revolutionised hard rock tunnelling. However, the success of rock drilling machines drew attention to another shortcoming of the drill-and-blast tunnelling method – that of the rapid blunting of the cutting edge of the drill steel bits. In Section 2.2 it was remarked that even during the manual system of drilling shotholes, the drills soon became blunt and a considerable team of men or boys was employed in returning these to the blacksmith for resharpening and supplying the miners with freshly sharpened drills; in America these boys were called 'nipper boys' (see Figure 2.1) and in English tunnels they were often the sons of the miners.[1] With the introduction of compressed air rock drilling machines, with their hard hitting capability, the drill steel bits became blunt even sooner than before. At first the solution to the problem was to organise the resharpening operation on a production-line basis with a large grinding shop established outside the tunnel portal, large stocks of drill steels, and a fleet of special rail cars which carried only drill steels in and out of the tunnel.[2] Disadvantages of this solution were that it was cumbersome, got in the way of other tunnelling operations and that the constant regrinding of the steels eventually reduced their length so that they were no longer usable. Concurrently, attempts were made to extend the life of the drills by using improved and harder steel and steel alloys in their manufacture.[3] Detachable drill bits which could be discarded or resharpened were also tried, but even as late as 1940, after drilling only 2 ft of hole a steel would be blunted. These then were only palliative measures; what was needed was a drill bit that was hard enough to match the performance of the drilling machines. For this the tunnelling industry had to wait until 1945 when the sintered tungsten carbide drill bit was introduced. In this Chapter the origin and development of this innovation is examined, and some of the problems that had to be overcome before it could be introduced to tunnelling practice are discussed.

Before doing so, however, it is helpful to consider briefly the process of percussive drilling of rock because it explains why it is important that the tool should have a durable sharp tip. When the tool is first struck, a high state of stress is induced at the point of contact of the tip with the rock, the rock fails in compression and a crushed zone is formed immediately beneath the point of contact. The tip of the tool can then penetrate some distance below the rock surface until the rock on either side of the chisel undergoes shear or tensile failure and small chips of rock are detached. Following a second blow another crushed zone is formed, leading to the detachment of further chips, and so on for successive further blows. The dust produced during percussive drilling is the material from the crushed zone, and it can be seen that a minimum of dust together with a maximum of rock chips indicate efficient drilling. As the tip of the tool becomes blunt it becomes less capable of penetrating and a higher proportion of the available energy is spent in creating the crushed zone and correspondingly less on producing rock chips – thus the drilling process becomes less efficient as the tool loses its edge. Therefore, not only must a drill steel tip be capable of taking a sharp edge, but ideally the tip should retain a sharp edge over a long period of drilling.

4.1 Discovery of tungsten carbide

The story of tungsten carbide begins with the French chemist Ferdinand Frédéric Henri Moissan, born in Paris on 28 September 1852.[4] He was appointed Professor of Inorganic Chemistry at the Sorbonne in 1900 and was a Nobel Prizewinner in 1906; he died in Paris on 20 February 1907. Moissan is well known for a number of advances that he made in chemistry, particularly for his work on metals, but the one that concerns us here was his invention in 1892 of the electric furnace which made possible the study of chemical reactions at high temperature.

Before the invention of the electric furnace the highest temperatures that could be reached by fuel furnaces were 1700–1800 °C and these were restricted to various technical processes. In the laboratory, temperatures above 1600°C could not be attained except by use of the oxyhydrogen blowpipe which achieved 2000 °C but was suitable only for heating small samples. Moissan's electric furnace, which utilised the heat of the electric arc produced by an electric current of 1000 A at 70 V, was capable of reaching temperatures of up to 3500 °C.[5] The invention of the electric furnace opened the way for experiments on the reactions of elements at very high temperatures. Moissan exhibited his electric furnace at the Académie des Sciences in December 1892.

One of Moissan's principal uses for the electric furnace was to prepare synthetic carbides of metals, and in 1895 he prepared a compound from tungsten and carbon having the chemical composition W_2C. (Although using the chemical notation of the time Moissan described it as Tu^2C.) He noted that the compound was iron-grey in colour and was very hard,[6] giving the first clue that tungsten and carbon could unite to form a hard substance.

In 1898, P. Williams, working in Moissan's Laboratoire des Hautes Etudes at the Ecole Supérieure de Pharmacie in Paris, prepared tungsten carbide having the chemical composition WC (called TuC at the time) which we will see is the compound that was to become so important later on. Williams' method of preparing tungsten carbide was to heat, in the electric furnace, an intimate mixture of tungstic oxide, carbon and iron. On cooling, an ingot was obtained which consisted of a mixture of tungsten carbide, iron, graphite and other impurities. The tungsten carbide was separated out by treatment with hot hydrochloric acid followed by magnetic and density separation. Williams studied the physical and chemical properties of the new compound. He observed that it was a grey powder which, when examined under the microscope, was seen to consist of opaque cubic crystals. It had a specific gravity of 15.7 and he noted that it was very hard, being able to scratch quartz easily.[7] In fact, tungsten carbide is one of the hardest substances known, having a hardness of 9 on Mohs' Scale (diamond = 10).

Tungsten carbide, however, suffers from a serious practical disadvantage: it is far too brittle for any industrial use. Before its desirable property of extreme hardness could be exploited, the problem of its brittleness had to be overcome. The solution was found by sintering the material.

4.2 Sintered tungsten carbide

If tungsten carbide powder is mixed with a small amount (about 10 per cent) of iron, nickel or cobalt, and the mixture is pressed into blocks which are then heated to 1500 °C in a reducing atmosphere of hydrogen, a material is obtained which has the hard property of tungsten carbide but is no longer brittle. The process is known as 'sintering' and the new material is called sintered tungsten carbide.[8] If before being sintered, the powder blocks are preheated to only 900 °C in hydrogen, they become only about as hard as chalk and can be readily cut and ground to any required shape; this property is most important because the material is far too hard to be shaped by normal machining processes after sintering.

In the 1920s the need for a hard die material to take the place of the very expensive diamond dies used to draw tungsten wire for incandescent electric lamp filaments led Karl Schröter of the Osram Studiengesellschaft, Berlin, to seek a substitute.[9] In January 1914 H. Voigtländer and H. Lohmann had described how to make dies and tools from pure sintered tungsten carbide which were extremely hard and resistant to wear but too brittle for use in industry. Schröter discovered that the addition of a little iron, nickel or cobalt facilitated sintering and gave a product that was no longer brittle and this was a decisive step in the development of usable sintered tungsten carbide products. The firm Friedrich Krupp of Essen took over the production of this material and in 1927 introduced hard metal alloys made from sintered tungsten carbide containing 6 per cent cobalt; these were marketed under the name 'Widia' from the German *wie Diamant* meaning 'like diamond'.[10] In 1928 the manufacture of sintered tungsten carbide dies and tools began outside Germany under patent, but for many years the new material was used only in the mechanical engineering industry for machining and forming metals. To give an idea of its importance to this industry, the production of sintered tungsten carbide in Germany rose from 1 ton per month in 1930 to 40 tons per month in 1944.

A detailed and complete description of the whole industrial process of manufacturing sintered tungsten carbide, starting with the extraction of tungsten from its ore right through to the production and testing of the final product has been given.[11] In the modern method of manufacture of tungsten carbide, metallic tungsten powder and carbon black are mixed together in atomic proportions and then heated in carburising furnaces at 1500–1600 °C.

4.3 Development of sintered tungsten carbide drill bits

The idea of using rock drills with tungsten carbide tips was first tried in Germany in the late 1920s, but the results were disappointing because of brittleness of the inserts and difficulties in brazing the bit in the steel, and nothing further was done. The idea, however, was taken up in Sweden in the early 1940s as a result of co-operation between Eric Ryd of Atlas Diesel, the firm which manufactured compressed air rock drilling machines, and Hans Hermann Wolff of Luma, a firm which made electric light bulbs. Atlas designed the sintered tungsten carbide bits and Luma manufactured them; after fitting them to drill steels they were then tested in the Atlas test mine. Later on Luma's tungsten carbide manufacturing operation was taken over by the firm Sandvik but Atlas and Sandvik continued with the joint development work, Wilhelm Haglund of Sandvik taking on the job of producing a reliable product and of brazing it to the steel. By 1947 these difficulties had been overcome and the now world-famous 'Coromant' drill steels tipped with sintered tungsten carbide came onto the market.[12]

Throughout the present-day tunnelling industry, wherever drill-and-blast tunnelling is being carried out, carbide tipped drills are used for drilling the blastholes. It seems unlikely that any other material will replace tungsten carbide for this purpose in the foreseeable future.[13] In this connection it should be noted that diamond-tipped drills are used only for rotary core drilling and not for the much faster percussive drilling which is the type of drilling used for shotholes in drill-and-blast tunnelling. Diamond would be too brittle for the tip of a percussive-type drill bit.

One further problem (touched upon above) that had to be solved before sintered tungsten carbide could be used to tip drill steels was to find a way of firmly attaching the drill bits to the steels. In the event two solutions were found.[14] Chisel-shaped sintered tungsten carbide inserts could be brazed into a milled groove in the end of the drill steel using silver or bronze as the brazing material, as shown in Figure 4.1, or sintered tungsten carbide buttons could be fitted into holes drilled in the bit body being held in place by heat shrinking or cold pressing, as shown in Figure

4.2. Both these techniques give a bond between carbide and steel that is able to withstand the operational stresses of drilling over a long period of time. The bits with the button inserts are for use in hard or abrasive rocks while the bits with chisel-shaped inserts are used for drilling in softer or less abrasive rocks.

4.4 Use of sintered tungsten carbide drill bits

The introduction of drill rods with sintered tungsten carbide bits had a profound effect on the logistics of hard rock drilling. In 1945, when the new sintered tungsten carbide bits were introduced in place of the forged

Figure 4.1 Drill steel bit with sintered tungsten carbide tip

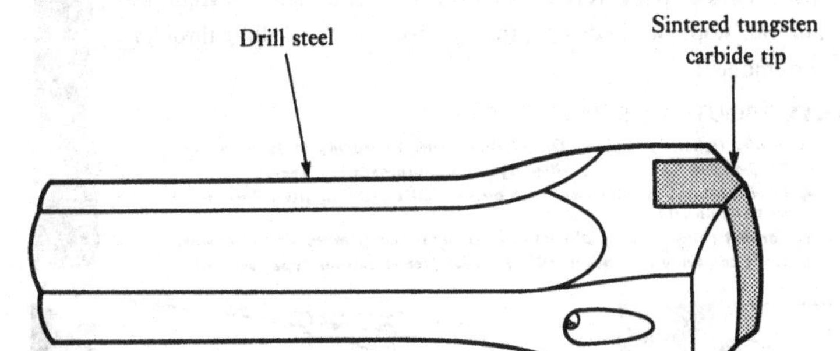

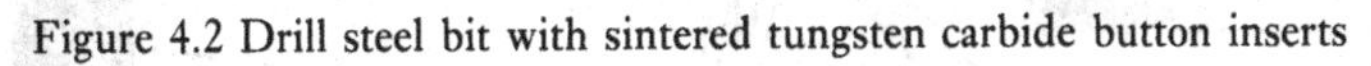
Figure 4.2 Drill steel bit with sintered tungsten carbide button inserts

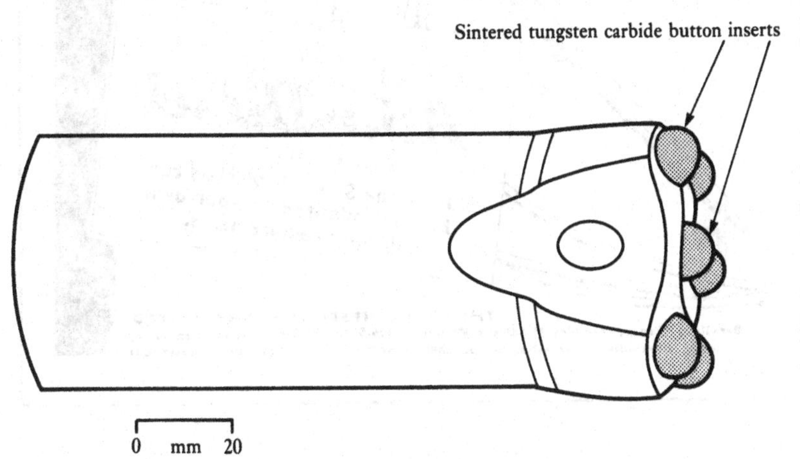

Figure 4.3 Advertisement for the Swedish Method (1951)

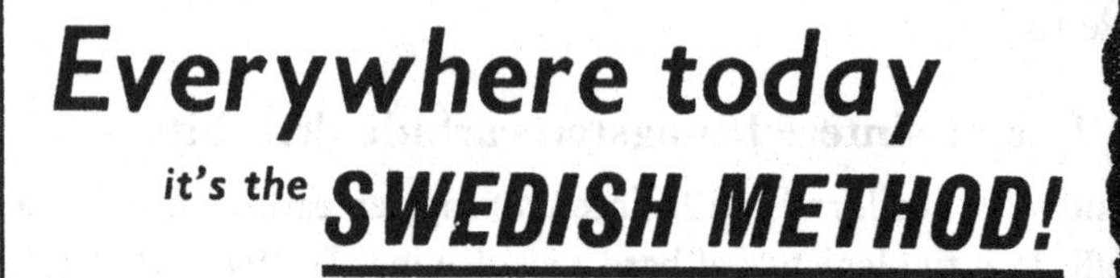

In mining, tunnelling and quarrying—in every job, all over the world, that calls for the drilling of rock—they're swinging away from the costly, wasteful ways of the past, and turning to the time-saving, money-saving Swedish Method. It's a revolution in drilling practice. A revolution nothing can stop—while drilling through rock means drilling through hours of money!

FACTS ABOUT THE SWEDISH GEAR

- *25,000 lightweight Atlas Diesel rock drills operating in 35 countries.*
- *500,000,000 feet of rock drilled by Atlas equipment last year.*
- *First company to utilize one man pusher drills with Tungsten Carbide tipped jumpers.*
- *Largest company specializing exclusively in compressed air equipment.*
- *Only company to convert 100% from drifter to pusher type rock drills.*

THE ATLAS DIESEL COMPANY LIMITED
Beresford Avenue, Wembley, Middlesex. Wembley 4426-9. *Service Depots:* MANCHESTER, LEEDS, GLASGOW, BIRMINGHAM, NEWCASTLE, BRISTOL, CARDIFF, NOTTINGHAM, BELFAST, DUBLIN, JERSEY (C.I)

steel chisel bits formerly used, 65 of the new drill rods replaced the 10 tons of drill steel that had been in use before. Although the tungsten carbide bits had to be reshaped periodically by grinding, the average life of one of the new drill rods was 500 ft before being worn down. Since that time they have been improved even more, the average life of a drill rod in hard rock increased by 1960 to 2000 ft with a drilling rate of 2 ft per minute.[15]

A full description of the properties and performance of sintered tungsten carbide drilling bits has been given.[16] Test drillings using both steel drills and carbide tipped drills in the same rock showed that in granite, quartz and sandstone, all hard and abrasive rocks, the carbide tipped drills gave a 30 times prolongation of tool life over steel drills. With softer or less abrasive rocks the improvement in performance was even better, rising to 75 times for slate.

The long life and rapid drilling rate of drill steels tipped with sintered tungsten carbide made possible the introduction in 1946 of what became known as the 'Swedish Method' of drilling blastholes. The firm of Atlas Diesel produced a lightweight compressed air rotary percussion drilling machine mounted on a pusher leg and fitted with a drill rod having a carbide bit. This was operated by one man and its high output was entirely due to the fact that the bit had to be sharpened far less often and the number of drill rod changes was much reduced.[17] The Swedish Method enabled the rapid development of tunnelling and underground construction in hard rock that took place in Sweden just after the Second World War, and from there the technique soon spread abroad. Figure 4.3 shows an advertisement used for marketing the Swedish Method in Great Britain in 1951 which makes a feature of the tungsten carbide tipped drill rods (called 'jumpers' in the advertisement text) utilised.[18]

The success of the sintered tungsten carbide drill tip in prolonging the working life of the drill rod brought with it another problem, for it was soon noticed that some of the rods fractured because of fatigue stresses after they had been in use for some time. In the case of the Coromant drill steels referred to earlier, the problem was solved by shot-peening the entire surface of the drill rod to prevent the formation of fatigue cracking starting from the surface. The steel was treated against corrosion as well since some fatigue fractures were initiated at corrosion sites in the metal. Coromant drill steels have come into wide use and are available with both the chisel type of sintered tungsten carbide insert shown in Figure 4.1 and the button type (Figure 4.2) as well as a cross type which is a variation of the basic chisel form.[19]

The successive introduction of the compressed air rock drilling machine, nitroglycerine explosive and the tungsten carbide tipped rock drill brought the drill-and-blast method of hard rock tunnelling to near perfection. This was particularly so when the use of multiple-drill jumbos[20] allowed a number of shotholes to be drilled at once under the control of a single miner. Figure 4.4 shows how improvements in the compressed air rock drilling machine coupled with the introduction of the tungsten carbide drill bit increased drilling speeds in Swedish granite from about 5 m per hour to 40 m per hour. The introduction of heavier jumbo-mounted drills almost doubled this rate, bringing drilling speeds up to 70 m per hour. The last item in the Figure, the introduction of the hydraulic rock drilling machine, increased the drilling speed still further to 90 m per hour, but with it we are looking ahead, this innovation being the subject of the next Chapter. Also shown on Figure 4.4 is the range of the Swedish Method referred to earlier.

4.5 Sharpening tungsten carbide

In spite of the extreme hardness of tungsten carbide, drill steel inserts made of the sintered material do become worn after extensive drilling. It is interesting that different kinds of rock produce different patterns of wear: for example, granite and gneiss give rise to mainly frontal wear so that the tip becomes blunt whilst quartzite produces mainly peripheral wear so that the corners get rounded off. The cause of this difference in behaviour is not known but probably lies in relative differences in the strength, hardness and abrasiveness of the different rocks. However, if either type of wear becomes excessive the drill bit must be resharpened and restored to its original profile. On first consideration, bearing in mind how hard tungsten carbide is, it would seem that this might be impossible, but that is not the case for sintered tungsten carbide inserts which can be reshaped using grinding wheels made from vitrified bonded silicon carbide. This is because while tungsten carbide is very hard, silicon carbide is in fact a little harder and so is capable of being used to machine tungsten carbide. Most manufacturers provide criteria for judging when an insert has reached the stage at which it should be resharpened, although on large worksites a fixed routine for regrinding, for example after a certain number of holes have been drilled, may be practised after preliminary trials have been made to determine the rate of wear. For the chisel-shaped type of rock drill bit shown in Figure 4.1 the edge angle of the insert should be reground to 110° when it becomes worn. Chisel-shaped rock drill bits can be reground with conventional style grinding

Figure 4.4 Increases in speed of drilling

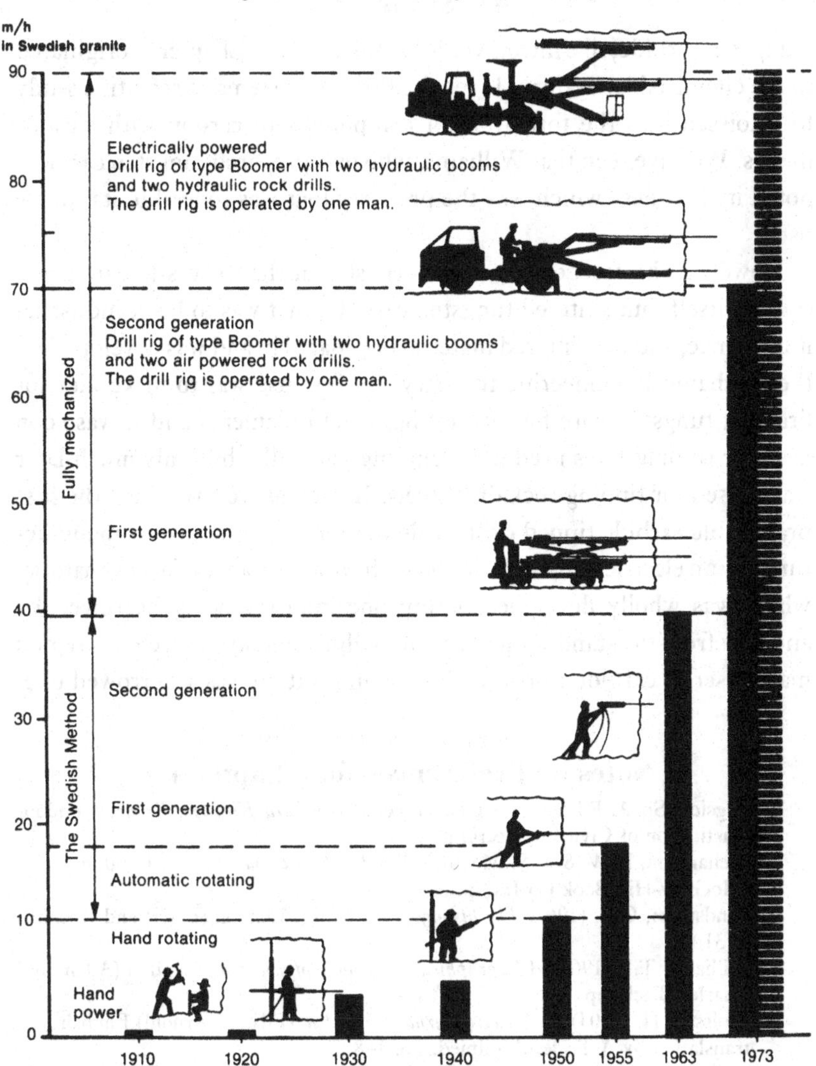

wheels but these are, of course, unsuitable for regrinding the button type of rock drill bit shown in Figure 4.2. Button bits are reshaped in either of two ways. The first utilises a special machine which has a grinding wheel having a peripheral concave groove shaped to the profile of the individual button inserts. The second uses a small grinding tool having a concave face. With both methods some of the steel body of the bit is ground away at the same time to leave the insert standing proud again.

4.6 Summary

Tungsten carbide, like nitroglycerine considered in Chapter 3, originated in the chemists' laboratory. It resulted from a systematic scientific study by Moissan into the formation of compounds of carbon with various metals. We have seen that Williams, who first prepared tungsten carbide, noted its hardness which was the property that was to lead to its future use.

However, the foregoing account has shown that it was not tungsten carbide itself, but sintered tungsten carbide, that was to be of industrial importance, and the sintered material originated in and was developed by the mechanical engineering industry. Its first use was to make dies for drawing tungsten wire for electric light bulb filaments and it was soon used for tipping tools used in machining generally, but only much later was it used for tipping rock drill steels. In fact, as we have seen, the first practicable carbide tipped drill steels were made using inserts manufactured by an electric lamp firm. Thus we have an instance of a new material which was wholly developed within one industry being borrowed by another: from the standpoint of the tunnelling industry, which was to be a major user of carbide tipped drills, the innovation was a borrowed one.

Notes and references for Chapter 4

1 Pugsley, Sir A. Editor (1976) *The works of Isambard Kingdom Brunel*. London (Institution of Civil Engineers) p. 45.
2 Richardson, H.W. & R.S. Mayo (1941) *Practical tunnel driving*. London (McGraw-Hill Book Co Inc) p. 324.
3 Sandström, G.E. (1963) *The history of tunnelling*. London (Barrie and Rockliff) p. 310.
4 Williams, T.I. (1969) *A biographical dictionary of scientists*. London (Adam and Charles Black) pp. 371–2.
5 Moissan, H. (1904) *The electric furnace*. London (Edward Arnold) English translation by A.T. de Mouilpied, pp. 1–8.

6 Moissan, H. (1896) Recherches sur le tungstène. *Comptes rendus hebdomadaires des séances de l'académie des sciences (Paris)*, **123**, pp. 13–16.
Smithells, C.J. (1952) *Tungsten: a treatise on its metallurgy, properties and applications*. London (Chapman and Hall Ltd) p. 283.
The element tungsten has now the chemical symbol W after wolfram its alternative name.
7 Williams, P. (1898) Sur la préparation et les propriétés d'un nouveau carbure de tungstène. *Comptes rendus hebdomadaires des séances de l'académie des sciences (Paris)*, **126**, pp. 1722–4.
8 Street, A. & W. Alexander (1960) *Metals in the service of man*. Harmondsworth (Penguin Books Ltd) p. 208.
9 Wulff, J. Editor (1942) *Powder metallurgy*. Cleveland, Ohio (American Society for Metals) pp. 20–1.
10 Dawihl, A. (1955) *A handbook of hard metals*. London (HM Stationery Office) pp. 1–2.
11 Hester, A.S., A.D. Mitchell & R.W. Rees (1960) Tungsten carbide. *Industrial and Engineering Chemistry*, **52**, (2), pp. 94–100.
12 Gårdlund, T, I Janelid, D Ramström & H Lindblad (1974) *Atlas Copco 1873–1973*. Stockholm (Atlas Copco AB) English translation by D. Jenkins, pp. 151 and 278–9.
13 Hester, A.S. *et al* (1960) *op cit*, p. 100.
14 Sandvik AB (1977) *Rock drilling manual*. Sweden (Sandvik AB) 2nd edition, p. 20.
15 Sandström, G.E. (1963) *op cit*, p. 311.
16 Shwarzkopf, P. & R. Kieffer (1960) *Cemented carbides*. New York (The Macmillan Co) p. 267.
17 Gårdlund, T. *et al* (1974) *op cit*, p. 287.
18 *Ibid*, p. 328.
19 The Coromant range of drill steels with sintered tungsten carbide inserts is described in:
Sandvik and Atlas Copco AB (1976) Choose your weapon. Sweden (Sandvik and Atlas Copco AB).
Illustrations showing the stages of manufacture of sintered tungsten carbide inserts for rock drill bits are given in:
Wolff, H. (1956) Hårdmetall för bergborr. *Bergslaget*, **3**, pp. 18–24.
A review of present-day rock drill bits is given by:
Pearse, G. (1984) Rock drill bits. *Mining Magazine*, **150**, (2), pp. 143–53.
20 Rubber-tyred and rail-mounted drilling jumbos are illustrated in the following manufacturer's handbooks:
Atlas Copco AB (1980–1) *Atlas Copco guide book of underground equipment*. Stockholm (Atlas Copco AB) pp. 16–21.
Atlas Copco AB (1975) *Atlas Copco manual*. Stockholm (Atlas Copco AB) pp. 207–39.

Hydraulic rock drilling machines

It is a question of hydraulics you see, and came within my own province.

By the 1960s the drill-and-blast method of tunnelling in hard rock appeared to have reached a state where little further improvement seemed possible. The compressed air rock drill, using drill steels tipped with tungsten carbide, and mounted on large drilling jumbos, together with the use of dynamite and other modern explosives for blasting, made possible rapid tunnel driving rates of up to 100 ft per week. However, drilling machine manufacturers, not satisfied with this state of affairs, turned their attention to ways of producing even faster rates of tunnel advance. Consideration of the existing drill-and-blast system showed that the tunnel driving rate could only be increased if the drilling rate could be increased, and this could not be done with the compressed air rock drill because the machine was found to have reached the limit of its development. This problem and its solution – the hydraulic rock drilling machine – are the subjects discussed in this Chapter.

Modern compressed air rock drilling machines are similar in principle to the early machines described in Chapter 2 except for one important difference. In the early machines, it will be recalled, the drill rod was attached directly to the reciprocating piston in the main cylinder of the drilling machine. In 1865, however, George Low in England took out a patent for a hammering drill in which the drill steel was kept stationary and a separate piston struck it in the manner of a hammer. Henry Sergeant patented a similar drill in America in 1884 and since then all compressed air rock drills have been of the hammering type.[1] The force that the piston of a compressed air rock drill can exert on the drill rod is determined by the air pressure supplied and by the cross-sectional area of the piston: the size of the force is given by multiplying these parameters together.

Theoretically, compressed air rock drills could be improved to any desired extent by simply increasing either or both of these parameters. In practice, however, there are limitations on the amount to which both air pressure and cross-sectional area of piston can be increased. Compressed air rock drills are normally operated with air at a pressure of 7 atm. This figure is not usually exceeded because of the danger to the miners of a rapidly expanding blast of air should a supply hose burst or split or should a connection come apart, and this effectively limits any improvement to drilling machine performance by raising the working pressure.

Attempts to increase drill performance by increasing the cross-sectional area of the piston met with another problem. This is that if the piston diameter greatly exceeds that of the drill steel shank, then there is a poor transfer of energy from the piston to the drill steel; ideally the piston should be of the same diameter as the shank. This is illustrated in Figure 5.1.[2] When the piston is much larger in diameter than the drill rod, the stress wave in the drill rod resulting from one blow of the piston is as shown in Figure 5.1(*a*). The drill mechanism is designed so that the amplitude of the stress wave does not exceed some value *A* in order to preserve the life of the drill steel. However, when a long piston the same diameter as the drill rod is used, then the stress wave in the drill rod is as shown in Figure 5.1(*b*). It can be seen that, for the same amplitude *A*, the area under the stress wave is much greater than it was for the large-

Figure 5.1 Comparison of stress waves from different pistons

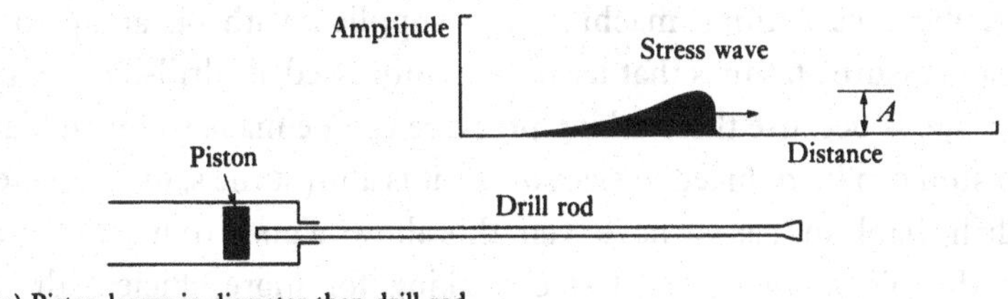

(*a*) Piston larger in diameter than drill rod

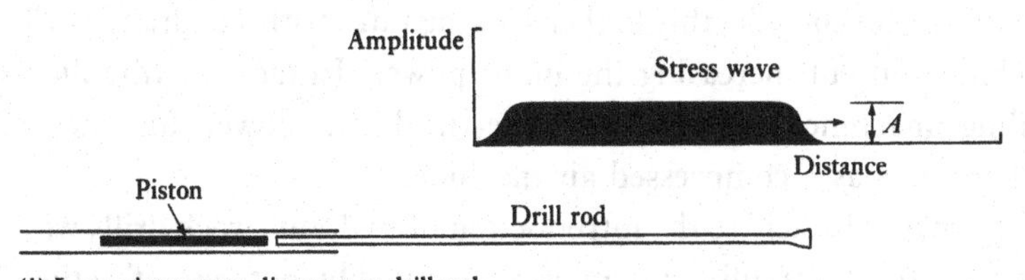

(*b*) Long piston same diameter as drill rod

diameter piston considered previously. This means that in the second case more energy per blow is being transferred to the drill rod and thus more energy is available to drill the hole. Increasing the piston diameter, therefore, results in less and less efficient drilling, and this effectively limits improvements in drilling machine performance by increasing the piston diameter.

Therefore the two parameters, working pressure and piston diameter, that governed the force that a rock drilling machine could apply to the drill rod, were both incapable of being increased further with a machine run on compressed air, leading to the conclusion that the compressed air rock drilling machine had reached the limit of its development.

The late 1960s, therefore, saw the remarkable situation wherein the compressed air rock drill, the first innovation in drill-and-blast tunnelling, the machine which alone had made the driving of long rock tunnels possible, was now itself the limiting element in the drill-and-blast system that was holding up further progress.

The solution to this problem was found in the introduction in the early 1970s of the hydraulic rock drilling machine. This is identical in principle to the compressed air drill except that a liquid (hydraulic oil) is used as the working fluid instead of a gas (compressed air). However, very important consequences follow from this change in working fluid. The first of these is that because compared with air hydraulic oil can be considered incompressible a burst in the system is not hazardous, and therefore very much higher working pressures can be used without danger to the miners. Hydraulic rock drilling machines are supplied with oil at up to 250 atm, a pressure 36 times that used for compressed air drills. The second advantage is because the working pressure can be made so much higher, the piston can be reduced in size so that it is almost the same diameter as the drill shank and, as we have seen, this allows a good transfer of energy from the piston to the drill steel making for more efficient drilling. Thirdly, because of the higher efficiency of a hydraulic system, it is possible to get considerably higher striking rates with a hydraulic drilling machine without increasing the input power. In fact, a hydraulic rock drilling machine requires only one-third the power for the same performance as a compressed air machine.

The main obstacle to the introduction of hydraulic rock drills was the fact that the hydraulic system is very quickly put out of action by contamination – microscopic particles of rock dust in the hydraulic oil can jam vital components such as valves. Because compressed air rock drills are almost immune from this problem being able to cope with grit, sand and rock dust without any effect on performance, the miners and other members of the tunnel workforce who are involved with the operation of the drills have to be retrained carefully to make connections and carry out maintenance without contaminating the oil. In spite of the seeming difficulty of doing this in a tunnel environment, hydraulic rock drills by the mid-1980s were superseding compressed air rock drills for major tunnel construction, hydraulic rock drills being now offered by nine manufacturers of rock drilling machines.[3]

To illustrate the increased power of the hydraulic over the compressed air rock drill a modern machine, the Atlas Copco COP 1038 hydraulic rock drill, will shortly be examined (Section 5.3), but before doing so a digression will be made to look at a much earlier machine which utilised the hydraulic principle – the nineteenth-century Brandt hydraulic rock drill.

5.1 The Brandt hydraulic rock drill

The first hydraulic rock drilling machine was invented by a German engineer, Alfred Brandt (1846–98), in 1876. When a junior engineer on the St Gotthard Railway, Brandt had been struck by what seemed to him a serious waste of energy involved in the use of water power to compress air to drive the compressed air rock drilling machines being used to excavate the St Gotthard Tunnel. Brandt had the idea of utilising the water power directly, thereby avoiding the energy loss involved in compressing the air. He first developed a rock drill of the percussive type similar to the compressed air rock drills except that it was driven by water pressure instead of compressed air, but this was unsuccessful. The second drill he designed was entirely diferent; the machine held the drill steel against the rock with a force of 12 tons while revolving the drill steel at about 4 revolutions per minute.[4] Both the thrust force on the drill and the rotation of the drill steel were provided by hydraulic power, and the waste water from the hydraulic rotary motor was used for washing the borings out of the drill hole. The Brandt drill also differed from other drills in that the drill steel was tubular with a head having hardened cutting teeth and so produced a core of rock which was periodically broken up and removed.

The mode of operation of this drill will now be examined.[5]

The Brandt hydraulic drill is shown in Figure 5.2.[6] Two machines were mounted side by side on a heavy iron thrust bar about 1 ft in diameter. The thrust bar was, in turn, mounted via a pivot on a very neat, small, rail-borne carriage, the mass of the drilling machines being counter-balanced by a weight at the rear end (see upper left in Figure 5.2). Because of this arrangement, the shotholes bored by the drills radiated in direction from the axis of the transverse thrust bar as shown in the lower drawing of the Figure. The feed of the rotary cutting tool was achieved by direct water pressure in a large cylinder, the piston of which automatically retracted when the water supply was stopped. The rotary action of the drill was achieved in the following way. Above the main body of the drilling

Figure 5.2 Brandt's hydraulic drill (1882)

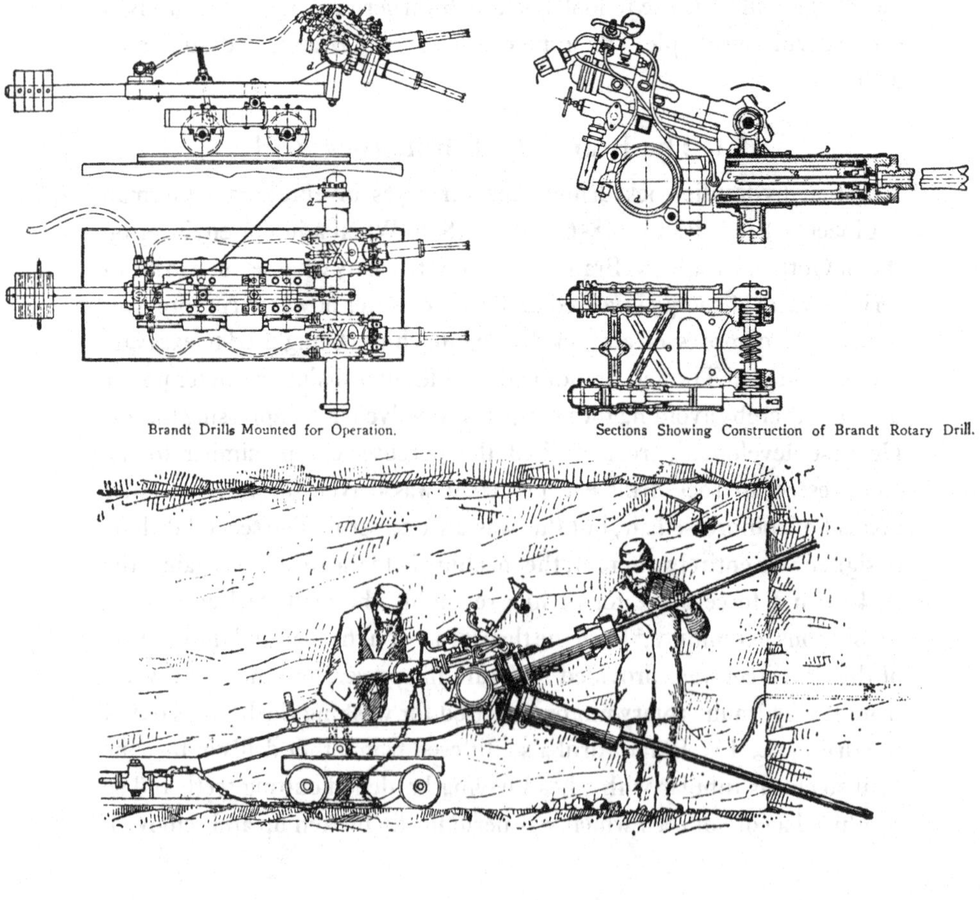

Brandt Drills Mounted for Operation.

Sections Showing Construction of Brandt Rotary Drill.

machine were mounted two cylinders of a hydraulic engine (see upper right in Figure 5.2), the pistons of which turned a small shaft which carried a worm gear. This worm gear meshed with a worm wheel which was mounted concentrically on the hollow main rotary drive shaft of the machine. This drove the drill rod and the cutting head, which was attached to the drill rod by a quick-fitting screw thread. Thus, rotation of the shaft by the hydraulic engine caused the drill rod and cutter to rotate. The two cylinders of the hydraulic engine were connected by cross waterways (well shown in Figure 5.2, upper right) so that the piston of one acted as the valve of the other. The speed of rotation of the drill rod was varied to suit the hardness of the rock, it was usually 5–7 but could go up to 10 revolutions per minute. As mentioned before, the drill rod was tubular, having an outside diameter at the cutter head of 3 in and a bore of 1½ in; the depth of hole bored was normally about 4½ feet. Usually, ten or eleven shotholes were bored for a 6½ ft by 9½ ft heading. The machine operated on water supplied at a pressure of 50–200 atm and the water consumption was 1.8–2.5 m^3 per hour. The hydraulic engine developed 8–15 hp and ran at 200–300 revolutions per minute. A Brandt drill can be seen today in the Bergbau Museum in Bochum, Federal Republic of Germany.[7]

During a trial in hard dolomite in the Sonnstein Tunnel in Austria, the Brandt hydraulic drilling machine drilled 12 ft of hole in 2½ hours, and using only one machine the tunnel heading was advanced 6 ft in 24 hours.[8] This drill was an immediate success and was used to drive the heading on the Swiss side of the later Arlberg Tunnel. On the subsequent Simplon Tunnel, the Brandt drill was used exclusively, four machines being used in each heading. A remarkable average rate of tunnel progress of 96 ft per week was achieved. In November 1898, while the Simplon tunnel was still being driven, Brandt died suddenly at a time when his rock drilling mahine seemed destined to be the one used for all future long hard rock tunnels. In the event, the Brandt drill seems not to have been used again after driving the Simplon Tunnel, being superseded by the Ingersoll type of compressed air drill, one of which, ironically, had been tested on one of the cross passages of the Simplon Tunnel. Some details of these tunnels and the drills used are given in Table 5.1.

In view of the promise shown by the Brandt drill and its success in use on the Sonnstein, Arlberg and Simplon Tunnels, it is worth examining why it virtually disappeared from tunnelling work after the Simplon

Table 5.1. *Nineteenth-century Alpine tunnels and drills used to drive them*

Tunnel	Length (miles)	Date	Rock drilling machines used	Type of rock drilling machine
Mont Cenis	7½	1857–1871	Sommeiller	Compressed air
St Gotthard	9¼	1872–1882	Dubois–François	Compressed air
			McKean	Compressed air
			Ferroux	Compressed air
Arlberg	6½	1880–1884	Ferroux	Compressed air
			Brandt	Hydraulic
Simplon	12¼	1898–1906	Brandt	Hydraulic

Tunnel. During the time the Brandt drill was in use, average daily drilling outputs for a single drill were: 22 ft in gneiss, 21 ft in granite, 19 ft in sandstone and conglomerate, and 13 ft in coal mining. The Ingersoll drill which, as we have seen, replaced the Brandt drill was described by H.S. Drinker as 'one of the best ever invented'.[9] It was a percussive drill and not a rotary drill, and was driven by compressed air, so was in the mainstream tradition of the compressed air rock drilling machines described in Chapter 2. The Ingersoll drill was patented in December 1871 and introduced in 1872 for driving the Musconetcong Tunnel in New Jersey, USA. As well as being a highly effective rock drill it also proved to be trouble-free, running for months with little or no repairs. The following average daily outputs were obtained for a single drill: 25–40 ft in syenite (Musconetcong Tunnel), 140 ft in marble, 40 ft in granite, and 50 ft in granite and feldspar ore. Comparison of these rates with the ones for the Brandt drill given above shows that the Ingersoll drill was capable of easily twice the output of the Brandt drill in comparable rock types. We need look no further for a reason why the Brandt drill was replaced by the Ingersoll drill; although the new hydraulic rotary drill was a remarkable innovation it was less effective than the existing compressed air percussive type. This highlights a fairly obvious but nevertheless important principle, namely that to be successful an innovation must work better than the thing it is intended to supersede. Modern experience is that rotary core drilling, even with diamond tipped drill bits, is far slower than percussive drilling, and has not, therefore, found any application in drilling shotholes for drill-and-blast tunnelling. There can be no doubt that the Brandt drill was superseded for the same reason.

Before leaving the Brandt drill it is interesting to note that although this type of rotary hydraulic drilling machine was not used again for drill-and-blast tunnelling, the principle much later reappeared in another guise; for in 1970, a hydraulically powered rotary drilling machine with a coring drill bit was introduced for underground prospecting in mines.[10] For this application the ability to recover an intact core of rock is far more important than speed of drilling.

5.2 Modern hydraulic rock drills

The history of modern hydraulic rock drills is so close to the present time that little has been written. Also, manufacturers are, at this stage, more interested in the current and future developments than those of the recent past. Apart from prototype machines made by Dormans of Stafford in the 1920s and by Richard Sutcliffe Ltd for the National Coal Board in 1962, the first working hydraulic percussive rock drills were those pioneered by the French firm of Montabert.[11] This company, a major drilling equipment manufacturer situated in Lyon, took the lead in producing and marketing hydraulic drills. Their first drill (the H50) was introduced in 1969 and by 1973 they had produced three different models, designated H50, H60 and H100.[12] At the end of 1973 they had sold over 130, some of which had been in service for over three years. The Seelisberg Motorway Tunnel in Switzerland, constructed between 1971 and 1977, was driven using Montabert drills.[13] Excavation was by drill and blast and the shotholes were drilled to just over 4 m depth by a Montabert jumbo carrying seven hydraulic drills. The jumbo took about one hour to drill the 105 holes required for a blast in the hard siliceous limestone rock. Figure 5.3 shows a modern Montabert rail-mounted hydraulic drilling jumbo;[14] the view is of the front of the machine as seen looking back down the heading from the tunnel face. It consists of a rail-borne carriage on the front of which are attached four hydraulically-controlled articulated arms, often called 'booms'. Each arm carries at its end a long rigid bar on which a Montabert hydraulic rock drilling machine is mounted and which incorporates the feed mechanism for the drill, usually a hydraulic chain feed, and a support for the drill steel. The jumbo chassis carries a group of hydraulic pumps, driven by a 30 kW electric motor, for supplying the hydraulic oil to the drilling machines, their feed mechanisms and the booms. The jumbo shown in Figure 5.3 is of suitable size for driving a highway tunnel, but smaller and larger ones are manufactured as well. It is interesting to compare the modern Montabert jumbo with the nineteenth-century drill carriages shown before in Figures 2.4 and 2.9.

At the same time that Montabert were developing their hydraulic drill, another French drilling manufacturer, Secoma, also of Lyon, decided to develop a fully hydraulic rotary percussive rock drill and by 1968 a prototype was ready for testing. Developments were also taking place in other countries. By 1970 the American firm Ingersoll–Rand had field tested a prototype hydraulic drilling machine and by 1973 another American firm Gardner–Denver had marketed prototype hydraulic drills. In Sweden by 1973 Atlas Copco had field tested a hydraulic drill and in Germany by 1974 Krupp had developed an all-hydraulic drill. All these firms were manufacturers of compressed air rock drills who realised that the future lay with a change to hydraulics.[15]

To sum up then, by October 1974 six major European and two American drill manufacturers had each developed a version of the new hydraulic rotary percussive rock drilling machine. All the development

Figure 5.3 Montabert drilling jumbo (1980)

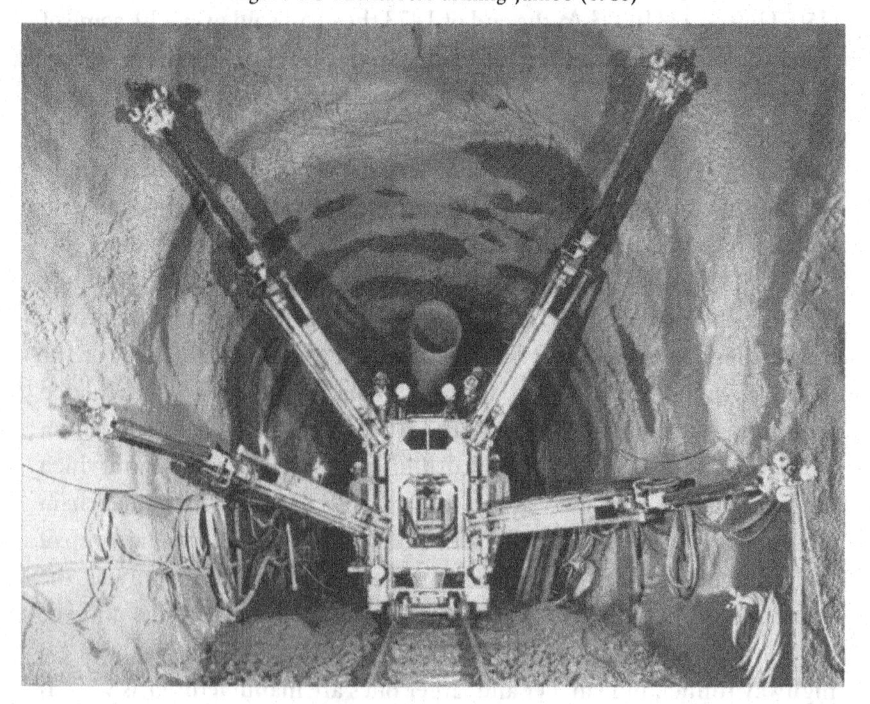

work for these drills took place 'in house', that is within each company's engineering design and development department. One exception to this was that the French firm Secoma co-operated with Birmingham University in the development of their hydraulic drill. However, apart from this, the development of the hydraulic rock drill was entirely industry-based.

5.3 Atlas Copco hydraulic rock drill

The Atlas Copco COP 1038 hydraulic rock drill,[16] first introduced in September 1973, is typical of the early generation of modern hydraulic rock drilling machines and is shown in Figure 5.4. The drill was developed by Atlas Copco's chief engineer Vigg Romell. It has a long slim piston 1, whose diameter is the same as that of the drill steel shank 3, thus ensuring maximum transfer of energy from the piston to the drill rod. It operates on hydraulic oil supplied at a pressure of 150–250 atm, and the impact rate can be varied from 2300–3600 blows per minute by means of a three-position adjuster 2, heavier impacts being given at the lower rate and lighter impacts at the higher rate. There is also a mechanism for varying the rotation rate of the drill rod, and an external water flushing system 4. A shock absorber 5, built into the heart of the drilling machine, dampens the reflected shock wave travelling back up the drill steel and thereby extends its life by reducing fatigue stresses. When first put to use, the COP 1038 was found to be capable of drilling holes in Swedish granite at a rate of 1.5–2 m per minute compared with a maximum rate of about 1 m per minute for compressed air drills.[17] However, a faster drilling speed was not the only advantage: the hydraulic drill showed significant energy savings and a lower working noise level compared with compressed air drills. There was also a better atmosphere in the tunnel because of the elimination of the fog that is sometimes produced by the exhaust of compressed air drills. The weight of the COP 1038 hydraulic rock drill is 145 kg, but this is no disadvantage because it is designed to be used on a jumbo, mounted on a boom which incorporates the feed mechanism which is consequently not shown in Figure 5.4. The length of the drilling machine shown in the Figure is just under 1 m.

Before being placed on general sale, the COP 1038 was tested for a year in the Boliden Mining Company's mine at Laisvall, Sweden, where two of the hydraulic drills were mounted on a twin-boom jumbo. The rock was

Figure 5.4 Atlas Copco hydraulic rock drill (1978)

hard sandstone and drilling rates of 1.4 m per minute were achieved when drilling 50 mm diameter, 4 m deep shotholes – this rate of penetration being 50–60 per cent higher than when using conventional compressed air machines. A total of some 200 000 m of drilling was carried out during this year's test. We will now consider the performance of the drill when driving an actual tunnel, the Megget Tunnel, and at the same time compare it with the performance of a compressed air drill in the driving of a similar tunnel, the Wyresdale Tunnel.

5.4 Wyresdale and Megget Tunnels

The Wyresdale and Megget Tunnels were of similar size, were driven in somewhat similar hard rock conditions by the drill-and-blast method and were constructed by the same contractor, Edmund Nuttall Ltd. The main difference in the jobs was that on the Wyresdale Tunnel compressed air rock drills were used while on the Megget Tunnel hydraulic rock drills were used. The Megget Tunnel was in fact the first civil engineering application of hydraulic rock drills in Great Britain.

The Wyresdale Tunnel forms part of a scheme, the purpose of which is the bulk transfer of water from the River Lune near Lancaster to the River Wyre, some 13 km to the south. The tunnel is $6\frac{1}{2}$ km long and passes under the Bowland Forest range of hills, having been excavated through hard sandstones, siltstones and mudstones. The tunnel was constructed between 1975 and 1978.[18]

The Megget Tunnel was constructed to carry water from the Megget valley dam to the Manor valley near Peebles in Scotland. The tunnel is almost 8 km long and was driven beneath the Greenside Law and Black Law hills, the drive being predominantly in hard sandstone. The tunnel was constructed between 1978 and 1980.[19] Table 5.2 compares the construction details of the two tunnel drives. It can be seen that construction details of the two tunnels are very similar except for the type of rock drilling machines used. Although rock conditions and the primary support required were somewhat different in the two tunnels, the contractor considered that the much better advance rate at Megget was largely due to the superior performance of the hydraulic rock drills over the compressed air rock drills used at Wyresdale. And it should be noted that two hydraulic drills gave a better tunnel advance rate than four compressed air drills.

Table 5.2. *Construction details of the Wyresdale and Megget Tunnels*

Construction detail	Wyresdale Tunnel	Megget Tunnel
Rock type	Sandstone, siltstone, mudstone	Mainly sandstone
Length	6.6 km	7.8 km
Profile	3 m horseshoe	3 m horseshoe
Rock drills used	4 Atlas Copco BBD 90	2 Atlas Copco COP 1038
Type of drill	Compressed air	Hydraulic
Tunnel advance rate	52 m per week	75 m per week

The superior performance of the hydraulic rock drilling machine at Megget over that of the compressed air machine at Wyresdale is just one example of the improvement that the hydraulic drill brought to the tunnelling industry. As hydraulic machines gradually became more widely used the speed of drilling shotholes rose by up to 30 per cent (last item in Figure 4.4), and with it the speed of tunnel advance.[20]

5.5 Summary

The hydraulic rock drilling machine was developed in the 1970s by most of the major manufacturers of rock drills because of a realisation within the industry that the compressed air rock drilling machine had reached the technical limit of its development. Replacing compressed air by hydraulic oil as the working fluid allowed the power and efficiency of the drilling machine to be greatly increased. This innovation did not depend either directly or indirectly on any scientific discovery or new principle. The transmission of power by hydraulics had been known for a very long time – as we have seen the Brandt hydraulic drill was invented in 1876. However, the Brandt drill being a rotary drill, was not the precursor of modern hydraulic rock drills which are of percussive type.

The development of modern hydraulic drills was made as the result of applied research carried out within the rock drilling machine industry. As R.L. Bullock, a reviewer of developments in hydraulic drills, says:

> The development of this line of drills has evolved through slow but in-depth research into all of the various components that must transmit energy from the hydraulic fluid to the rock.[21]

Coming from within the industry, the hydraulic rock drilling machine is a very good example of an innovation based on advances in the technology itself, and not due to an independent scientific discovery or any other external agency.

As will be seen later in this book (Chapter 11), full-face tunnelling machines for use in hard rock became a practical proposition from the 1950s onwards, thereby providing an alternative to the drill-and-blast method. Although there is no evidence that this had any direct bearing on the development of hydraulic rock drilling machines, the potential challenge of the tunnelling machine may have been in the back of the drill manufacturers' minds as an added incentive to produce a successor to the compressed air rock drilling machine that would have a much superior performance.

Notes and references for Chapter 5

1 Sandström, G.E. (1963) *The history of tunnelling*. London (Barrie and Rockliff) p. 293.
2 Edmunds, P.L. (1978) Could hydraulics help your drilling? *Mine and Quarry*, July/August, pp. 31–8.
3 Martin, D. (1979) Hydraulic drills twist the tail of pneumatics. *Tunnels and Tunnelling*, **11**, (10), pp. 14–21.
4 Sandström, G.E. (1963) *op cit*, pp. 181–2.
5 The Brandt hydraulic rock drilling machine is surprisingly well documented. Descriptions of it are given by:
Prelini, C. & C.S. Hill (1901) *Tunnelling: a practical treatise*. London (Crosby Lockwood and Son) pp. 102–7.
Lucas, G. (1926) *Der Tunnel: Anlage und Bau*. Band 2. Berlin (Verlag von Wilhelm Ernst und Sohn) pp. 49–55.
Andreae, C. (1948) *Les grands souterrains transalpins*. Zurich (S A Leeman Frères and Co) pp. 11–13.
Anon (undated) *Cinquantenario de traforo del Sempione*. Milano (Museo Nazionale della Scienza e della Tecnica Leonardo da Vinci) pp. 22–41.
6 Stauffer, D.M. (1906) *Modern tunnel practice*. London (Archibald Constable and Co Ltd) pp. 271–3.
7 The Bergbau Museum kindly provided the author with full technical data on their specimen of the Brandt hydraulic drill (Card No 622).
8 Drinker, H.S. (1893) *Tunnelling, explosive compounds and rock drills*. New York (John Wiley and Sons) 3rd edition, pp. 245–7.
9 *Ibid*, p. 264.
10 Anon (1970) New diamond drill cuts labour, adds speed. *Engineering and Mining Journal*, **71**, (Dec), pp. 85–6.
11 Bullock, R.L. (1974) Industry-wide trend toward all-hydraulically powered rock drill. *Mining Congress Journal*, **60**, (Oct), pp. 54–65.
Anon (1973) French drilling equipment. *Mining Magazine*, (June), pp. 481–3.

12 Described in the manufacturer's brochure:
Montabert SA (1973) Marteaux perforateurs hydrauliques H50, H60, H100. Lyon (Montabert SA).
13 Hone, A. (1975) Seelisberg Motorway Tunnel. *Tunnels and Tunnelling*, 7, (2), pp. 63–6.
14 Described in the manufacturer's brochure:
Montabert SA (1980) Brochure No 5–2.500. Lyon (Montabert SA).
15 Bullock, R.L. (1974) *op cit*, p. 56.
16 The Atlas Copco COP 1038 hydraulic drill is described in more detail in the manufacturer's brochure and handbook:
Atlas Copco AB (1978) Hydraulic rock drilling. Sweden (Atlas Copco AB).
Atlas Copco AB (1975) Hydraulic tunnelling. Sweden (Atlas Copco AB). Unpublished, but a copy is available in the Library of the Institution of Civil Engineers, London.
17 Edmunds, P.L. (1978) Hydraulics versus pneumatics. *Tunnels and Tunnelling*, **10**, (6), pp. 49–52.
18 Boyd, J.L. & J.R. Stacey (1979) Wyresdale tunnel: design, the contract and construction. *Tunnelling 79*, London (The Institution of Mining and Metallurgy) pp. 338–46.
19 Martin, D. (1978) Megget water tunnel breaks new ground. *Tunnels and Tunnelling*, **10**, (10), pp. 13–15.
20 A number of case histories of tunnels driven using hydraulic rock drilling machines are given by:
Ottosson, L. & T.I. Cameron (1976) Hydraulic percussive rock drills – a proved concept in tunnelling. *Tunnelling 76*. London (The Institution of Mining and Metallurgy) pp. 277–86.
21 Bullock, R.L. (1974) *op cit*, p. 55.

Tunnelling shields

Our Shield and Defender – the Ancient of Days.

The foregoing Chapters have dealt with innovations in methods of driving tunnels through hard rock. But in the nineteenth century the demand also arose for methods of driving tunnels through the gravels, sands, silts and clays that formed the subsoils of many cities. The problems of doing this were very different from those of rock tunnelling, and it is the purpose of this Chapter to discuss the outstanding innovation which made soft ground tunnelling a practicable proposition – the tunnelling shield. Before dealing with the tunnelling shield, an account will be given of how the need for tunnels in cities arose in the nineteenth century. The account will use the example of London, because it was there during the early and middle years of the century that the need for tunnels for a variety of purposes made itself felt in a pressing way. In the description that follows, most of the localities mentioned are shown on the map given in Figure 6.1. Although London is here taken for consideration it should be borne in mind that the problems that arose in London were also experienced in other Victorian cities to a greater or lesser extent. This has been discussed by Asa Briggs who has described, for example, how in the Australian city of Melbourne, a system of metropolitan sewerage was put in hand in the late 1890s that was similar to the main drainage scheme for London of the 1860s.[1]

6.1 The need for soft ground tunnels

During the nineteenth century, London experienced a rapid increase in population. This was principally because of the more general migration of people from the countryside to the towns that was a concomitant of other social and economic changes taking place in Great Britain at the same time, although natural increase also contributed to it. Greater London's rapid population growth during the century can be illustrated by the approximate figures given in Table 6.1. The increase in population of London and other cities created the need for more services to cope with their basic requirements of water, sanitation and transport as well as placing a strain on the existing already inadequate arrangements. The consequences of the deficiencies in water supply and sewerage are well known and resulted in the public health legislation which is such a feature of Victorian improvement.

The cholera epidemics of 1832, 1849 and 1854, which were spread by raw sewage being discharged into the Thames while at the same time drinking water was being pumped from it, finally forced the city to implement a system of drainage works. A Metropolitan Board of Works was set up in 1855, its main function being to design and construct a system of sewerage which would prevent all the sewage of the metropolis from passing into the Thames anywhere in the vicinity of the city. Because all the existing sewers discharged into the Thames, the principal element in London's main drainage scheme was the construction of five interceptor sewers running parallel to the Thames, three built on the north and two on the south side of the river, which intercepted the old sewers and allowed their contents to be diverted downstream where they

Figure 6.1 Localities in London mentioned in Chapter 6

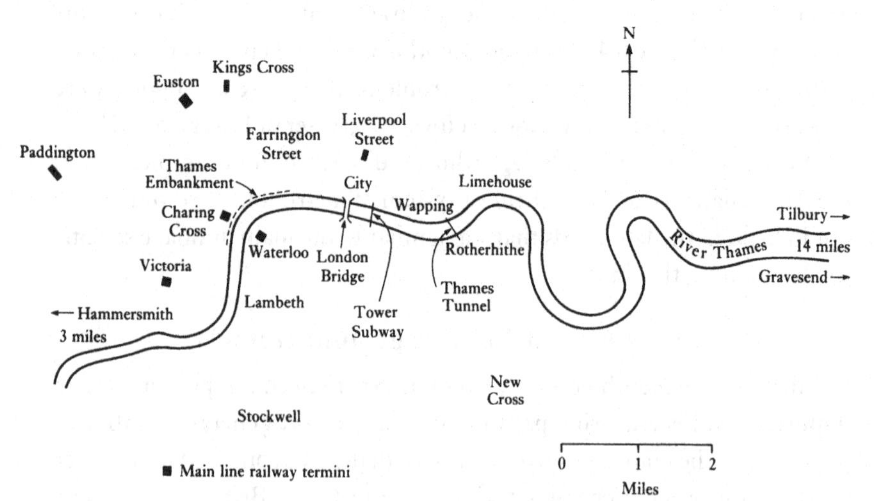

Table 6.1. *Growth in London's population during the nineteenth century*

Year	Population
1800	800 000
1820	1 400 000
1840	2 000 000
1860	2 800 000
1870	3 200 000
1900	5 000 000

could be discharged into the Thames far below the built-up area of the city. This work, 82 miles of brick-lined sewer, built in or tunnelled beneath a major city, constitutes a major achievement of civil engineering. It was carried out by Joseph Bazalgette (1819–91), the Board's Chief Engineer, and was completed in 1865, subsequently proving to be completely successful in operation.[2]

Bazalgette's interceptor sewers were constructed partly by the cut-and-cover method and partly by true tunnelling. The cut-and-cover method of construction consists of excavating a trench to the desired depth, building the tunnel structure in the bottom and then backfilling the trench. The interceptor sewers were constructed from brickwork, stock bricks being used for all but the invert which was lined with Staffordshire blue bricks. The thickness of lining varied from 9 in to 2 ft 3 in depending on the size of the sewer. The sewers varied in dimensions from about 4 ft by 3 ft up to 12 ft by 10 ft, getting larger as they went downstream. The usual cross section was egg-shaped with the narrow end down, but some were circular. The following description of a section of sewer constructed by tunnelling at Upper Mall, Hammersmith, is typical.[3] A hand-excavated heading, 10 ft high by 7½ ft wide, driven at a depth of 20 ft, was lined with timber along its walls and roof to support the ground. In this was constructed a brick-lined sewer of egg-shaped cross section, 4 ft by 2 ft 8 in internal dimensions. When the sewer had been constructed the space between the outside of the brickwork and the inside of the timbered heading was backfilled with spoil from the excavation. Some of the tunnels constructed by this method were as deep as 60 ft below the surface.

Figure 6.2 Uses of the underground at Charing Cross (1867)

As well as a sewerage system, the people of London needed to be able to get about and this was becoming increasingly difficult in streets crowded with horse cart, omnibus and cab traffic, pedestrians, street traders, itinerant salesmen and other colourful characters of the Victorian city street scene.[4] Furthermore, the main line railway termini, built between 1836 and 1874, were all sited well apart from each other (see Figure 6.1), adding to the transport problems of the city. Also, the people living on the outskirts of the metropolis needed to get into the centre daily to work. Underground railways were the solution to this problem, and the first of these, indeed the first underground railway in the world, the Metropolitan Railway Company's line from Paddington to Farringdon Street, was completed in 1863.[5] Much of this line was built only a few feet below the ground surface by the cut-and-cover method of construction. This proved immensely disruptive, resulting in chaos at the surface with streets obstructed, buildings in danger of collapse and with the need for sewers, gas and water mains to be diverted. Later, when the District Railway was built in 1869, the railway tunnel was constructed at the same time as a trunk sewer and the Thames Embankment (now called the Victoria Embankment) in order to gain the maximum advantage from the inevitable surface disruption. A contemporary print[6] reproduced in Figure 6.2 shows these works at Charing Cross, and illustrates the increasing use being made of underground space in nineteenth-century cities. The numbered items in Figure 6.2 are:

1 Subway designed to accommodate gas pipes, water pipes, telegraph wires and 'other underground apparatus of a great city'.
2 Northern low level intercepting sewer for London's main drainage system; 7 ft 9 in in diameter.
3 District Railway built by cut and cover. Now the District Line of the London Underground.
4 Pneumatic railway for transmission of letters, newspapers and small packets. This tunnel was intended to connect Charing Cross and Waterloo Station, but was in fact never built.

The cut-and-cover method of constructing underground railway tunnels proved unacceptable because of the scale of the surface disruption involved, and later lines, the 'tubes' were constructed by true tunnelling methods using shields. These provided the capital with the now familiar

Figure 6.3 Tunnel driving in soft ground (1840)

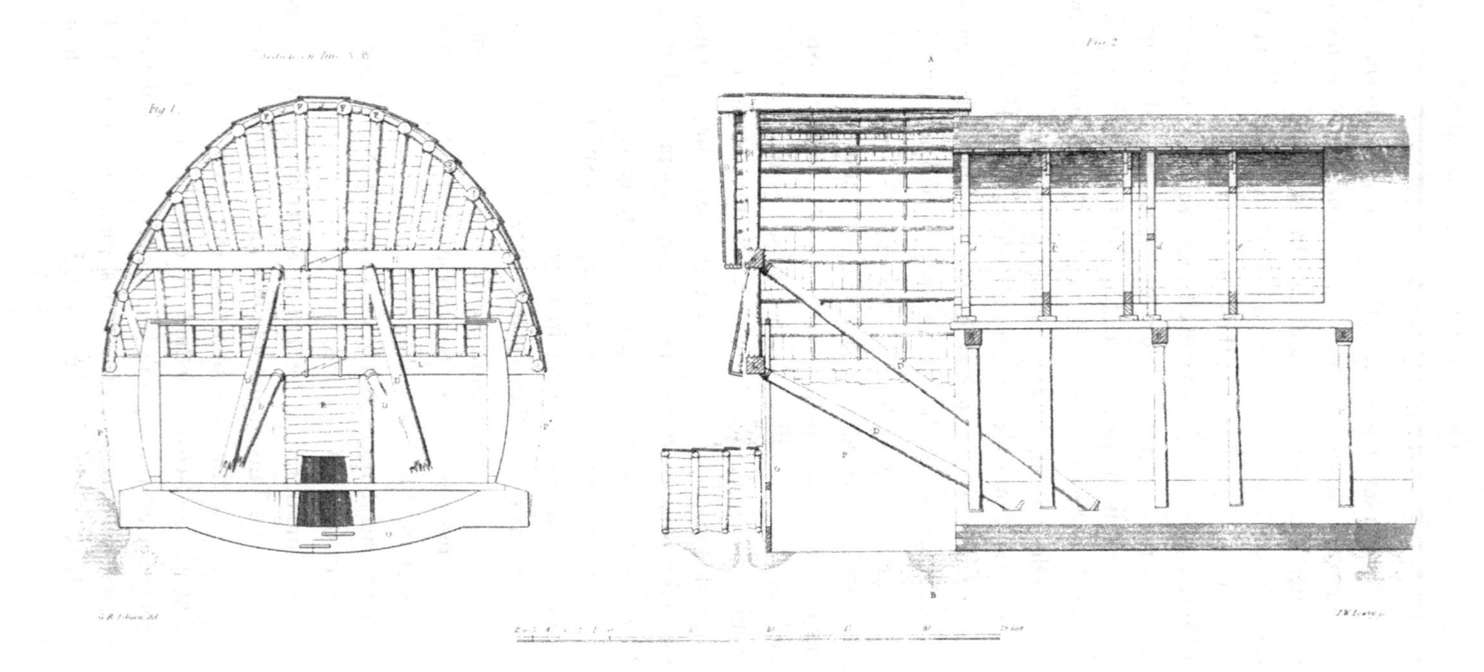

network of the London Underground without any construction chaos at the surface at all.

Other uses of the underground were made, for example tunnels were provided for gas and water pipes and telegraph lines, and for a brief period, for the pneumatic transport of parcels (see Figure 6.2). Also, below London Bridge the river Thames was a major barrier to north–south communication because bridges could not be built there on account of the need to keep the river unobstructed for navigation. As we shall see, this provided the incentive for three attempts at a tunnel crossing, one of which was successful. But before looking at these, a brief account will be given of the traditional method of driving tunnels in soft ground and some of the problems associated with it.

6.2 The traditional method of soft ground tunnelling

Before the 1850s the method of driving any tunnels that were needed through soft ground was essentially the method that had been developed some half a century before during the canal building period and which reached the peak of its refinement during the 1830s with the construction of some of the very early railway tunnels. Different methods evolved independently in different countries, but they were all the same in principle and the one which will be described here is the 'English Method'. The basic principle of all the methods consisted of supporting the ground during excavation with an elaborate temporary structure made of timber, while masons or bricklayers followed behind constructing a permanent stone or brick lining. Figure 6.3 shows the English Method as used on the Blechingley Tunnel in Surrey.[7] An upper main heading was driven ahead of the tunnel and this was supported by heavy timbers called 'crown bars' (F in the Figure), one end of each of which rested on the brick lining while the other end was supported on timber props G and H. Sometimes a smaller bottom heading was driven at the same time, as shown in Figure 6.3, to speed up the work. The miners then worked on either side, widening out the upper main heading, supporting the ground, if required, with poling boards. As the work progressed, further bars were installed until the whole length was supported over the full width. The lower half of the tunnel was then excavated, the sills K and L being progressively propped down to the tunnel floor as they were undercut. Massive shores D and D′ kept the sills securely in place against the ground at the face. Excavation was by teams of miners and the tools used were picks and shovels, or if in clay, spades. The spoil from the excavation was loaded into small skips which were pushed down the tunnel by men or hauled by horses. On a large railway tunnel, the work force, including those engaged on surface operations, might be as great as 4000 men and 300 horses. When the excavation of a length of tunnel, usually about 12 ft, was completed, the bricklayers then built the lining which was supported on timber centering *a, b, c, d, e* until the mortar had set. Miners remained in the tunnel to remove the bars, which were to be used again, as each section was supported by brickwork, an operation known as 'drawing the bars'. Some 5500 bricks per foot of advance were required for a large railway tunnel. The whole cycle of operations involved in soft ground tunnelling by the English Method has been reconstructed and described by Harding,[8] and the English Method and the other contemporary soft ground tunnelling methods used on the Continent have been described by G.E. Sandström.[9] It can be seen that, like the traditional method of hard ground tunnelling described in Section 2.2, the traditional method of soft ground tunnelling was slow and labour intensive; in addition it required elaborate timbering to provide temporary support before the lining was installed. Another shortcoming was that the method could be used only in soft ground *above* the water table or in impermeable strata.

During the nineteenth century tunnel driving in soft ground was revolutionised by some extremely important innovations. These were (i) the tunnelling shield, *c* 1820 (this Chapter), (ii) compressed air tunnelling, *c* 1830 (Chapter 7), and (iii) prefabricated tunnel lining, *c* 1870 (Chapter 8). The introduction of these measures completely changed soft ground tunnelling: before advance rates were typically 3 ft per day but afterwards advance rates of 13 ft per day were commonly achieved.

6.3 Early attempts to tunnel beneath the Thames

The first attempt to tunnel beneath the Thames was that proposed by Ralph Dodd (1756–1822) in 1798 for a crossing between Gravesend and Tilbury.[10] The tunnel was to be circular in cross section with an internal diameter of 16 ft suitable for use by carriages, 900 yd long and to pass about 20–30 ft under the river bed. Finance for the project was raised, an Act of Parliament obtained and trial borings were made. In 1800 the works were commenced by sinking a 10 ft diameter shaft to the west of Gravesend, from the bottom of which it was intended to drive the proposed tunnel. In December 1802, when the shaft had reached a depth of 85 ft below Thames high water level, the project was abandoned. The reason for this was that in spite of the facts that the shaft had been lined and a steam pumping engine had been installed, ground water continued to flood into the works. Well over half the sum of money raised for the whole project had been spent on sinking the shaft and in 1803 the proprietors wound up the affair.

Although Dodd's tunnel scheme ended in such premature failure, it did teach a lesson that should have been heeded for the future. This was that anyone considering a subaqueous tunnel or a tunnel beneath the water table would have to make adequate provision for dealing with the inflow of ground water that would inevitably be encountered. We shall see that this theme recurs in the history of soft ground tunnelling.

A second attempt at a Thames tunnel was proposed in 1802 by Robert Vazie, a Cornish mining engineer, who planned to drive a tunnel across the river from Rotherhithe to Limehouse, some two miles below London Bridge. The Thames Archway Co, as the backers of the scheme designated themselves, was formed in 1805, an Act of Parliament was obtained and work commenced in the same year with the sinking of a brick lined shaft 11 ft in diameter, from a site near Lavender Lane in Rotherhithe. When the shaft had reached a depth of 42 ft ground water flooded in and the Company ordered Vazie to suspend operations. The whole of the money that had been raised for the project had been spent. More finance was raised and Vazie and his workforce of Cornish miners went back to work. The shaft, now at a reduced diameter of 8 ft for economy, was sunk a further 34 ft. Vazie was now ready to start driving the actual driftway beneath the river. At this stage the Company lost confidence in Vazie and Richard Trevithick (1771–1833) the Cornish engineer, already well known as a steam engine and locomotive designer, was called in. Trevithick's initial appointment was as assistant to Vazie, but before the tunnel drive commenced he was promoted to chief engineer. Vazie, protesting, was made Trevithick's assistant, but this arrangement did not last long as Vazie eventually left in October 1807.

Trevithick started driving the driftway in August 1807. The dimensions of the driftway were 5 ft high, 3 ft wide at the bottom and 2 ft wide at the top. To stop the tunnel from falling in the miners followed the standard mining practice of shoring the sides and roof with timber. A steam engine at the bottom of the shaft worked a pump, powered a primitive ventilation system and pulled small muck wagons from the face of the heading where excavation proceeded by hand digging. Illumination was by candle.

Working in this way, Trevithick and his team of Cornish miners advanced the driftway more than 950 ft from the shaft by January 1808, in spite of repeated ground water problems. However, at the end of January an inrush of sand followed by a torrent of water completely flooded the workings. Clay placed on the bed of the Thames above the breach partially sealed the tunnel and within a week the workings were pumped out and the miners were back at work. By February they had advanced the driftway to 1059 ft, only 117 ft short of its final intended length, and Trevithick estimated that completion of the drive would take place in 10–12 days time. However, the tunnel again flooded and the directors of the Thames Archway Co would not advance any more money for putting in hand Trevithick's proposal for sinking a caisson in order to recover the works. The driftway was never completed, the Company went out of business and the tunnel was allowed to fall in and become destroyed. Contemporary opinion was that the Thames driftway episode cast doubt on the practicability of making a tunnel beneath the river at all.

In October 1980, the Science Museum Library acquired Trevithick's own original plan and description of the tunnel (Archive No MS 207/6), to a scale of 25 ft to 1 in, showing the state of the works in February 1808 and the proposed caisson by means of which Trevithick hoped to complete the drive to the north side of the river. It illustrates how the driftway was not a full-sized tunnel but a small pilot tunnel with a temporary timber lining, and also shows the large scale of the caisson works that would have been necessary to recover the tunnel workings after the last flooding.

The attempt to drive the Thames driftway had demonstrated how hazardous soft ground tunnelling was, particularly if carried out under a river or beneath the water table. Also, it should be noted, even if Trevithick had succeeded in completing his tunnel, it would have been of little use as it was. It was too small for any practical purpose and the timber lining would not have lasted long in the wet soil conditions. To have been of any use it would have been necessary to remove the timber lining, open the tunnel out to a larger size, and instal a permanent durable and watertight lining. To have done all of this would probably have been a more hazardous undertaking than the original drive. Trevithick was aware of this because on one occasion he refused to open out a section of the driftway to the full size of 16 ft high by 16 ft wide when asked to do so by one of the directors of the Company.[11]

This early attempt at driving a soft ground tunnel showed that the following items were needed: (i) a means of supporting the ground and protecting the miners during excavation of the tunnel heading and erection of the lining, (ii) a means of driving the tunnel from the outset at the full required size instead of driving a pilot tunnel which would have to be opened out afterwards, and (iii) a means of excluding ground water from the works. The rest of this Chapter discusses how some of these needs were met by the invention of the tunnelling shield, and shows that the origin of the shield sprang from a remarkable source.

6.4 The Thames Tunnel

Some ten years were to pass after the failure of the Thames driftway project before another Thames tunnel was proposed. The site chosen, between Rotherhithe and Wapping, was very close to that of Trevithick's driftway and it is interesting to digress for a moment to see why this location was so popular for a tunnel. At the time, because of working arrangements in the Port of London, freight and passengers often had to be conveyed from one bank of the river to the other. The passengers could be taken across by the 350 Thames Watermen who operated between Rotherhithe and Wapping, and it is estimated that up to 3700 passengers crossed daily by this means. But the cargo had to be taken by wagon and cart on congested streets via London Bridge. With all the potential traffic, a tunnel at this particular locality was, therefore, an attractive proposition. A tunnel beneath the Thames could not be built, however, until a means of tunnel driving in soft ground with safety had been devised, and to do this the tunnel shield was invented by Marc Isambard Brunel, a brief biographical note of whom will now be given.

Brunel was born in the village of Hacqueville in Normandy, France, on 25 April 1769. He was at first trained for the priesthood, but having no vocation for it and displaying drawing and mathematical abilities instead, was entered for service in the French Navy. In 1793, following the French Revolution, Brunel had to leave France because of his royalist sympathies. After some time in the United States, he came to England in 1799. Brunel soon occupied himself with setting up a block-making plant, the first of a number of enterprises in which great engineering skill was coupled with lack of business acumen. Also in 1799 he married Sophia Kingdom, who he had met earlier in France, and in 1806 their third child and only son Isambard Kingdom Brunel[12] – destined to become an engineer whose fame was to eclipse that of his father – was born. The elder Brunel's greatest project, and the one to concern us here, was the Thames Tunnel which he worked on from 1823 to 1843. In 1840 he was knighted. He died on 12 December 1849.

In 1823, at the urging of I.W. Tate, one of the promoters of the old Thames Archway Co, Brunel and some influential friends formed the Thames Tunnel Co with the object of driving a tunnel beneath the Thames between Rotherhithe and Wapping, and on 24 June 1824 an Act of Parliament for 'Making and maintaining a tunnel under the River Thames' received the Royal Assent. Work began in March 1825 with the sinking of a shaft 50 ft in diameter and brick lined throughout. The site of the shaft was $\frac{3}{4}$ mile west of Trevithick's abandoned driftway. Within eight months the shaft had been sunk to its final depth, a reservoir had been constructed beneath its base and steam driven pumps had been installed.

In November 1825, Brunel's shield, built by Henry Maudslay (1771–1831) in his Lambeth works, had been assembled at the base of the shaft and was ready to commence the tunnel drive. The great shield crept slowly forward at a rate of about eight feet per week and by May 1826 the tunnel had been advanced northward 100 ft. Brickwork was erected close behind the shield to form the twin horseshoe section passageways, each 14 ft wide by 15 ft high and capable of taking carriage and pedestrian traffic. By August 1826 some 200 ft of tunnel had been driven and the tunnel now extended under the Thames. From then on tunnel driving conditions became progressively worse with soft silt and water frequently running in at the face. Finally there were two very severe inundations resulting in complete flooding of the works, the first in May 1827 and the second in January 1828 during which six miners lost their lives. These were sealed by the same method Trevithick had used, that is to say by tipping clay on to the river bed from a barge positioned above the breach. These events, as indeed has the whole story of the Thames Tunnel, have been fully described both by contemporary[13] and later[14] writers; the episodes of the flooding of the tunnel and its recovery rank amongst the heroic epics of civil engineering. Because the ground has been so well covered, it is only necessary to describe the remaining events briefly here.

After the recovery of the works following the second inundation, the funds of the Company were exhausted and it was decided to stop work and block up the frames of the shield pending an appeal for further funds. A debenture scheme was floated, but not enough money was subscribed. In August 1828 the bricking-in of the shield was completed and a large mirror was placed at the end of the heading. The arch was stuccoed and lit with gas lamps and became a sort of novelty attraction, visitors paying for admission to see the tunnel.

Following the flooding of the tunnel in January 1828, the Tunnel Committee considered employing Charles Blacker Vignoles (1793–1875) as chief engineer in place of Brunel and in June 1829 Vignoles put a plan and model of his proposals for completing the tunnel on view in the Company's offices.[15] Little has survived of these except for the fact that Vignoles intended to save money by dispensing with the shield. Both Brunel and his son criticised these proposals severely and in the event the Board of the Company decided in June 1830 that if the tunnel were to be completed, they would not adopt Vignoles' plan but would continue with that of Brunel. It was as well for Vignoles and the miners that his scheme for the Thames Tunnel was not put to a practical trial, because as we have already seen, it was difficult enough to drive the tunnel *with* a shield and to have tried to do so without one would have been to court disaster.

In November 1834 the Government announced that it would advance money for the completion of the Thames Tunnel. By November 1835 the old shield was removed and in February 1837 a new shield, built by Rennies, was installed and commenced driving. Tunnelling conditions were as before, with wet silt and water an ever-present hazard, and to these was added the influx of gas which caused collapse and blindness of the workmen. In October 1840 a shaft was sunk on the Wapping shore to receive the tunnel, and in November 1841 the shield at last penetrated the brickwork of the Wapping shaft. Before the tunnel was opened the shield had to be removed, staircases had to be installed in the shafts and the archways paved, whitewashed and lit with gas. All this was done and the tunnel was finally opened to the public in March 1843.[16] This was shortly followed by a Royal visit.[17]

The tunnel proved popular for a while as a pedestrian crossing, but the Company would not approach the Treasury for the finance to construct the carriage ramps – which were never built. By 1846 it was clear that the tunnel was not a commercial success. From then on the tunnel fell on sad times. First it became the site of fairs and stallholders, but later, by the 1860s it had degenerated into a doss house and haunt for footpads and prostitutes. In 1865, however, a happier future was in store when the East London Railway bought the tunnel and from 1869 onwards ran trains through it from Liverpool Street to the south coast. Today the tunnel forms part of London's Underground system, being an essential part of the East London Section of the Metropolitan Line carrying the line south of the river to New Cross. The original engine house, designed by Brunel, which once housed the steam engines used for pumping water from the finished tunnel still stands close to the tunnel shaft at Rotherhithe; it has recently been restored by the Brunel Exhibition Project.[18]

6.5 Brunel's tunnelling shield

During the years following the failure of the Thames driftway project, Brunel, together with others, gave thought to methods of driving a tunnel beneath the Thames. Brunel concluded that there was no hope of successfully building a tunnel unless the face and the part of the tunnel in front of the masonry or brick lining were somehow supported. One day, while he was at Chatham Dockyard, Brunel noticed a piece of ship's timber which had been bored by shipworms, and in one of the holes there was a living example of the organism. He examined it closely together with the hole it had created, and this examination and the observations that Brunel made gave him the idea of the tunnelling shield.

The shipworm, *Teredo navalis*, is not really a worm at all but a mollusc.[19] It has become adapted to the specialised needs of a life spent burrowing through floating driftwood, ship's timbers and the wooden parts of marine structures. The body of the adult *Teredo* is long and worm-like, with a small shell confined to one end of the body. The shell consists of two valves, the outer surfaces of which are covered with rows of sharp ridges. The shipworm grasps the end of its burrow by suction using an organ called the foot, while the valves of the shell rasp at the wood surrounding the end of the burrow. The front end of the animal can turn through 180° during this process so that a burrow of circular cross section is bored through the wood. As the creature advances through the wood the body elongates while the shell end increases in size; a tapered burrow is thereby formed in the wood which can be up to one foot in length. As the shipworm bores through the wood it also lines its burrow with a hard layer of calcium carbonate. The shipworm in its burrow is shown in Figure 6.4.

This remarkable organism seemed to Brunel to possess the features he was looking for in a tunnelling method: the delicate tissues of the shipworm were completely protected, its head by the strong boring valves of the shell and its body by the lining of its burrow that it secretes as it advances. Brunel resolved to imitate this natural tunneller. With hindsight we can afford to be a little critical of the shipworm as a model of a soft ground tunnelling system. Boring through wood, a self-supporting medium, is more akin to boring through rock than through soil. The shell of *Teredo* does not support the wood during boring, and the calcium carbonate lining does not support the burrow – it probably serves to protect the body of the animal from being bored through by another shipworm. Nevertheless, *Teredo* gave Brunel the idea of the tunnelling shield.

Brunel's tunnelling shield for driving the Thames Tunnel has been described many times, for example, in References 12–14 previously cited. The following technical description is based on that given by Harding[20] but with some modifications. Figure 6.5 is one of the few surviving contemporary engineering drawings of Brunel's shield; it shows a vertical section through one of the frames of the shield. The shield was 38 ft wide

Figure 6.4 The shipworm in its burrow

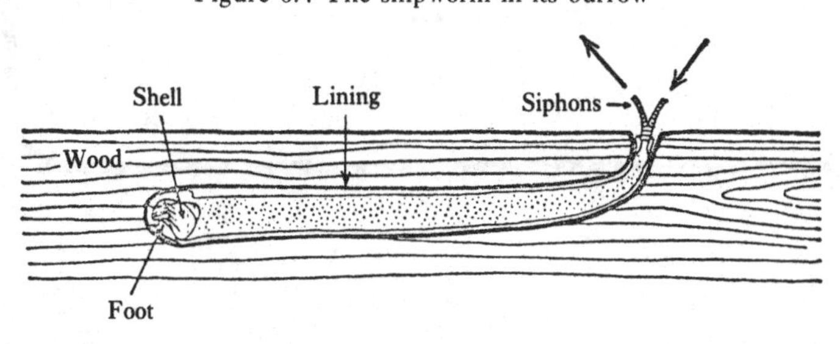

by 22 ft high and consisted of twelve vertical cast-iron frames, side by side, each with three compartments, one above the other. On the tops and sides of the outer frames and on the tops of all the others, cast-iron staves were placed which worked on rollers; they were moved forward as excavation proceeded in order to protect the roof and sides while the whole face of the

Figure 6.5 Brunel's Thames Tunnel shield (1826)

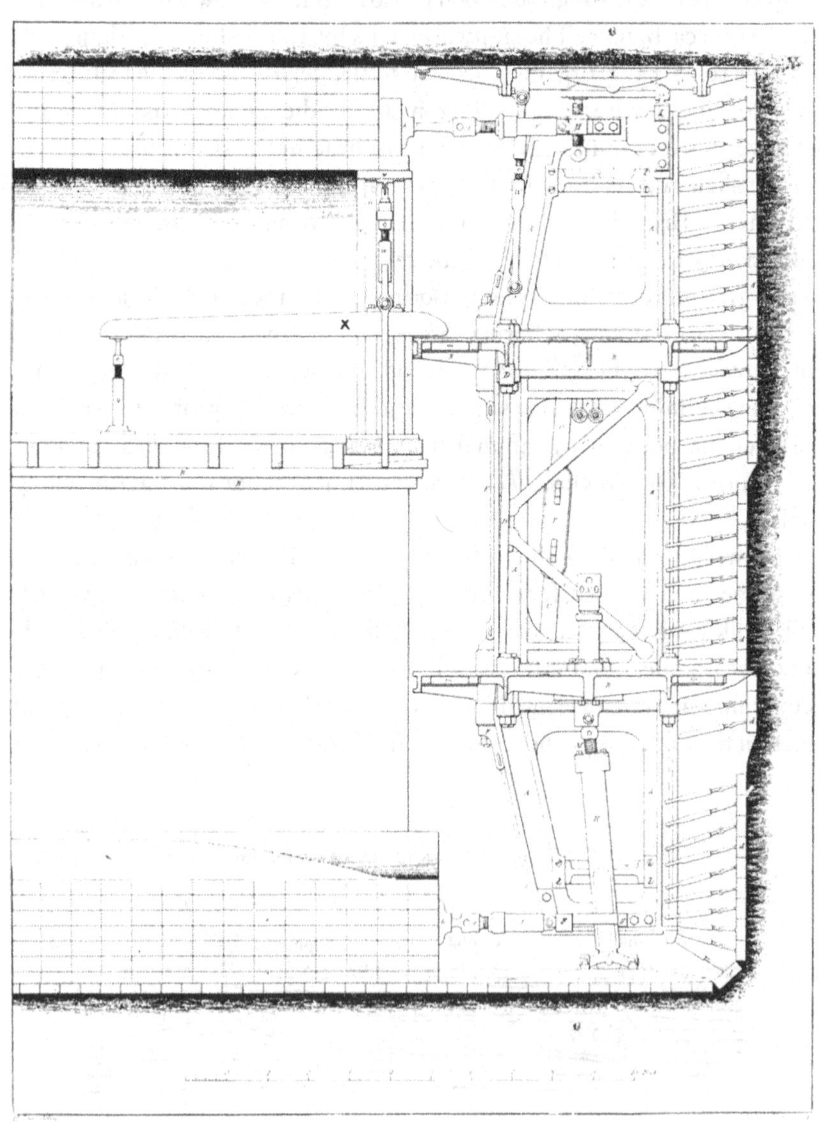

tunnel was supported by timbering. This consisted of horizontal poling boards, each 6 in high, 3 in thick and 3 ft long. Each frame thus had its own panel of boards, which were held against the ground by light screw jacks at each end. The boards could be removed one at a time from the top downwards and the ground in front carefully excavated. Each frame could be advanced separately by means of heavy screw jacks acting against the brick tunnel lining which was built close behind the shield. The shield, therefore, consisted of the equivalent of 36 small headings being driven almost simultaneously. The original method of working was to advance alternate frames $4\frac{1}{2}$ in so that there would never be a gap of more than $1\frac{1}{2}$ in between the ends of the poling boards. As each board was removed the ground was dug out and the board was replaced and strutted off the adjacent frames so that when the face had been 'taken down' the frame was free to be pushed ahead. When this had been done the boards were strutted to their own frame once more.

The tunnel was lined with bricks, the arch being six rings of bricks thick. A flat face to the brickwork was required for the shield jacks to press against. The spandrels were filled with brickwork and the arches between the two carriageways, which were an architectural feature of the tunnel, were cut out after the shield was well ahead so as not to weaken the centre wall on which some of the jacks had to press. The shield towed a timber stage for the bricklayers. The brick lining arch was turned on an iron semicircular rib carried on a long cantilever arm (X in Figure 6.5), one end of which rested on the floor of the upper compartment of the shield and the other on a screw jack on the stage.

In June 1975, for the Brunel Thames Tunnel 150th Anniversary Exhibition at the Institution of Civil Engineers, London, a full-size wooden replica of one of the compartments of a frame of the tunnelling shield was made; this now forms part of the Brunel Exhibition Project's collection. Two scale models of Brunel's shield, one in a model of the Thames Tunnel and the other an enlarged detail of three of the frames, were made for the Hall of Civil Engineering in the Museum of History and Technology of the Smithsonian Institution, Washington DC, USA.[21]

In January 1984 a remarkable collection of original drawings and designs by Brunel came to light. The owner of these is Mrs Marguerite Brunel Hurst, Brunel's great, great grand-daughter. They were restored and mounted by the Ashmolean Museum, Oxford, and put on public exhibition for the first time. The collection included drawings of the Thames Tunnel amongst which was an item of direct relevance to this book – a cardboard model of the tunnel and the shield. A photograph of this model is shown in the Frontispiece.

The model clearly shows some technical details of the shield. The many poling boards supporting the face together with their light screw jacks can be seen, together with the heavy screw jacks at the top and bottom of the frame which were used for advancing it. The screw jacks at the bottom of the frame are fitted with rectangular plates, probably to spread the load when jacking off the fresh brickwork of the lining. The men visible in the two upper cells are each holding a brush and a trowel, so it seems they are brick-layers putting finishing touches to a completed section of the lining. The staves at the top of the frames and the shoes at the bottom are also clearly shown. The other point of interest in the model is the detail of the brick lining that it reveals. We can see how the brick arch was constructed of six rings of brick and how there was a massive amount of brickwork required to fill in the spandrels, although it is probable that not all the infilling was done as neatly as shown in the model. Finally we can note a visitor being shown the tunnel, this being a common practice during its construction. It has not been possible to establish whether the shield depicted in the model is Maudslay's original shield or Rennies' later one – hence the two dates given in the caption.

Our admiration for Brunel, his tunnelling shield and the construction of the Thames Tunnel should not inhibit a considered engineering judgement of the shield. We need to consider how well it met the criteria set out at the end of Section 6.3. It can be seen that the cast-iron top and side staves supported the roof and walls of the excavation, while the poling boards supported the face (Figure 6.5). The need to protect the miners from a collapse of ground was therefore well provided for. The shield also met the second need, that of driving the tunnel at full size. However, the third need, to exclude ground water from the works was not met, and water flooding into the tunnel proved an ever-present major hazard during the whole drive as we have seen. Also, some criticisms of the shield can be made. The first of these is that the frames and staves were made of cast-iron and these cracked from time to time and had to be replaced.[22] A second is that the frames frequently became 'deranged' during the drive[23]

– that is gaps opened up between them and they went off line and level. Thirdly, much brickwork was wasted in converting a large rectangular opening into two smaller horseshoe-shaped carriageways (see Frontispiece). Finally, the shield had a painfully slow rate of advance, the greatest weekly advance being 14 ft, while 9 ft per week was a good average rate.

Because of the enormous cost of the Thames Tunnel project, the long time taken to complete it and the almost overwhelming problems encountered, it was not for a further 25 years that another tunnel under the Thames was undertaken. Nevertheless, despite these difficulties Brunel had demonstrated a potentially practicable method of soft ground tunnelling. Brunel's Thames Tunnel shield, however, with its multiplicity of parts and cumbersome method of working, was destined to be the first and last of its kind and its importance lies in the demonstration of the soundness of a principle – that of supporting the ground over the whole face of a drive – rather than in the detailed arrangements whereby it did this.

6.6 Brunel's patent

Brunel's patent specification of 1818 begins with a statement in which he clearly identifies the main need required to overcome the problem of subaqueous tunnelling in soft ground:

> The chief difficulties to be overcome in the execution of tunnels under the bed of great rivers lies in the insufficiency of the means of forming the excavation. The great desideratum, therefore, consists in finding efficacious means of opening the ground in such a manner that no more earth shall be displaced than is to be filled by the shell or body of the tunnel, and that the work shall be effected with certainty.[24]

An examination of Brunel's patent shows that it, in fact, covers the design of two shields, neither of which is the shield he actually used to drive the Thames Tunnel. The first shield described in the specification is a cylindrical shield about 14 ft in diameter, divided across the horizontal diameter by a working platform into an upper and a lower half. Each half was further divided into five vertical cells each of which accommodated a miner and each of which had its own set of poling boards to support the face. Each cell could move independently, pushed forward by a hydraulic ram which reacted against a frame erected further back in the completed section of tunnel. The lining, which was to be of cast-iron segments, was erected under protection of a tailskin at the back of the shield. The hydraulic rams were specified to be double-acting so that they could also pull the reaction frame forward from time to time as the tunnel progressed. The whole arrangement has a thoroughly workmanlike appearance and looks much more practicable than the cumbersome rectangular shield Brunel eventually used in 1825 when he came to drive the Thames Tunnel. Maybe the cylindrical shield with independently moving cells, each of a different size and shape, was too difficult to manufacture whereas a rectangular shield in which all cells were identical was easier to make. Whatever the explanation, the first shield in Brunel's patent of 1818 is much more like tunnelling shields as they eventually developed than is his Thames Tunnel shield. Also the cast-iron segmental lining is clearly shown and its safe erection described whereas the Thames Tunnel was constructed with the less safely built and more time-consuming brick lining.

The second shield described in the patent specification is also a cylindrical shield, but quite different in design and operation from the first, being like a large pair of concentric augers; it is generally considered to be impracticable. However, it is of interest to us here because in the preamble Brunel states that its design is based on *Teredo navalis*. Now in fact the design of the first shield is more like that of the shipworm than is that of the second shield; the design of the second shield is more like that of an auger – as Brunel himself acknowledged by sometimes calling it an auger as well as a *Teredo*. Therefore it must be the case that the shipworm provided the idea for the concept of the tunnelling shield in general, and not just for the design of the second one mentioned in the patent specification.

6.7 The Tower Subway

The next tunnel to be driven under the Thames was the Tower Subway.[25] This tunnel was driven from the bottom of a shaft, 10 ft in diameter and 60 ft deep, sited at the corner of Lower Thames Street in Tower Hill on the north bank on the Thames to the Vine Lane end of Tooley Street on the south side of the river – a distance of 1350 ft. The engineer for the tunnel was Peter William Barlow (1809–85) but the man in charge of the actual drive was James Henry Greathead (1844–96). Greathead was born in

Cape Colony, South Africa, but was sent to England, aged 15, in 1859 in order to complete his education. In 1864 he was working under the tutelage of Barlow, who by then was established as an engineer of repute with an office in London. The Tower Subway tunnel was constructed in 1869 and was circular in cross section, 7 ft in diameter and lined with cast-iron lining segments which were bolted together. To drive the tunnel a tunnelling shield was used which will be described shortly. The 900 ft long section under the river was completed in 14 weeks and the whole tunnel completed in under a year, almost without incident. Barlow had located the level of the Tower Subway over 22 ft below the river bed in the firm stratum of London Clay which lay below the soft alluvial deposits. By doing this he had placed the tunnel in an almost perfect medium for shield tunnelling. Because of this and because the shield used was so much simpler to operate than Brunel's, the works proceeded at a pace of startling rapidity compared to the Thames Tunnel, although it must be borne in mind that the excavated face area of the Tower Subway was only one-twentieth of that of the Thames Tunnel.

When it was first opened, passengers were conveyed through the Tower Subway in small cylindrical cars holding 12 persons each that nearly filled the tunnel. The cars were drawn on 2 ft 6 in gauge rails by cables operated by small steam engines situated in the shafts. There were also steam-operated lifts in the shafts. Later, the scheme not paying its way, the tunnel was converted to a pedestrian walkway and the lifts in the shafts were replaced by spiral staircases. The pedestrian tunnel was finally closed to traffic in 1898 when the nearby Tower Bridge was opened. The Tower Subway was then purchased by the London Hydraulic Power Co and two hydraulic power mains were installed in it. It was also used to accommodate two large-diameter water mains for the Metropolitan Water Board. In December 1940, during an air raid on London, a bomb fell in the river near Tower Pier and made a crater in the river bed very close to the tunnel. The shock of the explosion cracked the cast-iron lining longitudinally and the lining closed like a spring over itself, reducing in diameter at the point nearest the bomb to about 4 ft. During the course of repair work the opportunity was taken to examine the condition of the 70 year old tunnel. The cast-iron lining was found to be in excellent condition and very few places were found where the grouting had not been entirely satisfactory.[26]

6.8 The Barlow–Greathead shield

In 1862 Barlow was resident engineer on the construction of Lambeth Bridge, the piers of which were constructed by sinking cast-iron cylinders vertically into the London Clay. This gave him the idea of a tunnel shield that would advance as a whole, not in parts as Brunel's did, and with which tunnels might be driven as easily as the bridge pier cylinders had been sunk. In 1864 and 1868 Barlow patented his idea of a circular tunnelling shield.

The tunnelling shield for driving the Tower Subway was designed by Greathead, but it was obviously inspired by Barlow's ideas. The upper illustration in Figure 6.6 shows a drawing prepared in Greathead's office of the Tower Subway shield. The Barlow–Greathead shield consisted of a cylinder 7 ft 3 in in diameter made from ½ in thick iron plate. At the front (right on the drawing) was a cast-iron ring to which was bolted a diaphragm made of wrought-iron plate having a rectangular opening in the middle extending to within a few inches of the top to permit the passage of workmen and materials. Just in front of the diaphragm was fitted a sharp circular cutting ring to assist in penetrating the ground. The cylindrical portion of the shield supported the radial pressure of the ground while the diaphragm supported the face; if required the doorway in the diaphragm could be closed by dropping across it 3 in wide boards, the ends of which could be slotted in the channel bars on either side of the doorway. The inside of the shield was fitted with six 2½ in diameter screw jacks (three are shown in the drawing) which enabled the shield to be jacked forward off the tunnel lining. The method of working was as follows. The London Clay was found to be stiff enough to be readily excavated without need of support from the diaphram and so three miners worked in front of the shield digging out the clay and passing it back through the door in the diaphragm. The shield was then jacked forward 18 in by means of the screw jacks (see drawing at bottom left of Figure 6.6) which were operated by hand. Following the shove forward of the shield, a permanent lining made from cast-iron segments was erected (see drawing at bottom right of Figure 6.6).

This was the first time a permanent lining of cast-iron had been used for a tunnel, and was in itself a major innovation in soft ground tunnelling. Not only could the segments be placed and bolted up more speedily than a masonry or brick lining could be built, but the cast-iron, unlike the other linings, could immediately bear the reaction of the shield jacking screws. The cast-iron lining for the Tower Subway consisted of 6 ft $7\frac{3}{4}$ in internal diameter rings 18 in long, each of three segments and a key piece. The

Figure 6.6 The Tower Subway shield (1869)

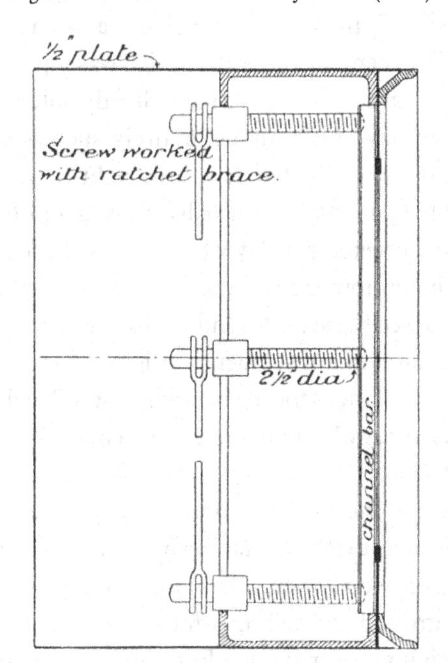

thickness of the cast-iron was $\frac{7}{8}$ in. A ring of segments was erected and bolted to the previously built ring under cover of the overlapping rear tailskin of the shield after the jacks had been withdrawn to their starting position. When the shield was jacked forward off the latest ring of lining segments, a small annular gap was left between the outside of the lining and the excavated surface of the clay corresponding to the thickness and clearance of the skirt. This was about one inch and was filled with a grout made from Blue Lias cement and water which was injected into the gap by a workman using a hand syringe. We have already noted that this grouting was later found to have been almost entirely successful.

Each working day was divided into three eight-hour shifts and in each shift two rings of segments were usually erected and bolted up. In this way, six rings were erected per day which corresponded to a 9 ft advance of the tunnel. The bottom segment of a ring was installed first, then the two side segments (see Figure 6.6.) and finally the ring was closed with the key piece.[27] The three main segments weighed 4 cwt each and the key piece 1 cwt. The iron flanges for the bolt holes were 2 in deep. Lighting for the workmen was by candle. The tunnel was reported to be absolutely dry during construction.

The importance of the Tower Subway tunnel cannot be overestimated for it has a significance extending far beyond the tunnel itself. Barlow and Greathead had given a convincing demonstration of a safe and practicable method of soft ground tunnelling which has served as the model for tunnel construction all over the world right up to the present day. The cylindrical tunnel shield, the segmental lining and the grouted annulus are all essential features of soft ground tunnelling today. If Greathead's miners could be brought back to a soft ground tunnel under construction today they would recognise these essential features in a modern tunnel drive – the only difference is that today machinery has replaced the old hand methods of excavating the clay, advancing the jacks, injecting the grout etc.

Following completion of the Tower Subway, Greathead went on to drive the tunnels of the City and South London Railway, constructed between 1884 and 1890, using a shield of similar design to the Tower Subway shield but having the larger diameter of 11 ft 3 in.[28] During this work Greathead perfected the shield tunnelling system that had been so successfully used on the Tower Subway, notably inventing the grout pan, a device which enabled the linings to be quickly and effectively grouted by using compressed air pressure to force the grout behind the segments. Between 1896 and 1907 the deep tube tunnels of the London Underground were driven, using what had become known as 'Greathead shields', the level of the tunnels being selected so that the drives were as far as possible in London Clay.[29] The idea of the cylindrical tunnel shield quickly spread abroad and other cities soon followed London's example in providing themselves with underground railways.[30] However, Greathead was to see nothing of this great increase in underground construction, nearly all of it carried out by the shield tunnelling method he had pioneered based on Barlow's original idea, for he died in 1896.

6.9 Barlow's patents

Examination of Barlow's patent of 1864[31] shows that it is concerned with two matters, the first being the working of railways which does not concern us here and the second being 'In constructing tunnels for railways to pass under rivers or in tunnelling beneath towns.' To achieve this objective Barlow's patent specification describes a tunnelling shield which consists of a cylinder of wrought-iron or steel fitted at the front with eleven vertical and three horizontal cutting plates having sharpened edges. The method of operation envisaged was for the whole cylinder to be jacked forward and for the earth which came in between the cutting plates to be removed. Cast-iron segmental lining was to be erected inside the cylinder and, after the shield had been jacked clear, the space between the lining and the ground was to be filled with fluid cement grout. Barlow further specified that poling boards could be used at the face if the ground was weak.

Barlow's patent of 1868[32] is simply an addendum to his patent of 1864. It specifies that instead of fitting the cylinder with the array of cutting plates it should be fitted with a diaphragm having a door that can be closed. The patent of 1864, as modified by that of 1868, therefore, specifies the Barlow–Greathead shield that was actually used to drive the Tower Subway. The shield in Barlow's patent specification of 1864 is similar to Brunel's first shield of his patent specification of 1818 in that the shield is cylindrical and that the lining is of cast-iron segments and is erected under protection of the shield, but differs in that Barlow's shield was jacked forward as a whole whereas Brunel's was to be jacked forward

Figure 6.7 The St Clair River Tunnel shield (1886)

one cell at a time. Both shields have poling boards to support the face, but in Barlow's 1868 patent these are replaced by a diaphragm.

It is interesting to note that in his patent specification of 1864, referring to the role of the diaphragm, Barlow said 'If water rushes in the door can be closed and air pressure applied to the tunnel'. Barlow is here envisaging an additional function of the diaphragm. Not only is it to prevent a run in of loose or soft ground at the face but also to be used in conjunction with compressed air to stop water entering the tunnel from the face. The shield would then be jacked forward to stop the leak and the air pressure could then be taken off. We will be considering the use of compressed air in Chapter 7.

Before leaving the tunnelling shield, we will briefly consider an early application of it abroad by looking at the St Clair River Tunnel on the border between Canada and the United States, which was driven by means of a very large Barlow–Greathead-type shield.

6.10 The St Clair River Tunnel shield

The St Clair River Tunnel was constructed between 1888 and 1891 to provide a rail link between Canada and the United States from Sarnia, Ontario, to Port Huron, Michigan, beneath the wide St Clair River. The man appointed as chief engineer for the tunnel was Joseph Hobson, who was also chief engineer of the Canadian Grand Trunk Railway. To drive the tunnel Hobson designed a huge shield 21 ft 6 in in diameter. Although it was based upon the principles of the Barlow–Greathead Tower Subway shield, the St Clair River Tunnel shield was three times the diameter, and also twice the diameter of the shield Greathead used on the City and South London Railway. A contemporary print of the St Clair River Tunnel shield is shown in Figure 6.7.

The shield followed the Barlow–Greathead design in having a cylindrical body fitted with a diaphragm with closable doors. However, because of its large size it was provided with two working platforms so that miners had access to the whole face and in this respect it was similar to Brunel's shield. It was driven forward by jacking off the lining using 24 hydraulic rams distributed around the periphery of the shield. After each advance the rams were retracted and a ring of cast-iron lining segments was erected under the protection of a tailskin at the rear of the shield. Again, this was similar in principle to the Tower Subway shield.

However, because of the huge size of the St Clair River Tunnel shield, a manually-operated counter-balanced erector arm was used for placing the segments into position[33] instead of manhandling them as in the small Tower Subway. This invention of the erector arm for the St Clair Tunnel shield completes the full development of the shield tunnelling system. No other innovation was necessary and the only significant improvement to the present day has been the substitution of machinery for hand operations.

Figure 6.7 shows virtually all the important operations in a tunnelling shield. The miners are seen excavating the face through the doors in the diaphragm; there were usually about 12 miners working at the face. The cylindrical shield with its hydraulic jacks is shown and the new lining erector arm is shown placing a ½ ton lining segment under the cover of the shield's tailskin – men are waiting to receive it and bolt it into position. Two men are also shown caulking the joints in the completed lining a little further back in the tunnel. Finally, in the invert of the tunnel, mule-drawn rail-mounted muck cars are shown about to dispose of the spoil from the face. A word of explanation is required as to why the print bears the inscription 'The Beach hydraulic tunnelling shield'. This is because the early cylindrical tunnelling shields used in America were called 'Beach shields' after Alfred Ely Beach (1826–96) who was the editor of the journal *Scientific American*, a successful inventor and a patent attorney. Although Beach was not a civil engineer, he designed a shield to drive the 8 ft diameter Broadway Subway in New York in 1869 – the same year that the Tower Subway was constructed. R.M. Vogel considers that because of his position Beach must have been aware of Barlow's patents for tunnelling shields and that it is therefore very unlikely that Beach invented the tunnelling shield independently.[34] And it is called a 'hydraulic shield' because it was jacked forward with hydraulic rams.

As well as spreading to the United States, shield tunnelling was quickly taken up elsewhere abroad, particularly in France. It became adapted to suit different kinds of ground conditions and various technical refinements were introduced. These developments have been described by W.C. Copperthwaite,[35] by R. Legouëz[36] and by R. Phillipe.[37] By the turn of the century shield tunnelling was an established and proven technique which could be undertaken with confidence for soft ground and subaqueous tunnels. In 1897 a shield was made which incorporated a completely mechanical arrangement for excavating the ground at the tunnel face, but the story of this development belongs to Chapter 9. Shields were not always circular in section: on the Clichy Sewer Tunnel in Paris elliptical and semi-elliptical shields were used.[38]

Finally, on the St Clair River Tunnel, it can be noted that as well as the shield that has been described, compressed air was used in the tunnel, and this makes a convenient link with Chapter 7 which will deal with the subject of compressed air tunnelling.

6.11 Summary

In this Chapter we have seen how during the nineteenth century, demographic, social and economic changes gave rise to the need for a method of driving tunnels in soft ground beneath cities and under rivers, and we have followed the invention and development of the tunnelling shield – the innovation that satisfied this need. We will now consider its origin.

There is no doubt as to the origin of the first tunnelling shield. Brunel's idea of a tunnelling shield clearly resulted from his encountering the piece of keel timber bored by shipworms whilst he was on a visit to Chatham Dockyard. Examination of the shipworm and its burrow gave him the basic idea for his shield and the method of driving a tunnel with it. In the origin of Brunel's tunnelling shield we therefore have a clear instance of an innovation deriving from an external idea. We should note, however, that at the time of Brunel's encounter with *Teredo navalis* he had been pondering the problem of tunnelling beneath the Thames so that the external idea, when it came, impinged upon a prepared mind.

It might be argued that Brunel's shield, whilst being a remarkable innovation in itself and also being the precursor of the tunnelling shield, is not the true original model of the tunnelling shield as it in fact developed. A case could be made for considering the Barlow–Greathead shield as the first true tunnelling shield and for this reason its origin should also be considered. We have seen that Barlow got his idea for a tunnelling shield from his experience with sinking the cylinders for the piers of Lambeth Bridge – the tunnelling shield was simply one of these bridge cylinders turned on its side. This then would seem to be borrowing an idea from another technology. But there is more to a tunnelling shield than a cylinder, and the Barlow–Greathead shield incorporates features such as the diaphragm and jacks that have exact counterparts in the poling boards and jacks of Brunel's shield. And of course Barlow would have been well aware of all the details of Brunel's shield when he came to design his own shield. Brunel's shield, therefore, has a better claim for being considered as the first true tunnelling shield.

We must also take account of the fact that the first shield driven tunnel in the United States, the Broadway Subway, was constructed at the same time as the Tower Subway, and consider the origin of this tunnelling shield. We have noted that Beach must have been aware of Barlow's earlier shield patents and would certainly have known of Brunel's shield. The idea of the independent invention of the shield by Beach is therefore highly improbable.

To conclude: in Brunel's shield we have the first tunnelling shield, and it is clear that the origin of this innovation was an external idea. This is not to belittle the achievements of Barlow and Greathead whose contributions were in making Brunel's basic idea into a practicable method of tunnelling.

Notes and references for Chapter 6

1 Briggs, A. (1971) *Victorian cities.* Harmondsworth (Penguin Books) pp. 283–4.
2 Sheppard, F. (1971) *London 1808–1870: the infernal wen.* London (Secker and Warburg) pp. 279–84.
3 Bazalgette, J.W. (1864–5) On the main drainage of London, and the interception of the sewage from the River Thames. *Minutes of Proceedings of the Institution of Civil Engineers,* **24**, pp. 280–314.
4 Seaman, L.C.B. (1973) *Life in Victorian London.* London (B.T. Batsford Ltd) pp. 79–88.
5 Sheppard,F. (1971) *op cit,* p. 141.
6 Anon (1867) The Thames Embankment. *The Illustrated London News.* **50**, (1432), pp. 632–4.
7 Simms, F.W. (1896) *Practical tunnelling.* London (Crosby Lockwood and Son) 4th edition, revised and greatly extended by D.K. Clark, *passim.*
8 Pugsley, Sir A., Editor (1976) *The works of Isambard Kingdom Brunel.* London (Institution of Civil Engineers) pp. 46–8.
9 Sandström, G.E. (1963) *The history of tunnelling.* London (Barrie and Rockliff) pp. 114–31.
10 James, J.G. (1976) Ralph Dodd, the very ingenious schemer. Paper presented to the Newcomen Society for the study of the History of Engineering and Technology, 14 January, pp. 8–9.
The above paper contains far more detail about the Gravesend tunnel than the abridged version that was published:
James, J.G. (1974–6) Ralph Dodd, the very ingenious schemer. *Transactions of the Newcomen Society,* **47**, pp. 161–78.

11 Trevithick, F. (1872) *Life of Richard Trevithick, with an account of his inventions*. Volume 1. London (E. and F.N. Spon) p. 261.
12 Rolt, L.T.C. (1976) *Isambard Kingdom Brunel*. Harmondsworth (Penguin Books).
13 Law, H. (1845) *A memoir of the Thames Tunnel. Part 1. From the commencement of the works to their suspension in 1828*. London (John Weale).
Beamish, R. (1862) *Memoir of the life of Sir Marc Isambard Brunel*. London (Longman, Green, Longman and Roberts).
14 Noble, C.B. (1938) *The Brunels – father and son*. London (Cobden-Sanderson).
Lampe, D. (1963) *The tunnel*. London (George G. Harrap and Co Ltd).
Clements, P. (1970) *Marc Isambard Brunel*. London (Longmans Green and Co Ltd).
15 Vignoles, K.H. (1982) *Charles Blacker Vignoles: a romantic engineer*. London (Cambridge University Press) pp. 34–8.
16 Anon (1843) Opening of the Thames Tunnel. *The Illustrated London News*, **2**, (48), pp. 226–8.
17 Anon (1843) The Queen's visit to the Thames Tunnel. *The Illustrated London News*, **3**, (65), p. 75.
Anon (1843) Royal visit to the Thames Tunnel. *The Illustrated London News*, **3**, (66), p. 96.
18 Brunel Exhibition Project (1980) *Brunel's tunnel and where it led*. Rotherhithe (Brunel Exhibition Project).
19 There are many descriptions of the shipworm, probably because of its economic importance as a pest of the wooden hulls of ships and the timbers of marine civil engineering works. Some of these are:
Figuier, L. (1872) *The ocean world*. London (Cassell, Petter and Galpin) New edition revised by E. Perceval Wright, pp. 321–6.
Woodward, S.P. (1910) *A manual of the mollusca*. London (Crosby Lockwood and Son) Reprint of the 4th edition, 1880, pp. 506–7.
Woodward says 'The operation of the *Teredo* suggested to Mr Brunel his method of tunnelling the Thames'. Footnote to p. 507.
Yonge, C.M. (1971) *The sea shore*. London (Collins: The Fontana New Naturalist) pp. 204–8.
20 Pugsley, Sir A., Editor (1976) *op cit*, pp. 30–3.
21 Vogel, R.M. (1964) Tunnel engineering – a museum treatment. *Contributions from the Museum of History and Technology*, Paper 41. Washington DC (Smithsonian Institution) pp. 203–39.
22 Law, H. (1845) *op cit*, pp. 52 and 61–2.
23 *Ibid*, pp. 55 and 58–9.
24 Brunel, M.I. (1818) Forming tunnels or drifts under ground. *British patent* No. 4204. London (Great Seal Patent Office).
25 Greathead, J.H. (1896) The City and South London Railway; with some remarks on subaqueous tunnelling by shield and compressed air. *Minutes of Proceedings of the Institution of Civil Engineers*, **123**, Paper No. 2873, pp. 39–73, and Discussion, pp. 74–111.
26 Harding, H.J.B. (1945) Emergency repair to the Tower Subway, London, after air-raid damage. *Journal of the Institution of Civil Engineers*, **25**, pp. 73–9.
27 Anon (1869) The Thames Subway at the Tower. *The Illustrated London News*, **55**, (1564), p. 440.
28 Greathead, J.H. (1896) *op cit*, pp. 39–73.
29 Jackson, A.A. & D.F. Croome (1962) *Rails through the clay: a history of*

London's tube railways. London (George Allen and Unwin Ltd) *passim*.
30 Nock, O.S. (1973) *Underground railways of the world*. London (Adam and Charles Black).
31 Barlow, P.W. (1864) Constructing and working railways. *British Patent* No. 2207. London (Great Seal Patent Office).
32 Barlow, P.W. (1868) Constructing tunnels. *British Patent* No. 813. London (Great Seal Patent Office).
33 Vogel, R.M. (1964) *op cit*, p. 238.
34 *Ibid*, p. 227.
35 Copperthwaite, W.C. (1906) *Tunnel shields and the use of compressed air in subaqueous works*. London (Archibald Constable and Co Ltd) *passim*.
36 Legouëz, R. (1897) *De l'emploi du bouclier dans la construction des souterrains*. Paris (Libraire Polytechnique Baudry et Cie) *passim*.
37 Phillipe, R. (1900) *Le bouclier et les méthodes nouvelles de percement des souterrains*. Paris (Libraire Polytechnique Ch Beranger) *passim*.
38 Prelini, C. & C.S. Hill (1901) *Tunnelling: a practical treatise*. London (Crosby Lockwood and Son) pp. 253–6.

Compressed air tunnelling

Die Luft ist ein Körper.

Neither Brunel's Thames Tunnel shield nor the Barlow–Greathead shield met the third criterion for a safe method of driving a subaqueous tunnel, namely a means of excluding ground water from the works. This was not achieved until the introduction of compressed air tunnelling which is discussed in this Chapter together with some of the problems associated with it.

The principle on which the idea of compressed air tunnelling is based is a very simple one. It had long been known that air under pressure will balance a head of water, 1 atm being capable of supporting a water column some 30–34 ft high. Consequently if a tunnel was being driven beneath a river, the water surface level of which was 30 ft above the tunnel, then if the pressure of the air in the tunnel heading were to be increased by one atmosphere, the hydrostatic presure would be balanced and the tendency for water to flow into the works would be prevented.

When Brunel's Thames Tunnel suffered its two severe inundations in 1827 and 1828 (see Section 6.4) he was overwhelmed with suggestions as to how to recover the tunnel. Many of these were fanciful but the Swiss physicist Jean-Daniel Colladon (1802–93) wrote to Brunel and suggested that compressed air should be used in the tunnel to exclude water from the works. Colladon expressed surprise that Brunel had not himself thought of working in compressed air since, he suggested, 1 or 2 atm pressure would have been sufficient to keep the water out if the entrance to the tunnel had been closed with an iron door and a compressor had been installed to compress the air inside the tunnel. Colladon was at this time in frequent contact with Benjamin de Lessert, the French mining engineer, who was interested in the Thames Tunnel and was so struck by the idea that he got Colladon quickly to prepare a note on it which he did in January 1828[1] and this was sent to Brunel as mentioned above. Then in 1830 Thomas Cochrane patented a complete system of applying compressed air to a subaqueous tunnel under construction. References to Colladon's note and Cochrane's patent in Brunel's diaries and letters, quoted by Glossop,[2] show that he took these two suggestions seriously, and in fact, he drew both schemes to the attention of the Directors of the Thames Tunnel in 1831. In the event neither was used, it being thought now that the reason why Brunel did not use compressed air in the Thames Tunnel was partly that he wanted to show that his shield alone was adequate to ensure success, and partly because he may not have realised the full effects that compressed air would have had in improving driving conditions. Colladon's note has not survived so it is not known just how detailed his suggestion was, but Cochrane's patent is extant and because his ideas were so fully developed we can now examine them in detail. Before doing so, we should note that although Brunel put forward the proposals for compressed air tunnelling to the Thames Tunnel Co as being the ideas of Colladon and Cochrane, he may have had a half-formed notion of it himself, as indicated by his diary entry for 13 July 1830 (see Section 1.2):

> Why not make use of the tunnel even as a Diving Bell. The idea has frequently occurred to me, and Isambard has mentioned it in this connection – the more I think about it the more I find it practicable and the more important I think it.

On the 21 July he added:

> The tunnel bell can, nevertheless, be no more than an auxiliary to the shield.

By this last remark Brunel showed that he realised an extremely important fact, which was that however effective compressed air was in excluding water from the tunnel workings, the shield was still needed to support the ground and protect the miners. We shall see later in this Chapter how, in fact, an early attempt was made to drive a large-diameter tunnel using compressed air without a shield. Brunel's reference to using the tunnel like a diving bell will also be taken up again in Section 7.8 when the origin of compressed air tunnelling is considered.

In compressed air tunnelling it is standard practice to express the pressure in the tunnel as the pressure over and above normal atmospheric pressure, and this convention will be followed in this book. Thus, a pressure of 1 atm means that the pressure in the tunnel is 1 atm above the normal atmospheric pressure outside the tunnel. This is also sometimes called 'gauge' pressure because it is the pressure that is indicated by a pressure gauge located outside the tunnel but connected to the pressurised section of the tunnel.

7.1 Cochrane's patent

Cochrane was born on 14 December 1775, the eldest son of the ninth Earl of Dundonald.[3] He joined the Royal Navy in 1793 and served during the Napoleonic wars. Between 1806 and 1818 he was a Member of Parliament. He was then invited by Bernardo O'Higgins (1778–1842), leader of the Republic of Chile to organise and command the Chilean Navy, and by defeating the Spanish fleet in the Pacific was partly responsible for the liberation of Chile and Peru. He returned to England in 1828 something of a popular hero. He succeeded to the title of tenth Earl in 1831 on the death of his father and in 1832 was gazetted Rear Admiral of the Fleet. In 1829 he settled near Southampton where he cultivated an interest in science, engineering and invention, one outcome of which was the compressed air tunnelling system that he patented in 1830. Cochrane died on 31 October 1860 and was buried in Westminster Abbey.

Cochrane's patent specification of 1830 is a fully worked out scheme for applying compressed air to a tunnel under construction. The fact that the invention must have been inspired by the difficulties Brunel was having with his Thames Tunnel is indicated by the following statement by Cochrane in the specification:

> The essential properties of my Invention will be explained by the following description of an apparatus, for the purpose of sinking a perpendicular shaft at the shore by the side of a river, and then excavating a drift or tunnel from the bottom of that shaft to extend horizontally (or nearly so) beneath the bed of that river, being a similar undertaking to that which is now executing beneath the river Thames at Rotherhithe.[4]

A copy of the drawing attached to the patent specification is shown in Figure 7.1. AB is a column of cast-iron cylinders bolted together by internal flanges which is sunk by the usual method of shaft sinking. When

Figure 7.1 Cochrane's scheme for compressed air working (1830)

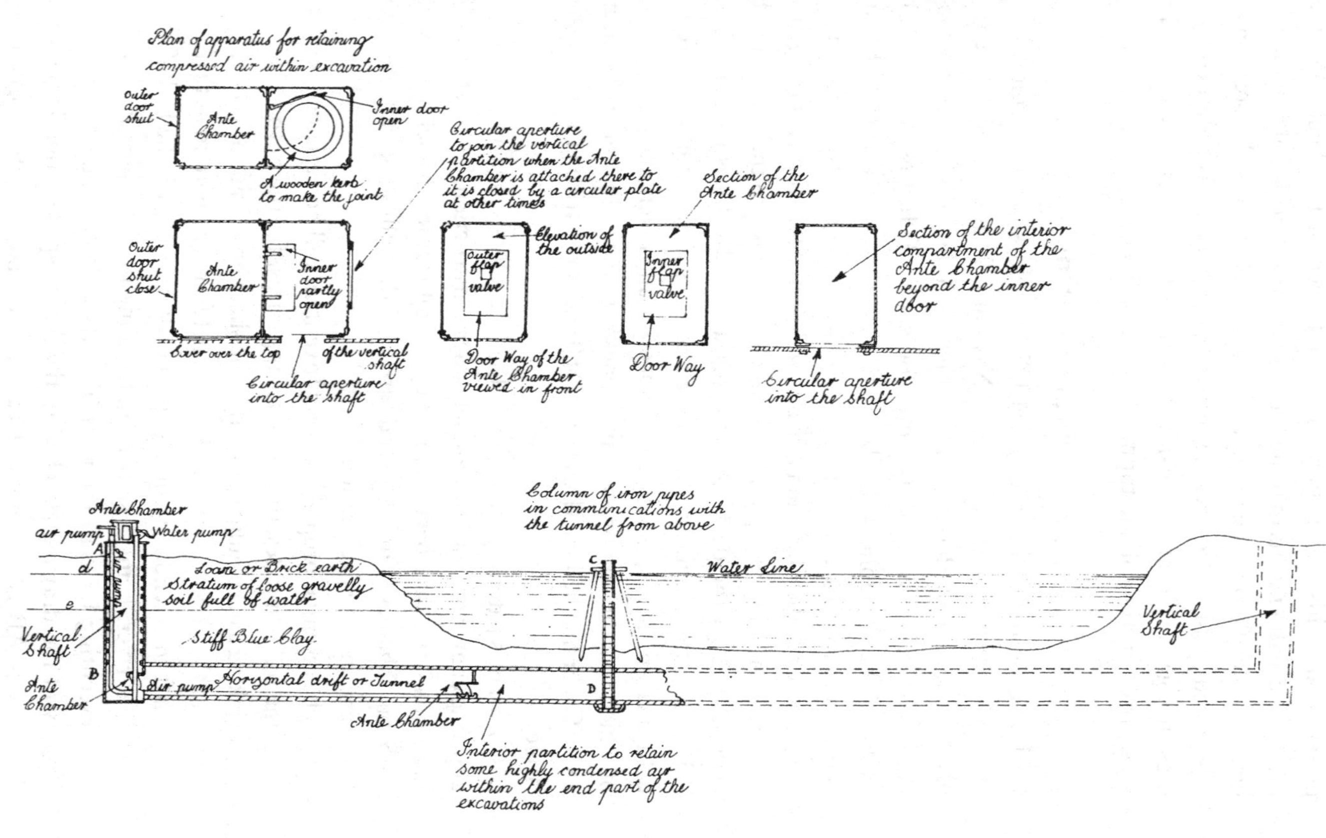

the shaft has been sunk to the required depth, the tunnel is driven horizontally from the base of the shaft as shown in the Figure. On the top of the shaft at A is fitted a compartment which Cochrane called an 'ante chamber', but which we would call today an air lock. The purpose of this is to allow workmen to pass from the outside where the air was at normal atmospheric pressure into the works which were under compressed air. Cochrane's ante chamber and a modern air lock are the same in principle, the main difference being that the flap valves in the doors have been replaced with stopcock-operated valves.

Cochrane provided another ante chamber at the bottom of the shaft at B, and yet another along the tunnel towards the working face. The purpose of these, he explained, was to allow workmen who were working in the shaft and the completed section of the tunnel to do so at lower pressures than those who were working at the tunnel face. Cochrane gave as an example a pressure of $\frac{2}{3}$ atm in the shaft, $1\frac{1}{3}$ atm in the first section of the tunnel and 2 atm in the heading; this would have given safe working conditions at the face for a head of 60 ft of water above the tunnel. Separate air supply pipes led from the air compressor on the surface to the different sections of the works. This idea is quite sound theoretically, but in practice has since been found to be an unnecessary complication to an otherwise simple arrangement; occasionally a higher air pressure has been provided at the face than in the rest of the tunnel by installing an additional air lock in the heading, this being called multi-stage compressed air working.[5] However, the idea of a single air lock in the tunnel rather than at the top of the shaft is one which is fairly common today.

So long as the tunnel face was reasonably close to the shaft, Cochrane envisaged that the muck would be passed out through the ante chamber at the shaft bottom and thence up the shaft to the surface. However, when the heading was far advanced he envisaged another method of mucking out. This was to instal a column of iron pipes as shown at CD in Figure 7.1. The pipe was to be sunk from the river, and when completed would extend from above high water level down to a chamber constructed by the side of the tunnel and communicating with it. The lower part of the chamber was to be fitted with a large cistern that was to be kept full of water and in which the lower end of the pipe was to be immersed. The pipe was to be of sufficient diameter to allow free passage up and down inside it of a bucket. The compressed air in the tunnel would have raised a column of water in the pipe and maintained it at a level depending on the air pressure in the tunnel. The earth excavated from the tunnel face was to be drawn up the pipe by filling the bucket and introducing it under water in the cistern beneath the open lower end of the pipe. The bucket would then be hauled up to the surface through the column of water standing in the pipe. Materials for constructing the tunnel lining were to be passed down in the bucket on the return journey. Although Cochrane did not make any mention of it, it is clear that the method of disposing of muck from the top of the pipe was to be by tipping it into a barge, because the drawing (Figure 7.1) shows the pipe column surrounded by a timber dolphin probably intended as a mooring as well as for protection of the pipe. The barge would also have been used for bringing in bricks or other materials to be passed down the pipe to the tunnel.

This scheme was ingenious, theoretically sound and would have permitted the flow of muck and material between the tunnel and the surface without involving the use of the air locks in the tunnel. However, it suffered from a grave practical shortcoming that would have made it potentially very hazardous to use. This was that it depended critically for safe operation on maintaining the water seal in the cistern at the bottom of the pipe. If for any reason, the water level fell below the bottom of the pipe, the column of water would have been blown out at the top and there would have been a sudden and catastrophic loss of air pressure in the tunnel. Because of its potential danger this method of muck removal has never been used on a tunnel as far as is known, although a somewhat similar arrangement was used in sinking one of the caissons for the Brooklyn Bridge.[6] The solution adopted in later compressed air tunnels was simply to make the air lock long enough to accommodate several muck cars and then to operate in the normal manner; on some large-diameter tunnels there has been room to instal a small man lock and a large muck lock side-by-side.[7]

In his specification Cochrane made no mention of the method of advancing the tunnel; the purpose of the compressed air was to 'counteract the tendency of superincumbent water (or of such superincumbent earth as is rendered semifluid by admixture with water) to flow by gravitation into such excavations'. Cochrane was not, of course, an engineer and left the actual method of excavating the heading to be decided by those who were. In fact, as we shall see, compressed air was at first used without a shield – with consequences that will be described.

It is important to note that Cochrane's specification covered both the sinking of the shaft under compressed air once the bottom of the shaft had passed below ground water level, and the driving of the tunnel under compressed air. His specification is therefore the source of innovation for both compressed air shaft sinking and compressed air tunnelling, but in fact the idea was used extensively in sinking shafts and caissons during the period 1840–70, long before it was first used in a tunnel.[8]

It was one thing to propose that compressed air could be used to exclude water from the works in shaft sinking or tunnel driving, and another to show how this could be done practically. Cochrane's specification is important because it did both. In order to be a practical proposition some means had to be devised for allowing men and material to pass from the outside at normal atmospheric pressure into the workings at a higher pressure without releasing the compressed air. As we have seen, Cochrane solved this problem by devising the air lock (which he called the ante chamber, see Figure 7.1). The air lock consists of a chamber having two doors, each door opening in the direction towards the higher pressure so that it cannot be opened until the pressure on either side is the same. To enter the workings the miners enter the air lock, close the outer door and then admit some compressed air from the workings using the valve in the inner door until the pressure in the air lock is the same as that in the working area. The inner door is then opened and the miners pass through. In coming out the process is reversed, the inner door being closed and the pressure in the air lock being allowed to fall to normal atmospheric pressure by releasing air using the valve in the outer door. The operation of the air lock is directly analogous to the operation of the canal lock: the air lock allows the passage of men or material from one air pressure to another in exactly the same way that a canal lock allows the passage of a boat from one water level to another. Air locks, based on the principle established by Cochrane, have been used in compressed air shafts and tunnels right up to the present day.

Although, as we have seen, Colladon's letter to Brunel predated Cochrane's patent it is unlikely Cochrane would have heard of this private communication and he should at least be credited with the independent invention of the principle. The air locks, which make the idea practicable, can certainly be credited to Cochrane.

Finally, although Cochrane's scheme for compressed air tunnelling was inspired by the problems of the Thames Tunnel, we have seen that in the event, the Thames Tunnel was completed using the shield alone without any use of compressed air (Section 6.4), and earlier in this Chapter we have noted the probable reasons for this. However, although there was to be no further loss of life due to flooding of the tunnel, if Brunel had used compressed air as well as the shield, many of the miners and bricklayers might have been spared permanent damage to their health because the use of compressed air would have undoubtedly prevented the influx of those noxious gases into the tunnel which caused much serious sickness, including blindness, during the second phase of the works. From contemporary scale drawings of the Thames Tunnel we know that to have balanced the water pressure at the mid-height of the shield, Brunel would have needed to have used compressed air at 2 atm pressure when the river was at high water, reducing to $1\frac{1}{3}$ atm at low water, showing that the air pressures required in the tunnel were not unrealistic, and in fact similar to those envisaged by Cochrane. Before putting compressed air tunnelling into practice, however, Brunel would have been faced by the problem that the then existing air compressors would not have been capable of delivering the volume of air needed for the Thames Tunnel at the required pressure.

7.2 The Kattendyk Tunnel, Antwerp

The first time compressed air working was used in a tunnel under construction was in 1879 when it was used by H. Hersant on the Kattendyk Tunnel in Antwerp. This tunnel was built to provide a drainage gallery associated with the construction of some dry docks for the Port of Antwerp. The tunnel was driven through fine sand at a depth of 30 ft below the ground surface and some 20 ft below the water table and it was, therefore, decided to use compressed air to ensure the safety of the works, a single air lock being installed at the top of the shaft.[9] The tunnel was lined with sets of four slightly curved cast-iron segments, bolted together with internal flanges, giving a cross section having the form of a rectangle with slightly convex sides; the internal dimensions were 5 ft high by 4 ft wide. This was the second tunnel on which cast-iron segmental lining was used. Because of the small size of the tunnel, no shield was used during excavation of the face which was advanced in lengths of 1 ft 8 in, corresponding to the width of a segment. We can note in passing that illumination was provided by Swann incandescent light bulbs with specially thickened glass – this is thought to be the first time electric light was used in a tunnel under construction. We will now see what happened when compressed air alone was used to drive a very much larger subaqueous tunnel.

7.3 The first Hudson River Tunnel

The first tunnel under the Hudson River was proposed in 1873 by DeWitt Clinton Haskin (1833–1900) in order to provide a rail link between New Jersey and New York City. Haskin was a wealthy mine and railway builder from San Francisco, who had been impressed by the effectiveness of compressed air used in the sinking of caissons for bridge piers. Haskin had the idea that a tunnel could be driven through the silt beneath the Hudson River using compressed air to exclude water from the works.[10]

Work began in 1874 with the sinking of a 38 ft diameter shaft on the New Jersey shore to a depth of 54 ft. Half-way down the shaft a horizontal air lock was built into the side from which the tunnel was started. From the air lock an unusual conical shaped enlargement was constructed down to the level of the tunnel proper. Both the conical enlargement and the tunnel itself were lined with wrought-iron rings made from thin plates bolted together using angle iron. But as the rings in the tunnel were completed they were in turn lined with bricks set with cement mortar. The tunnel proper was built as two single-line tubes of elliptical shape, 18 ft high by 16 ft wide in internal dimensions. The ground was a very soft water-saturated silt and the tunnel crown was some 30 ft below the mean water level of the river above. No shield or timbering was used to support the heading; instead Haskin decided to use compressed air only. Driving of the first of the tubes, the northern one, commenced in 1879 using compressed air at a pressure of 1.2 atm and in six months some 280 ft of tunnel had been driven. Work on the southern tube then commenced under an air pressure of 1.4 atm. In July 1880 work was being carried out in reconstructing the entrance, when a sudden loss of air occurred at the point in the roof of the tunnel where the permanent lining abutted the conical enlargement.[11] This event is described in more detail later, in Section 7.6. Just before it happened, air pressure in the tunnel was 1.1 atm, only one tenth of an atmosphere more than was needed to balance the hydrostatic head of water above at mean river level. A number of writers have quoted this incident to demonstrate the danger of using compressed air without a shield, but the loss of air occurred not at the face or in the heading where a shield would have been, but at the entrance to the tunnel; nevertheless it does show what might have happened at the face. A caisson was sunk to recover the entrance chamber, the roof of which had collapsed following the loss of compressed air, and work once more commenced on the tunnel drives. But in 1882 the company suspended operations, Haskin having run out of funds; at this time 1540 ft of the northern tube and 600 ft of the southern tube had been driven.

In 1889 work was resumed under the direction of British contractors who were advised by the eminent consulting engineer Benjamin Baker (1840–1907) and Greathead himself. They recommended that the tunnel be driven with a shield together with the use of compressed air, and that the tunnel should be lined with cast-iron segments of 18 ft 2 in internal diameter. This was put in hand and once the shield was started an advance rate of 10 ft per day was achieved. The silt was so fluid however, that the air pressure in the tunnel had to be raised to 2 atm to control it where it was exposed through the doorways in the diaphragm of the shield. In 1891 work was suspended again because of the promoter's lack of funds, but it was resumed in 1902 and the tunnel was completed in 1905. The use of a medical lock on this tunnel job will be discussed later, in Section 7.6.

7.4 The St Clair River Tunnel

The shield used for driving this tunnel was described in Section 6.10, but as well as a shield compressed air was used during some of the construction. The total length of tunnel driven by shield was 6000 ft and on the 2300 ft section under the St Clair River, compressed air also was used. The tunnel was lined with a cast-iron segmental lining of 19 ft 10 in internal diameter. The tunnel was driven from both sides of the river simultaneously and air locks were built in the headings when they arrived at the river banks. The ground consisted of very soft clay with pockets of gravel and sand. The water level of the river was up to 60 ft above the tunnel crown and compressed air at a pressure of between $\frac{2}{3}$ atm and just under 2 atm was used during the drive beneath the river.[12] In all respects the work proceeded smoothly, an average monthly rate of advance for both headings together of 455 ft being achieved, whilst the maximum rate of advance in a single heading was 15 ft per day – i.e. double the average rate. The St Clair River Tunnel drive demonstrated the soundness of the concept of using compressed air and the shield together: the compressed air safely excluding water from the workings while the shield allowed the soft wet clay at the face to be controlled. This combination set the pattern of construction for all subsequent subaqueous soft ground tunnels up to the present day.

7.5 The City and South London Railway

It will be recalled from Chapter 6 that the Metropolitan Railway and the District Railway, the first and second underground railways to be built in London were constructed by the cut-and-cover method. So much surface disruption was caused by their construction that no further lines were built by the cut-and-cover method in the metropolis. The next underground railway to be constructed in London, the City and South London Railway, was built by true tunnelling methods utilising a tunnelling shield, which was, for short lengths of the drive, supplemented by compressed air.

In 1884 an Act of Parliament was passed authorising the construction of an underground railway, called the London and Southwark Subway, between King William Street in the City and the Elephant and Castle in south London.[13] Between 1887 and 1893 further Acts were passed sanctioning extensions of the line and changing the name of the undertaking to the City and South London Railway. The line as built ran from the City to Stockwell (see Figure 6.1). Work started in 1886 with the construction of a working shaft in the River Thames at Old Swan Pier near London Bridge. The first objective was to construct the twin tunnels under the Thames because this part of the works was thought to be the most difficult of the whole project. The engineer for the scheme was Greathead and his plan was to drive the tunnels using a larger version of the shield that he had so successfully used to drive the Tower Subway (see Section 6.7). Also, as for the Tower Subway, Greathead proposed keeping the levels of the tunnels within the stratum of London Clay that had proved to be so perfect a tunnelling medium. In the event, no difficulty was encountered in driving the tunnels beneath the river. However, in driving the upper of the two tunnels northwards from the river the crown of the tunnel broke through the top of the London Clay into the overlying bed of water-bearing gravel and sand. The face of the shield was immediately closed and an air lock was installed in the tunnel close to the heading. The tunnel was then driven for about 150 ft under compressed air after which the shield again encountered a full face of London Clay. The compressed air was taken off and the remainder of the drive to the King William Street terminus was made under normal atmospheric pressure. During the period of compressed air working it was noticed that air escaped from the tunnel into the river showing that the water in the stratum of gravel and sand was in communication with the water in the river. This was confirmed by the observation that the air pressure in the tunnel automatically adjusted itself to that of the changing head of water in the river between high and low tide.

The most notable use of compressed air on this project was at the south end of the railway beneath the Clapham Road at Stockwell. Here, for a length of about 600 ft the two tunnels were driven through coarse gravel and sand under a head of about 35 ft of water. The tunnels were shield driven under normal atmospheric pressure to the point where the cover of London Clay had reduced to about 5 ft. Air locks were installed close to the heading, compressed air at about 1 atm was applied and the tunnel drives were continued using both the shield and compressed air. At Stockwell it was noticed that because the water pressure at the top of the tunnel was due to a head of 25 ft of water whilst that at the bottom was $36\frac{1}{2}$ ft of water, this difference corresponding to the diameter of the tunnel, it was difficult to adjust the air pressure so that neither a large amount of air escaped at the top nor a large amount of water flowed in at the bottom. This point will be returned to later. The rate of progress of tunnelling under compressed air in the gravel and sand was about 5 ft per day, compared with about 16 ft per day when driving through London Clay without compressed air.

The City and South London Railway illustrates how compressed air can be used to enable tunnelling to proceed through short lengths of water-bearing strata in a drive consisting predominantly of ground where ordinary shield tunnelling will suffice. The railway was, as we have seen, mainly sited in the London Clay, ideally suited to the use of the shield without compressed air. However, over short but critical lengths of the tunnel drive, water-bearing gravel and sand was encountered with which the shield alone would not have been able to cope. Compressed air used

Table 7.1. *Methods used to drive tunnels beneath the Thames in London*

Tunnel		Methods used	
Date of construction	Type (Pedestrian: P; Road: Rd; Rail: Rl)	Shield	Compressed air
1805–1808	Driftway	–	–
1825–1842	Thames Tunnel P	+	–
1869	Tower Subway P	+	–
1886–1890	City and South London Rl	+	–[a]
1892–1897	Blackwall Rd	+	+
1894–1898	Waterloo and City Rl	+	+
1899–1901	Greenwich P	+	+
1901–1906	Baker Street and Waterloo Rl	+	+
1904–1908	Rotherhithe Rd	+	+
1910–1912	Woolwich P	+	+

[a]Compressed air used occasionally elsewhere on the drive (see Section 7.5).

together with the shield enabled work to proceed with safety and allowed the line to be completed. The tunnels were lined with cast-iron segmental linings, which will be described in Chapter 8. The railway was opened to the public in 1890,[14] and by 1894 was carrying almost seven million passengers per year.

To show how important was the combination of compressed air working and the tunnelling shield, a list has been compiled of all the major tunnels constructed beneath the Thames up to 1920; this is shown in Table 7.1. The driftway and the Thames Tunnel have already been discussed. The remaining eight tunnels were all driven by shield and of these, six were driven using compressed air as well. The two on which compressed air was not used were both driven through London Clay – a material virtually impermeable to water – which explains why compressed air was not necessary. And this combination of the shield and compressed air was not only used for driving tunnels in London. Table 7.2 shows the major tunnels constructed beneath the Hudson and East Rivers in New York for the same period. Here, of eleven tunnels all were driven with compressed air and of these, ten were driven using the shield as well. And in the one tunnel on which the shield was not at first used, the first Hudson Tunnel, the tunnel was eventually completed using both compressed air and the shield. Before leaving Table 7.2 it can be remarked that the Ravenswood Tunnel is notable for having had used on it what is

Table 7.2. *Methods used to drive tunnels beneath the Hudson and East Rivers in New York*

Tunnel		Methods used	
Date of construction	Type { Gas: G, Rail: Rl }	Shield	Compressed air
1879–1882	First Hudson Rl	–	+
1889–1905	First Hudson Rl	+	+
1892–1894	Ravenswood G	+	+
1903–1906	Battery Rl	+	+
1902–1908	Hudson and Manhattan Rl	+	+
1905–1906	Hudson, Penn RR Rl	+	+
1904–1909	East River, Penn RR Rl	+	+
1905–1907	Steinway Rl	+	+
1914–1919	Old Slip Rl	+	+
1914–1920	Whitehall Rl	+	+
1916–1919	14th Street Rl	+	+
1916–1919	60th Street Rl	+	+

thought to be the highest air pressure used in American tunnelling, $3\frac{1}{2}$ atm having been used at times.

These experiences of compressed air tunnelling can be summarised as follows. The first compressed air tunnel, the Kattendyk Tunnel, was driven successfully using compressed air alone. This success is undoubtedly due to the tunnel's small size, only 5 ft by 4 ft in cross section. In fact it is interesting to note that the Kattendyk Tunnel was only a little bigger than Trevithick's driftway which was 5 ft by 3 ft, and which was almost successfully completed without either compressed air or shield. When compressed air alone was used to drive a large tunnel, the first Hudson Tunnel, the danger of using it without a shield was recognised, and the Thames Tunnel had earlier demonstrated the corollary – namely that the shield without compressed air was equally hazardous. Nearly all subsequent subaqueous tunnels were driven using a combination of compressed air and the shield (for examples see Tables 7.1 and 7.2) unless there was a special reason such as the tunnel being driven through a stratum so perfectly suited for tunnelling as the London Clay. Finally, the St Clair River Tunnel shows how the combination of compressed air and the tunnelling shield reached early technical perfection. The City and South London Railway demonstrated another vital role that compressed air tunnelling had to play; this was to provide a means of dealing with short lengths of water-bearing gravel and sand in an otherwise trouble-free shield drive through London Clay.

7.6 Disadvantages of compressed air tunnelling

Although the effectiveness of compressed air in keeping water out of subaqueous tunnels had been so convincingly demonstrated, there remain some disadvantages associated with its use. Three of these were discovered as soon as the method was used, but a fourth did not come to light until much later. These are (i) the danger of a 'blow', (ii) the risk to miners of 'the bends', (iii) the risk to miners of 'getting blocked', and (iv) the risk to miners of bone necrosis.

The danger of a 'blow' occurs when a subaqueous tunnel of fairly large diameter is being driven under compressed air, and the cause is the fact that with a large-diameter tunnel there are unbalanced pressures at the face. Consider the tunnel shown in Figure 7.2, where following the usual practice compressed air has been applied at a pressure required to balance the hydrostatic pressure at tunnel axis level. In the example shown in the

Figure 7.2 Unbalanced pressures at crown and invert of tunnel under compressed air

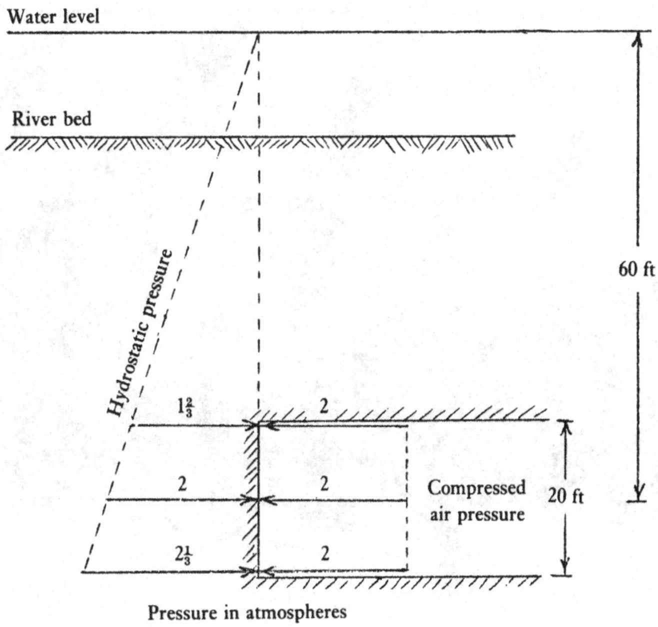

Figure, the tunnel is 20 ft in diameter and the tunnel axis level is 60 ft below the surface level of the water in the river above; in these circumstances a pressure of 2 atm needs to be applied in the tunnel. It will be seen that if this is done, then the air pressure at the top of the tunnel face, is $\frac{1}{3}$ atm higher than the water pressure and the air pressure at the bottom of the face is $\frac{1}{3}$ atm lower than the water pressure. There will therefore be a constant tendency for air to leak out at the crown of the tunnel and for water to leak in at the invert. If the tunnel drive runs in to very permeable ground, say sand or gravel, and the air finds a passage through to the surface, there may be a sudden escape of air – called a 'blow' – and a consequent sudden loss of air pressure which is followed by an inrush of water because the hydrostatic pressure is no longer balanced. In these circumstances the miners down their tools and rush to the air lock.

Early compressed air tunnels suffered a number of spectacular blows. During the driving of Haskin's tunnel under the Hudson River a blow occurred through the roof of the tunnel in July 1880 which has been referred to earlier. As soon as this happened the miners took refuge in the

Figure 7.3 Blow in Haskin's tunnel under the Hudson River (1880)

air lock, but one of the doors was jammed by falling earth and lining plates and twenty men were drowned. Figure 7.3 shows a contemporary print of this incident, the air lock is on the left. This tunnel is thought to be the first in which telephones were used; electric arc lights were also used and it is interesting to note that they are both shown in Figure 7.3, although the telephone wires must have been routed in a different way from that shown by the artist otherwise they would have been cut the first time the air lock door was closed!

Another remarkable instance of a blow occurred in 1905 on the Battery Tunnel being driven beneath the East River between Brooklyn and New York for the Rapid Transit Subway system.[15] Eight men were in the shield when the compressed air being used blew a hole through to the river bed 5 ft above the crown of the tunnel. While attempting to stem the blow by stuffing the hole with a bale of straw, the usual practice, a miner, Richard Creegan, was blown into the hole, through the silt of the river bed and upwards through a further 15 ft of water to the river surface. Some astonished longshoremen in a passing boat rescued him unharmed by the incident.

The danger to miners of 'the bends' was also experienced right from the start of compressed air tunnelling; it is the name given to a sickness which can strike tunnel workers during decompression after working in compressed air. The cause is as follows. During the period when a man is working in compressed air, because of the excess pressure the blood stream becomes saturated with more nitrogen than is usually found in the blood at normal air pressure. During decompression this excess nitrogen comes out of solution to form bubbles in the blood and in the tissues of the body, which give rise to the symptoms of the bends. The extremely severe pains of the bends are caused by a stoppage of the oxygen supply. The red cells in the blood carrying the oxygen fail to get past the nitrogen bubbles and the body tissues are denied the vital oxygen they must have – resulting in severe pain. The affected parts are very often the joints of the limbs which gives the sickness its name. If the nitrogen bubbles are liberated in the spinal chord paralysis can be caused, and if liberated in the veins and carried to the heart, can cause death. There were also other symptoms such as muscle pains, stomach cramp, swollen joints and headaches – all caused by nitrogen bubbles coming out of solution in different parts of the body. We now know that the solution to the problem of the bends lay in following three important rules: restricting working pressures to less than 3 atm, restricting the length of time men had to work in compressed air, and, most important of all, carrying out the decompression slowly enough for the nitrogen to come out of solution and be removed by the lungs without forming bubbles. The incidence rate for the bends is very low for pressures up to 1 atm, but rises steeply between 1 and 3 atm; the regulations for working in compressed air, therefore, become progressively more restrictive above a pressure of 1 atm.

But the cause of the bends was not known until it was discovered by the French physiologist Paul Bert (1833–86), Professor of Physiology at Bordeaux and the Sorbonne, in 1872 and attempts to deal with the bends before this and for some time afterwards were therefore unsuccessful; although Andrew H. Smith, the medical consultant to the New York Bridge Co, had thought that slow decompression might be the solution to the puzzling illness. Indeed, in 1872 Smith proposed a device which was in all respects the same as the medical lock that is now a standard feature on all compressed air tunnel jobs, a special chamber into which a miner suffering from the bends can be placed, quickly returned to tunnel pressure and then very slowly decompressed.[16] The bends had been known for some time before compressed air tunnelling, because of the earlier use of compressed air in caisson sinking where the workmen had suffered from this sickness. For example, a pressure of up to 3 atm was used to sink one of the caissons for the St Louis Bridge across the Mississippi in 1870. In fact its other name, 'caisson disease' is derived from this association. A very detailed account of the occurrence of the bends suffered by the workmen during the sinking of the caissons for the foundations of the piers of the Brooklyn Bridge in 1870–2 has been given by D. McCullough.[17] Cases of the bends became as frequent on compressed air tunnelling jobs as they were on caisson sinking, although the air pressures used in tunnelling were usually lower than those used in caisson sinking.

On the first Hudson River Tunnel described earlier, prior to 1889 deaths due to the bends were at the rate of 25 per cent of the men employed, and it is surprising that men would work in the tunnel at all. In 1889 when the British contractors took over the works, Ernest William Moir (1862–1933) installed a medical lock, constructed and used in the way suggested by Smith back in 1872. This was the first time a medical lock was used and it at once reduced the mortality rate to less than 2 per cent.[18] On the St Clair River Tunnel, also described earlier, the incidence of cases of the bends was reduced by replacing the 4 in diameter air release valve in the air lock by a $1\frac{1}{2}$ in diameter one because, as we now know, this had the desirable effect of increasing the time of decompression of the miners. As a curious footnote to the physiology of working in compressed air, on the St Clair River Tunnel it was discovered that mules were far better able to survive working in compressed air than horses.

During the construction of the City and South London Railway, on the short sections of compressed air working through the gravel and sand the men were observed to bear the effects of compressed air without ill effect, there being no cases of the bends at all. We now know that the reason for this is probably because of the relatively low air pressure of 1 atm that was used, although at the time Greathead believed that it was because of the purity of the air in the tunnel which was a consequence of the large volume of air supplied to make good that escaping from the face through the gravel and sand.[19]

A full account of the occurrence of the bends in early compressed air tunnelling, including a description of the measures taken by the State of New York to prevent or reduce the incidence of cases of the bends during the driving of the Hudson River and East River tunnels, has been given by B.H.M. Hewett and S. Johannesson.[20]

We have seen that the danger of the bends arises during decompression, and that the principal method of preventing it is to carry out the decompression very slowly. There is another hazard to miners but in this case one that occurs during compression, and this is the risk of 'getting blocked'. To understand it we must briefly consider the structure of the human ear. The outer ear is separated from the middle ear by the membrane of the ear drum. The middle ear is connected with the back of the nose by a narrow tube known as the Eustachian tube. When a person is subjected to increasing air pressure, unless the Eustachian tube freely admits air into the middle ear, the air pressure in the outer ear will become higher than that in the middle ear, the membrane of the ear drum will be pressed in and eventually rupture of the membrane will occur. This is not only very painful but may result in temporary or permanent deafness. By a swallowing movement or by holding the nostrils and blowing the nose some people can open the Eustachian tubes and equalise the pressure on either side of the ear drum. However, other people cannot do so and this constitutes a small but persistent hazard in the use of compressed air. Even miners who can clear their tubes in the manner described sometimes get blocked because a head cold or some other cause has resulted in an obstruction of the Eustachian tubes. The only remedy for getting blocked is to release the compressed air and return the affected person to normal atmospheric pressure.

The fourth disadvantage of compressed air tunnelling, bone necrosis, was discovered during the 1950s in both the United States and Great Britain. It is a disease to which miners who have spent a long time working in compressed air are prone and it is characterised by the death of parts of the bone. Bone necrosis, like the bends, is caused by nitrogen bubbles forming in the blood stream during decompression, but in this case the blockage occurs in the blood vessels supplying the bone. The bubbles prevent oxygen getting to the bone tissue and bone death results. The symptoms may not appear until the miner has left compressed air working for some time, typically 8–10 years, and therefore the association of the disease with the cause was not for many years recognised. The connection between bone necrosis and compressed air tunnelling was suspected in the 1940s, but it was not until studies were made of the medical history of the miners at the third Lincoln Tunnel, New York, constructed between 1953 and 1957, the first Dartford Tunnel built between 1957 and 1959, and the Clyde Tunnel built between 1958 and 1962, that the link became clear. No way of preventing the disease is known so far, but extending the decompression time is practised in an attempt to reduce its incidence.

Today, compressed air tunnelling is subject to regulations[21] aimed in particular at reducing the chances of miners contracting the bends (now termed 'decompression sickness'); the regulations also provide for radiological examination to detect the onset of bone necrosis and compression tests to eliminate those who cannot clear their tubes. The realisation that compressed air tunnelling was fraught with these dangers, in particular blows, the bends and bone necrosis, provided the impetus to seek a less hazardous alternative, and this has now been found in a further innovation – pressurised face tunnelling shields, the invention of which will be described in Chapter 9.

7.7 The diving bell

Before discussing the origin of compressed air tunnelling it is necessary to consider briefly the precurser of this innovation, namely the diving bell. Although the idea of the diving bell can be traced back into antiquity, it is generally agreed that the first practical working diving bell[22] was that constructed by Edmund Halley (1656–1742), better known as the celebrated astronomer (discoverer of Halley's comet) and Secretary of the Royal Society. Halley constructed his diving bell in 1690 and described it at a meeting of the Royal Society in 1716.[23] Halley's diving bell consisted of a hollow truncated cone made of wood, 3 ft in diameter at the top, 5 ft in diameter at the bottom and about 10 ft high. The internal volume of the bell was 60 ft^3. In the top a piece of strong clear glass was let in to give light and a stopcock was fitted to let out stale air. The bottom was open, but a bench was fitted around the inside for the divers who were to go down in the bell. A stage was suspended from the bottom of the bell which carried weights to take the bell down, and from the top the bell was suspended by a rope leading to the surface. Whilst submerged, the diving bell was supplied with fresh air periodically by lowering weighted barrels full of air down to it. The air was transferred from the barrel to the bell by means of a leather air tube connected up by one of the divers. Halley reported that using this apparatus, he had stayed for $1\frac{1}{2}$ hours at a depth of 60 ft.

Some time after this successful demonstration, diving bells gradually came into use for marine salvage and maritime civil engineering operations. Notable amongst the latter were the diving bells used by John Smeaton (1724–92) in 1790 and by James Rennie (1761–1821) in 1812 on the construction of underwater works at Ramsgate harbour.[24] Both these bells were made of cast-iron and were supplied with air through a pipeline from a force pump at the surface. Rennie's bell had some freedom of movement, being suspended from a trolley running on rails above.

During the construction of the Thames Tunnel (see Section 6.4) a diving bell, brought from the West India Docks, was used regularly after April 1827 to inspect the river bed and superintend the placing of clay over the breach following the inundation of the tunnel.[25] Both Brunel and his son made a number of descents; on one occasion Brunel's son was able actually to stand on one of the frames of the shield. On another occasion Roderick Impey Murchison (1792–1871), the Director of the Geological

Survey, went down in the bell and has left a vivid description of the experience.[26]

The pressure of the air in a diving bell is, of course, determined by the depth of the bell. As the bell sinks, the air inside it is compressed by the free water surface at the bottom of the bell so that the air pressure is always in precise equilibrium with the water pressure. In a tunnel no such automatic control is available and, as we have seen, the engineer must measure the height of water above the tunnel axis level and apply the requisite air pressure to balance it. In a diving bell the divers enter the bell at the surface under normal atmospheric pressure and are subjected to a gradually increasing air pressure as the bell sinks. Again, in a tunnel no such automatic process is available and, as we have seen, the miners have to enter the compressed air workings by passing through an air lock. In spite of these differences, the diving bell was a most effective demonstration that water could be excluded from a working area by air under pressure.

7.8 Summary

Cochrane's patent specification of 1830 is the source of both compressed air shaft sinking and compressed air tunnelling. And we have seen that, incorporating as it did the idea of the air lock, it was capable of immediate practical application. Compressed air shaft sinking began in 1839 and compressed air tunnelling in 1879. However, where did Cochrane get his ideas for compressed air working and the air lock? Cochrane makes no mention of the matter in his patent specification and Glossop, who made a close study of Cochrane's invention, said that in Cochrane's letters and papers he could find no evidence as to how Cochrane had thought of compressed air tunnelling. The following discussion is, therefore, to some extent speculative.

The circumstances are that Cochrane was by profession a naval officer, and moreover, one who had had much experience. It is, therefore, certain that he would have known of both the diving bell and the lock; he would have been familiar with the lock because it must have been a common sight on the canals of early nineteenth-century England to say nothing of his encountering it in ports and harbours, and he would have been familiar with the diving bell, by then common apparatus of marine salvage and underwater construction.

In 1829 we know that Cochrane had settled near Southampton and had applied himself to mechanical invention – surrounding himself with friends of scientific and engineering turn of mind. When the Thames Tunnel was flooded, a matter of great public interest, he applied his mind to the problem of subaqueous tunnelling and came up with the solution of compressed air tunnelling. It is reasonable to suppose that he would have drawn the ideas for this from his knowledge of the diving bell and the lock because we have shown that the diving bell embodies the principle of compressed air working and the canal lock is the same in principle as the air lock. This supposition, at least in respect of the diving bell, is supported by an entry that Brunel made in his diary for 7 March 1831 after having seen Cochrane's patent:

> Went to the Patent Office found that Lord Cochrane had taken out a Patent for some improvements in the Art of Mining – which is the scheme he has for tunnelling – this plan is to force air in the excavations as under a Diving Bell, in proportion to the head of water or the weight of ground that must be supported.

We have already noted Brunel's likening of compressed air tunnelling to the diving bell in the introduction to this Chapter, from which it is reasonable to infer that his own tentative idea of working under compressed air was suggested by the diving bell. Thus Cochrane devised a method of compressed air tunnelling which embodied two items, air under pressure in a tunnel and the air lock, which can be shown to have direct analogies with items with which he already would be familiar, air under pressure in a diving bell and the canal lock.

Notes and references for Chapter 7

1 Mallet, M.A. (1893) *Autobiographie de J.-Daniel Colladon.* Paris (Société des Ingénieurs Civils de France) p. 8.

2 The historian of the early use of compressed air tunnelling and shaft sinking, Rudolph Glossop, has published the following account which includes many quotations from Brunel's papers:
Glossop, R. (1976) The invention and early use of compressed air to exclude water from shafts and tunnels during construction. *Géotechnique*, **26** (2), pp. 253–80.

3 Stephen, L., Editor (1887) *Dictionary of national biography*, Volume 11. London (Smith, Elder and Co) pp. 165–75.

4 Cochrane, Sir T. (1830) Apparatus for excavating, sinking and mining. *British*

Patent No. 6018. London (Great Seal Patent Office).
5 Széchy, K. (1970) *The art of tunnelling*. Budapest (Akadémiai Kiadó) pp. 772–3.
6 McCullough, D. (1972) *The great bridge*. New York (Simon and Schuster) pp. 175–6.
7 Hewitt, B.H.M. & S. Johannesson (1922) *Shield and compressed air tunnelling*. New York (McGraw-Hill Book Co Inc) p. 223, Figure 101.
8 Glossop, R. (1976) *op. cit*, p. 266, Table 1.
9 Legouëz, R. (1897) *De l'emploi du bouclier dans la construction des souterrains*. Paris (Libraire Polytechnique Baudry et Cie) pp. 349–51.
10 Gilbert, G.H., L.I. Wightman & W.L. Saunders (1912) *The subways and tunnels of New York*. New York (John Wiley and Sons) pp. 7–9.
11 Copperthwaite, W.C. (1906) *Tunnel shields and the use of compressed air in subaqueous works*. London (Archibald Constable and Co Ltd) p. 160, Figure 99.
12 Vogel, R.M. (1964) Tunnel engineering – a museum treatment. *Contributions from the Museum of History and Technology*, Paper 41. Washington DC (Smithsonian Institution) pp. 203–39.
13 Greathead, J.H. (1896) The City and South London Railway; with some remarks on subaqueous tunnelling by shield and compressed air. *Minutes of Proceedings of the Institution of Civil Engineers*, **123**, Paper No 2873, pp. 39–73.
14 Anon (1890) The City and South London Railway. *The Illustrated London News*, **97**, (2690), pp. 579–80.
15 Sandström, G.E. (1963) *The history of tunnelling*. London (Barrie and Rockliff) p. 236.
16 McCullogh, D. (1972) *op cit*, p. 322.
17 *Ibid*, Chapters 13 and 14.
18 Copperthwaite, W.C. (1906) *op cit*, p. 172.
19 Greathead, J.H. (1896) *op cit*, p. 69.
20 Hewett, B.H.M. & S. Johannesson (1922) *op cit*, Chapter 16.
21 Construction Industry Research and Information Association (1973). A medical code of practice for work in compressed air. *CIRIA Report* 44, London (Construction Industry Research and Information Association).
22 Heinke, J.W. (1870) *A history of diving*. London (Heinke and Davis) pp. 10–11.
23 Halley, E. (1716) The art of living under water. *Philosophical transactions of the Royal Society*, *Abridged*, **6**, pp. 258–62.
24 Davis, Sir R.H. (1955) *Deep diving and submarine operations*. London (Siebe, Gorman and Co Ltd) 6th edition, pp. 610–1.
25 Law, H. (1845) *A memoir of the Thames Tunnel. Part 1. From the commencement of the works to their suspension in 1828*. London (John Weale) pp. 71, 75–8, 84–5, 102–4 and Plate 15.
26 Rolt, L.T.C. (1976) *Isambard Kingdom Brunel*. Harmondsworth (Penguin Books) pp. 53–4.

Prefabricated tunnel linings

Decus et tutamen.

Early tunnels in soft ground, for example those constructed by Bazalgette for London's main drainage, were lined with brick. Brick lining was also used by Brunel for the Thames Tunnel. However, as soon as the tunnelling shield came into general use a brick, or masonry, lining was seen to have two grave disadvantages. The first of these was that until the mortar used to lay the brickwork had set, the lining was unable to sustain the reaction of the jacks used to thrust the shield forward. The second was that until the mortar had set, the lining was unable to sustain the pressure of the ground which it would have to do as soon as it had emerged from the tailskin of the shield. These considerations were not much of a problem with Brunel's Thames Tunnel shield because, it will be recalled (Section 6.5), this shield was advanced one frame at a time, but with all subsequent shields which were advanced as a single unit, the matter became pressing. In order to take full advantage of the tunnelling shield, a lining had to be devised which would take the full thrust of the shield and the full overburden pressure of the ground as soon as it was erected.

As well as the two problems outlined above, the traditional brick tunnel lining had an intrinsic disadvantage of a more serious nature. It will be recalled from Section 6.2 that in the traditional method of soft ground tunnelling the tunnel was first temporarily supported by timbering and then a permanent brick lining was built by bricklayers on centring. When the mortar had set, the centring was struck, the crown bars were removed and the brick lining allowed to take the pressure of the ground. To do this, of course, the ground had to move down towards the tunnel and this in turn led to subsidence at the surface above the tunnel. Surface subsidence above a tunnel was of little consequence if the tunnel was being driven beneath open country, but in urban areas it could be disastrous – as the experience of the Metropolitan Railway was to show.

8.1 The Metropolitan Railway and the problem of subsidence

In Section 6.1 we have already noted that the Metropolitan Railway and the District Railway, London's first two underground lines, were built mainly by the cut-and-cover method of tunnel construction and that this left such a swathe of destruction in its path that it was not repeated again in the city. However, on the Metropolitan Railway there were three conventionally driven tunnels as well. These were the Campden Hill Tunnel, the Clerkenwell Tunnel and the Widening Tunnel which was parallel to the Clerkenwell Tunnel. These three tunnels were driven by the traditional method of soft ground tunnelling. During construction of the Widening Tunnel from November 1865 to May 1867 the resident engineer, Mr Morton, took progressive measurements of the settlement above the tunnel and these were:

Commencement to getting in top sill	¾ in
Completion of timbering	2 in
After striking back props	1½ in
Further settlement	3 in
Made up total of	7¼ in

These measurements were made in a section of tunnel in London Clay where the depth from the ground surface to tunnel invert level was 55 ft, and showed the considerable amount of subsidence occurring at the time of construction, but with time the ground moved further onto the tunnel lining and the subsidence continued. Baker remarked:

> Tunnelling through a town is a risky operation, and settlement may occur years after completion of the works. Water-mains may be broken in the streets and in the houses, stone staircases fall down, and other unpleasant symptoms of small earthquakes alarm the unsuspecting occupants.[1]

It was observed that the general draw of the ground was towards the working end of the tunnel and that fissures formed in the ground which ran along the surface parallel to the tunnel, these being so predictable that the resident engineer could forecast where a building or wall standing over the tunnel line would be most severely cracked; it was further noted that the slope of the fissures was $\frac{1}{2}$ (horizontally) to 1 (vertically) measured from tunnel invert as shown in Figure 8.1.[2] Several miles of fissures occurred during construction of the Metropolitan Railway and resulted in heavy extra expense to the contractors who assumed responsibility for the damage to adjoining property. And it was not only on the new underground railway tunnels that there were problems. When the London and North-Western Railway Company drove a tunnel at Primrose Hill it was observed that 'from end to end all the buildings were cracked in the neighbourhood of the tunnel'.[3]

Clearly this was a problem of such magnitude that it would have soon prevented the construction of tunnels beneath cities or built-up areas. Taken together with the disadvantages of trying to use a brick lining with a shield that were discussed above, the problem of the large subsidence over traditional soft ground tunnels led to the devising of prefabricated tunnel linings which could be used with the new tunnel shield and which would support the ground with a minimum of ground movement, this in

Figure 8.1 Damage caused by traditional tunnelling (1866)

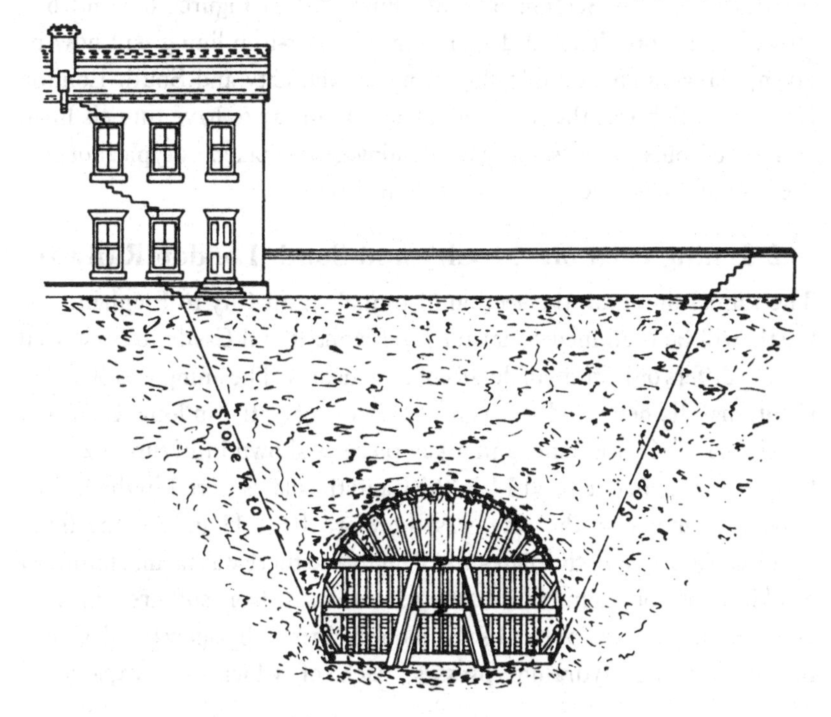

turn minimising subsidence and damage to property over the tunnel.[4] The purpose of this Chapter is to examine this innovation and its subsequent development, to discuss some of the problems associated with it and to indicate its likely origin.

The solution to all the problems of the traditional lining was a prefabricated tunnel lining which consisted of cast-iron segments which were bolted together to form a ring which in turn was bolted onto the previously built ring, giving a lining which had instant strength both longitudinally and transversely. This idea was around at the same time that the tunnelling shield was thought of, because, as described in Sections 6.6 and 6.9, cast-iron segmental lining was described in both Brunel's and Barlow's patent specifications for the tunnelling shield, being considered as an important part of the method. For example, Brunel says in his patent specification of 1818:

> The body or shell of the tunnel may be made of brick or masonry, but I prefer to make it of cast iron, which I propose to line afterwards with brickwork or masonry.[5]

A general idea of an early cast-iron segmental tunnel lining was given in the description in Section 6.8 and illustrated in Figures 6.6 and 6.7. However, a more detailed description of a cast-iron lining will now be given, using as an example the lining for the City and South London Railway, which was the first underground railway to have tunnels lined with cast-iron segments and whose lining can be taken as typical of that used on all subsequent tube tunnels in London.

8.2 Lining used on the City and South London Railway

The tunnel lining on the section of tunnels from City to Elephant and Castle will be examined;[6] that for the Elephant and Castle to Stockwell section differs only in small details of dimensions. The lining was 10 ft 2 in in internal diameter and was composed of rings 1 ft 7 in long, each ring consisting of six segments and a key piece as shown in Figure 8.2. The flanges were $3\frac{1}{2}$ in deep and $1\frac{3}{16}$ in thick and the plates were 1 in thick. The outside diameter of the ring was therefore 10 ft 11 in. All the holes required were cast in the plates and flanges and there was no machining of any kind done on them. The segments were made from soft grey pig-iron and were cast in compressed sand moulds formed by specially designed machines using hydraulic presses, each of which was capable of manufacturing moulds for 14 segments per hour. The segments were then dipped into a mixture of pitch and tar while hot to give them a protective coating.

The segments were fastened together by means of nuts and bolts passing through holes cast in the flanges. There were, therefore, two kinds of joints: those between individual segments in a ring being termed longitudinal joints and those between one ring and another being termed vertical joints. Details of these are shown in Figure 8.2. Soft pine packings $\frac{1}{4}$ in thick were placed between the flanges in the longitudinal joints, and in the vertical joints a rope of tarred hemp was placed between the flanges behind the bolt holes. Subsequently all the joints were pointed with cement.

Although the segments weighed $4\frac{1}{2}$ cwt each, and in spite of the fact that earlier Greathead had designed a segment lifting apparatus, on this project the lining rings were erected manually. A team of six miners placed the segments into their positions by hand, using for the upper two and the key piece a light temporary stage which was also used for bolting. Small pulley blocks were used for slinging the lower side segments. For larger tunnels, however, segment erecting arms were essential, as described in Section 6.10 and shown in Figure 6.7. The ring of segments

Figure 8.2 Cast-iron lining (1887)

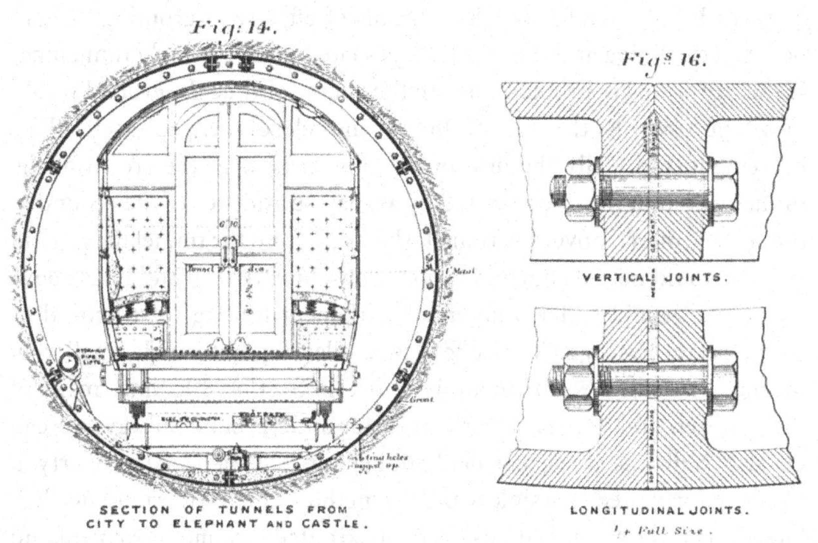

was erected within the tailskin of the shield; when complete the shield was shoved forwards off the ring with its hydraulic rams. When the ring emerged from the tailskin there was an annular gap 5½ in wide between the ring and the clay corresponding to the difference between the external diameter of the lining ring, 10 ft 11 in, and the excavated diameter of the tunnel cut by the shield, 11 ft 4½ in. This difference arises because of the thickness of the metal plate of the tailskin of the shield plus the necessary clearance between the lining ring and the tailskin.

We have seen that on the Tower Subway Greathead had the miners fill this gap between the lining and the ground with Blue Lias cement grout injected by means of a hand syringe. The repair work on the Tower Subway carried out after air-raid damage in 1940, referred to in Section 6.7, later showed that this grouting was satisfactory, but at the time Greathead was very concerned that this method of injecting the grout behind the lining might not be effective in filling the whole of the void. Because of this, for the City and South London Railway linings, he devised both a method of grouting behind the lining and the equipment for doing it. This system was so successful that it has persisted to the present day.

8.3 Greathead's grout pan

Before looking at Greathead's grouting apparatus and how it was used, it is worth briefly examining why a means of effectively grouting the gap between the lining and the ground was so important to shield tunnelling. If the gap were not grouted, then in the course of time the ground would move in towards the tunnel lining and close the gap itself. This movement, especially the downward movement over the crown of the tunnel would in turn be reflected by a small subsidence of the ground at the surface directly over the tunnel. But away from the tunnel the ground surface would not, of course, subside. Consequently any buildings above the tunnel would suffer differential settlement leading to structural or architectural damage. This effect is similar to, but much smaller in magnitude than, the settlement described in Section 8.1. Since many of the required tunnels, particularly those for underground railways were in densely built-up areas, if shield tunnelling had damaged property it would not have been possible to use the method where it was most needed. However, if the gap could be effectively grouted, ground movement and surface settlement would be minimised if not eliminated and there would be no obstacle to the use of shield tunnelling beneath cities.

The grouting apparatus devised by Greathead, which he called a 'grout pan', is shown in Figure 8.3. It consisted of a cylindrical vessel, capable of holding a pressure of 5 atm, through which was fitted a shaft carrying paddles which can be turned by a handle outside. The cement and water were put into the vessel through an opening at the top which was afterwards closed with an airtight lid. The grout was discharged from the bottom of the pan through a flexible pipe and nozzle by applying compressed air through the top of the pan. The grout pan was usually operated by two men; one man kept the grout mixed by continually turning the paddles and he also operated the air-admittance and grout-discharge valves V, while the other man handled the nozzle end of the grout pipe.

The gap behind the lining was grouted by inserting the nozzle into a hole in the segment which had been specially provided for this purpose (see Figures 8.2 and 8.3). Beginning at the lowest segment in the ring, grout was injected until it reached the hole above it; the lower hole was then plugged and the nozzle applied to the upper hole, and so on until

Figure 8.3 Grout pan (1887)

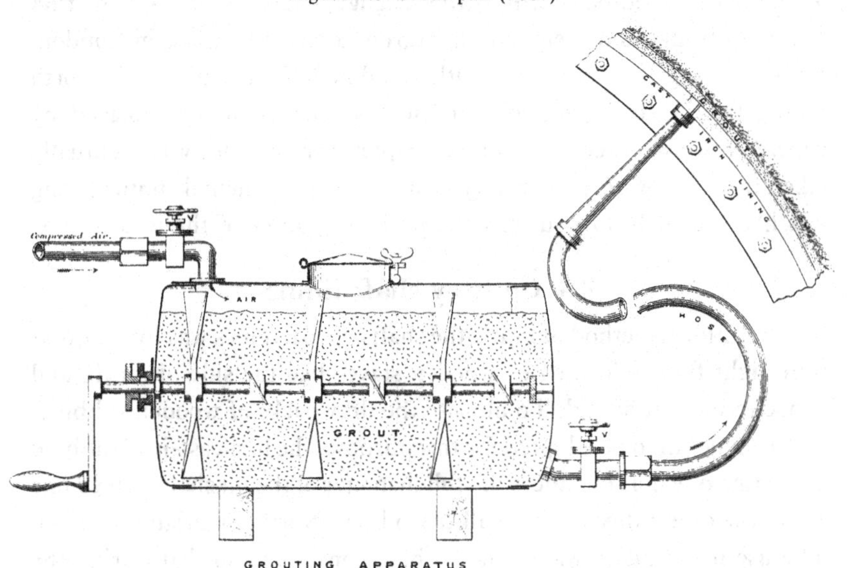

finally the highest hole, that in the key piece was reached. In this way the whole ring of segments was grouted, the full pressure from the grout pan being applied at the last injection.

Greathead carried out a number of experiments in making grout with different cements. He also tried adding sand and using different amounts of water. He concluded that for grouting tunnels in London Clay, a grout made with Blue Lias cement and water was the best. Because there was no point in having a grout that set harder than the clay itself there was no need to use the more expensive Portland or Medina cements, and the extra trouble of mixing in sand was not repaid by the small cost saving over cement. He also found that it was most important that there should not be an excess of water because this led to shrinkage of the grout. During the works for the City and South London Railway, the opportunity was taken to examine the effectiveness of the grouting when some 2000 ft of tunnels were enlarged to form stations. The iron linings were found to be completely encased and even adjacent cavities and cracks in the ground had been filled with grout.

The $3\frac{1}{4}$ mile long City and South London Railway tunnels, completed in four years, were a convincing demonstration that tunnels could be driven beneath city streets with little or no damage to property by a combination of shield tunnelling, cast-iron lining and effective grouting. These basic methods, with little change, were used to construct successfully all the subsequent underground railway tunnels in London. Before leaving the City and South London Railway lining it is worth noting that Moir suggested that the cast-iron could be replaced by concrete;[7] we shall see later in this Chapter how this idea was eventually taken up. Having seen what a typical cast-iron segmental tunnel lining was like, we shall now suggest the probable source of the idea.

8.4 Colliery shaft lining

The traditional method of lining the shafts of coal mines was by means of watertight frames formed of wooden staves like the side of a tub, and hence termed 'tubbing'. To obviate the drawbacks of wooden tubbing, cast-iron began to be adopted for this purpose, the first cast-iron tubbing consisting of whole cylinders the full diameter of the pit shaft, placed one above the other; this was at Walker Colliery, Northumberland, in 1795. The use of cylinders was found to be inconvenient so that during the construction of a shaft at the Percy Main Colliery, also in Northumberland, in 1796–9, John Buddle the elder[8] employed tubbing made of cast-iron segments 4 ft long by 2 ft high and bolted together with internal flanges.[9] However, it was found that in use the internal flanges of the tubbing were a nuisance because the containers used in drawing coal up the shaft tended to catch on them. Moreover, the flanges also broke up an otherwise smooth surface and impeded the free flow of ventilating air. Later on, therefore, segments with external flanges without bolts like those shown in Figure 8.4 were used and became the normal method of lining colliery shafts in water-bearing ground.[10] The shaft lining with external flanges could not be bolted because, of course, the miners, who were inside erecting it, could not get access to the outside where the flanges were. This did not matter for a shaft where a ring of segments was on its side, but when the idea was later used for tunnel construction in which a ring would be upright, bolting became desirable for stability during erection.

During the nineteenth century, when Brunel, Barlow and Greathead were designing their shields, it was undoubtedly the cast-iron segmental shaft lining for colliery pits that gave them the idea of the cast-iron segmental tunnel lining. But for the tunnel lining they reverted to the original idea of internal bolted flanges, this arrangement being more suited to tunnel construction for the reason given above. The cast-iron segmental tunnel lining was, therefore, the same as Buddle's original shaft tubbing with its internal bolted flanges.

8.5 The McAlpine lining

In his patent specification of 1874, referring to the tunnel lining, Greathead wrote:

> Instead of metal the casing may be formed of Voussoir segments of artificial stone, cement, concrete or other material previously moulded, and secured in their places by dowelling.[11]

However, it was many years before this suggestion was taken up, cast-iron remaining the preferred material from which to make prefabricated tunnel lining segments.

The first prefabricated concrete tunnel lining was that developed by the firm of Sir Robert McAlpine and Sons Ltd in the early 1900s but which was not used extensively until it was adopted as the primary lining for sewer tunnels in west Middlesex between 1931 and 1935. The lining was made up of segments made of precast concrete, each was 1 ft wide with tongue-and-groove joints on all four sides. After erection of a ring of segments, the inside of the vertical joint was pointed with cement mortar and a steel hoop was inserted into the joint. The next ring was then erected. In this way steel reinforcement was built into the lining of the tunnel, ring by ring, as it was constructed. The space behind the lining

Figure 8.4 Colliery shaft tubbing (1859)

was also grouted with cement through special grout holes in the segments. The lining was not used in conjunction with shield tunnelling and has only found limited application since its use on the West Middlesex Main Drainage Scheme;[12] a brief description is included here, however, because of its historical interest as the first of what was much later to become a prolific line of development.

The McAlpine lining has another historical claim to fame in that it was the first tunnel lining to be tested scientifically.[13] In 1911 a special purpose-made testing machine was constructed to test in compression an assembly of three complete rings. Unfortunately the lining was far stronger than the loading capacity of the machine and the tests were, therefore, inconclusive! It was intended that the testing machine would also be used to test tunnel linings of brickwork and cast-iron so that the strength of the new McAlpine concrete lining could be compared with these traditional linings. In the event these linings were also stronger than the capacity of the machine.

8.6 Precast concrete lining used on the Ilford tube

In 1936 the London Passenger Transport Board embarked on a programme of extending the tube railway system. In 1937 during the letting of the tunnelling contracts for these works, it was found that the demands of national rearmament had resulted in an acute shortage of cast-iron. Consequently, there were fears that interruption in or cessation of the supply of cast-iron lining segments might endanger the proposed programme of tunnelling works. Because of this, the Board decided to develop a precast concrete tunnel lining. This was done and the new lining was tried out on the Redbridge–Newbury Park section of the Ilford tube extension of the Central Line; the lining was designed by the notable bridge and tunnel engineer David Anderson (1880–1953), the Board meeting the cost of the experimental work.[14]

The precast concrete tunnel lining devised was generally similar in form to the cast-iron lining used exclusively in London's tube railway construction until this time. The rings, 1 ft 8 in wide, each consisted of six segments plus a key piece, the segments being bolted together through their longitudinal flanges as were cast-iron segments. Bolts through the vertical flanges connected ring to ring – again as in cast-iron lining. The thickness of the wall of the lining was 2 in compared with $\frac{7}{8}$ in for cast-iron and the flanges were much thicker than for cast-iron segments. They also differed from cast-iron segments in that there were four transverse ribs to give additional strength when the shield was shoved forward. Each concrete segment had two $1\frac{1}{4}$ in diameter grout holes through the wall just like the cast-iron segments. The precast concrete rings were of 12 ft 3 in internal diameter and 13 ft 1 in outside diameter, so that they were a little bigger than the Transport Board's standard cast-iron rings (12 ft internal diameter, 12 ft 10 in outside diameter) but the shields were easily adapted to accommodate the slightly larger concrete rings. The reason for this difference in size was to allow room for standard signal apparatus which was designed to nest into the cast-iron segments but which would not fit into the smaller space between the flanges of the concrete segments. Steel reinforcement, cast into the segments during manufacture, provided them with tensile strength which concrete by itself does not possess. Bituminous packing with a hessian matrix was used in the longitudinal joints while creosoted wood packing was used in the vertical joints. The concrete lining was not expected to withstand the rough handling that cast-iron does and modifications to the tunnel miners' accustomed heavy-handed methods of getting the segments into position and the rings into shape were therefore made. A complete concrete ring weighed 1.39 tons compared with 1.66 tons for a cast-iron ring, both rings being of the same width. The manufacture of the precast concrete segments was carried out by two manufacturers, each of whom used steel moulds. The concrete was compacted into the moulds by the use of either vibrating tables or shock tables – both methods gave satisfactory results. After manufacture the backs of the segments were heavily coated with bituminous emulsion to prevent possible sulphate attack from taking place when they were installed in the London Clay.

Some 8700 rings of precast concrete lining were used in the construction of a $2\frac{3}{4}$ mile section of tunnel on the Ilford tube, no special difficulties arising. The cost of construction, excluding the cost of the lining segments, was the same as when using standard cast-iron lining, and the same rate of progress in shield driving was made when using either type of lining. However, the cost of the precast concrete lining was nearly 25 per cent less than the cost of cast-iron lining. The Ilford tube experiment, which had been initiated because cast-iron had become a strategic material and was therefore in short supply for civil works, had demonstrated that there was a much more economical way – using precast concrete – of making a tunnel lining.

Figure 8.5 shows a section of completed tunnel on the Ilford tube. In the foreground is the Transport Board's standard cast-iron lining, and beyond is the precast concrete lining. From the Figure, and from the description given above, it is apparent that the precast concrete lining is a copy of the cast-iron lining made out of another material. Anderson had designed a lining to be made from precast reinforced concrete, but which had to be as like as possible the Board's existing cast-iron segmental lining so that the same shields, methods of erection, grouting and other operations could be used with little change. If he had been given a free hand to design a precast concrete lining without these constraints, he would, no doubt, have based a design on the properties of concrete instead of following almost exactly a design based on cast-iron. Referring to this type of lining W.H. Ward has commented succinctly:

> Clearly the reinforced concrete segment, which after all was merely an imitation of the traditional cast-iron form in another material, is a poor design.[15]

Figure 8.5 Lining on the Ilford tube (1939)

In spite of this stricture, precast concrete segmental lining with the segments bolted together as in cast-iron lining has continued to be used right up to the present day, although other more rational designs of precast concrete lining are available. One of these will shortly be described, but before doing so a brief discussion on the principles of tunnel lining design will be necessary.

Although the cast-iron segmental tunnel lining had such a long and successful application in soft ground tunnelling, a number of engineers had questioned the principles on which its design appeared to be based. For a tunnel of circular cross section in ground such as London Clay, A.W. Skempton had shown that the forces acting on the tunnel lining are exactly those arising from the overburden load.[16] If the overburden load were assumed to act on the tunnel as a uniform all round pressure there would be no bending moment in the lining. If this were so, then there would be no need for the segments to be bolted to each other except for convenience of erection, in fact structurally it would be better if they were not since, as Karl Terzaghi (1883–1963) pointed out, their flexibility would allow the lining to accommodate itself to any local variations in the state of stress in the clay.[17] To resolve this question, in 1955 the Building Research Station carried out a test on the lining of an old deep London tube tunnel.[18] The bolts on the longitudinal joints of three cast-iron rings were slackened, the segments in the rings having been previously fitted with strain gauges. The slackening of the bolts was found to have a negligible effect on the strain in the segments showing that the bolts were not providing any appreciable resistance to bending at the longitudinal joints. Bolting of the joints, therefore, it was concluded, served no useful structural purpose. It should be emphasised that this argument does not apply to pressure tunnels, the linings of which have to withstand an internal hydrostatic pressure. However, for all other tunnels, the realisation that the segments of a prefabricated lining need not be bolted together opened the way for new designs, one of which will now be considered.

8.7 The wedge-block lining

The wedge-block lining was developed from an earlier lining known as the Don-Seg lining. The Don-Seg lining was designed by H.J. Donovan in 1947 and manufactured by the firm of Kinnear, Moodie and Co Ltd in the early 1950s. The segments of the Don-Seg lining were made of unreinforced concrete, there were ten identical segments to a ring, each segment being of simple rectangular cross section but of tapering form in plan. The lining was erected by placing alternate segments directly against the clay with their wide ends against the last built ring. The ring was completed by jacking home the intermediate segments with their narrow ends towards the last built ring. No bolts were used at all, the lining being held in place by being tightly expanded against the clay. As the first Don-Seg tunnels were water tunnels,[19] a steel internal lining was sometimes placed inside the concrete lining and grouted with cement – this vitiated the structural function of the Don-Seg system, but the steel lining was thought necessary to ensure watertightness.

Early experience with the Don-Seg system showed that the lining had a number of serious shortcomings. These were that the amount of friction generated on the sliding surfaces of the segments when the intermediate set of five were being jacked home was very large resulting in high jacking loads being needed, all the segments had to be jacked completely home or the ring was not true and the lining became progressively more irregular, the dimensions of the segments had to be very accurate and consistent and the shield had to cut a tunnel of accurate and consistent diameter and shape. The wedge-block lining was designed to overcome all these difficulties.[20]

The wedge-block lining, designed by F. Tattersall in 1955, is shown in Figure 8.6. The lining is made up of unreinforced concrete segments of rectangular cross section. Of the eleven segments in a ring, nine are identical and have parallel sides. The two crown segments, however, have a taper on one side and the ring is closed by a tapered key or 'wedge-block' from which the lining takes its name. All segments are placed in their final position directly against the clay and the last built ring, and there is little movement of them during closure of the ring which is accomplished by jacking in the wedge-block. The wedge-block is deliberately made shorter than the other segments in the ring so that some variation in the diameter of the excavated tunnel or some variation in the size of the segments can be accommodated by varying the distance to which the wedge-block is driven home; the small space left is filled with concrete afterwards. The rings are usually erected by a gang of four men; up to tunnel axis level the segments are handled and erected with the aid of a winch but the remainder are lifted into position by hand, the men working on a removable timber stage. It should be noted that the wedge-block lining, like the Don-Seg lining before it, was not erected within the tailskin of the shield as were the bolted linings, but erected directly against the ground after the shield had shoved forward. This meant that these unbolted linings could only be used in ground like London Clay which was cohesive enough to stand unsupported until the whole ring was erected. The bolted linings, on the other hand, could be used in any sort of ground.

The wedge-block lining was first used on a short experimental length of the Metropolitan Water Board's Thames–Lee Valley tunnel, constructed between 1955 and 1959,[21] the main part of which was lined with Don-Seg lining. Subsequently, however, the wedge-block lining has been used instead of the Don-Seg lining in all the Board's tunnelling schemes. The

Figure 8.6 Wedge-block lining (1955)

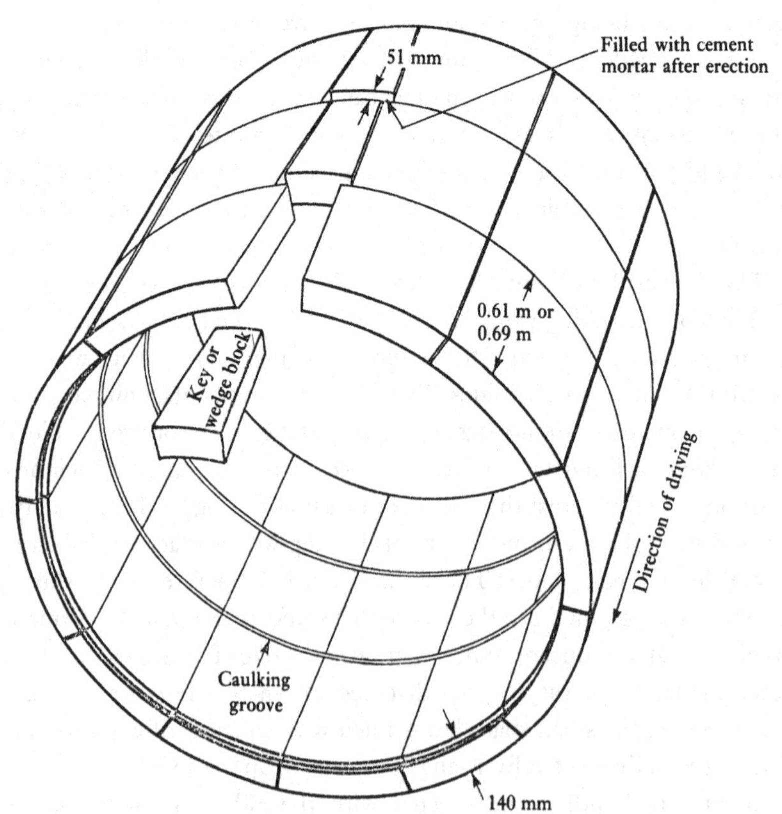

wedge-block lining is usually used to build tunnels of 8 ft 4 in internal diameter, this being the Board's standard size, but short lengths of tunnel in other diameters have been built. The internal steel liner which was at first used with the Don-Seg lining has been dispensed with altogether for concrete water tunnels in London Clay.

Because it is so quickly erected the wedge-block lining has enabled some outstanding rates of tunnel progress to be achieved. For example, a typical average erection time for one ring of wedge-block lining segments is 4 minutes, compared with 20 minutes for a ring of bolted segments. The record progress rate for a tunnel excavated with a mechanical digger shield and lined with wedge-block lining is 200 ft per day.

A comment can be made on the number of types of prefabricated concrete tunnel linings that have been designed. In the case of the cast-iron bolted lining, the basic type has remained unchanged from Greathead's original design for the City and South London Railway right down to the present; the only changes have been variations in dimensions and in details of the flanges and bolts. By contrast, there has been a plethora of types of precast concrete lining. In principle there are two concepts, the bolted segmental lining and the unbolted: the bolted lining has been exemplified here by the Ilford tube lining and the unbolted by the wedge-block lining. However there have been many variations of both the bolted and unbolted concrete segmental tunnel linings which it is not proposed to examine further here. Reviews of them have been given by Ward[22] and by R.N. Craig and A.M. Muir Wood,[23] the last being extremely detailed and well illustrated and containing historical notes on both cast-iron and concrete linings. The development of expanded linings has been reviewed by Donovan.[24] It is particularly interesting to see that an unbolted expanded concrete tunnel lining, similar in principle to the wedge-block lining but having two tapered keys instead of one, was used in 1972–6 on parts of the Jubilee Line of the London Underground whilst traditional-type bolted cast-iron lining was used on other parts,[25] showing that new and original methods are in use side-by-side on the same tunnel. This two-key wedge-block lining proved so successful that it was used again on the sections of drive through London Clay on the 6 km long extension of the Piccadilly Line to Terminal 4 at Heathrow Airport in 1983–4. The lining used is shown in Figure 8.7; the ring is 3.81 m internal diameter and consists of sixteen precast concrete segments together with two wedge-shaped concrete keys. The lining is erected and then expanded against the ground by driving in the wedge keys using two 15 tonne rams mounted in the shield.[26] It can be noted that the two wedge keys are in the more convenient lower side positions of the lining rather than in the crown as in the original wedge-block system (see Figure 8.6). As on the Jubilee Line, short sections of cast-iron lining were also built where the ground was unsuitable for the use of wedge-block lining.

Much has been made of the expanded tunnel lining, but it should be emphasised that the bolted precast concrete tunnel lining still has an important application where a tunnel is being driven through soft ground that is unsuitable for the expanded type of lining. The bolted lining has been considerably developed since the days of the Ilford tube where the

Figure 8.7 Two-key wedge-block lining (1983)

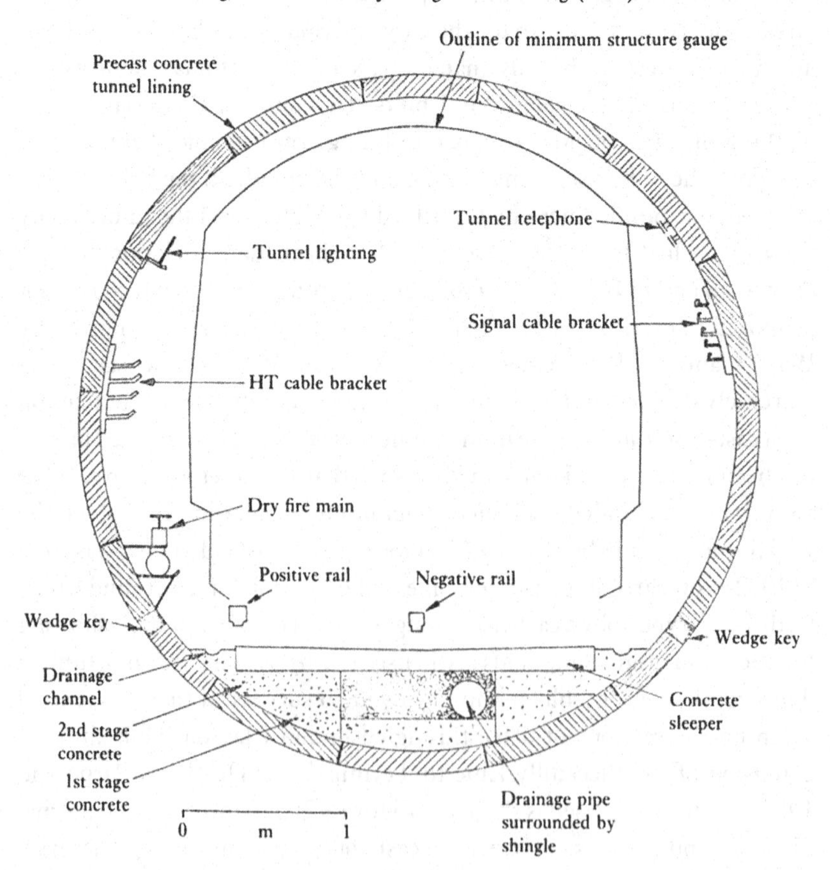

lining was simply a copy of the old cast-iron type. Modern bolted precast concrete tunnel linings are designed according to the principles of structural concrete design and stress analysis, and take into account the ground loads at the depth they are intended to be installed. In particular, the design is based on the concept of a flexible ring, the bolts at the joints being used only for erection purposes – the longitudinal joints are not intended to take bending moments. An example of a bolted precast concrete tunnel lining designed, manufactured and erected according to these criteria is the lining developed in 1965 for a 3 mile extension to the subway system for the Transit Commission in Toronto, Canada.[27] The ground was dense sand with little or no cohesion, for which the expanded type of lining would have been unsuitable. The tunnels were 16 ft internal diameter, the segments were 6 in thick, 2 ft wide and there were eight to a ring. The ring was erected in the normal manner within the tailskin of a shield, and grouted as soon as possible after the shield had shoved forward. In all, over $1\frac{1}{2}$ miles of the new extension were lined with the bolted precast tunnel lining; cast-iron lining was used on the other $1\frac{1}{2}$ miles, again showing that this old method was still being applied. The only problems encountered with the installation of the bolted precast concrete lining at Toronto were occasional damage to the lining when the shield was being shoved forward and some leaking of ground water into the tunnel. The first problem was quickly solved by replacing the broken segments, but the second problem was not so easy to solve although caulking of the joints was carried out as well as grouting. It was finally recommended that this type of lining should not be used below the water table.

An attempt has been made to extend the applicability of the expanded type of tunnel lining to ground in which it is at present unsuitable for use, by providing the whole lining with a sheath of polythene or similar flexible material.[28] The idea is that this sheath would provide the reaction normally supplied by the clay, in ground consisting of other, looser materials than clay. This idea, however, has not yet been given a practical trial, probably because the bolted lining already fills the need in a satisfactory manner.

Before leaving the subject of prefabricated tunnel linings an interesting parallel can be drawn with an aspect of the history of dams. For as N.A.F. Smith has pointed out, it was during the 1920s and 1930s that an understanding of the soil mechanics of earth dams was gained,[29] and one of the main contributors to this was the same Terzaghi who, as we have seen, also said that a flexible tunnel lining would adjust itself to the stress conditions in the ground. Terzaghi was active in the whole field of geotechnology and is regarded as the founder of modern soil mechanics.[30]

8.8 Summary

We have seen that the old hand-built brick or masonry lining proved unusable as soon as soft ground tunnels came to be driven beneath city streets because of the excessive subsidence that occurred at the surface when the ground was allowed to come down on the lining. And as soon as the tunnelling shield came into general use, the traditional lining was seen to have other operational disadvantages. To overcome all these problems cast-iron prefabricated tunnel lining was introduced and proved to be an immediate success. It is still in use at the present day, together with other forms derived from it and newer linings produced as a result of development within the industry.

It was undoubtedly the well-established cast-iron segmental lining for colliery shafts that inspired the idea of the bolted cast-iron segmental lining for tunnels, and this form became widely used and stayed in use right up to the present day because it was so well suited to erection within the tailskin of a tunnelling shield. The bolted cast-iron segmental tunnel lining was, therefore, borrowed from the mining industry. The bolted precast concrete segmental tunnel lining was clearly a copy in another material of the cast-iron lining and can, therefore, be considered as the development of an innovation rather than an innovation in its own right.

The unbolted precast concrete segmental lining, however, arose within the tunnelling industry itself. We have seen how very early on the McAlpine lining was introduced but that although used quite extensively on one tunnelling scheme it never became widely adopted by the tunnelling industry. It was dissatisfaction with the design principle of the bolted segmental lining that led engineers to devise unbolted linings made from precast concrete segments such as the Don-Seg and later the wedge-block systems. In this we see the application of an analytical approach as instanced by the contributions by Skempton, Terzaghi and the Building Research Station to the understanding of how tunnel linings behave.

Notes and references for Chapter 8

1 Baker, B. (1885) The Metropolitan District Railways. *Minutes of Proceedings of the Institution of Civil Engineers*, **81**, Paper No 2054, pp. 1–33.

2 Baker, B. (1881) The actual lateral pressure of earthwork. *Minutes of Proceedings of the Institution of Civil Engineers*, **65**, Paper No 1759, pp. 140–86.

3 Galbraith, W.R. (1885) Discussion on Metropolitan Railway Papers Nos 2054 and 2060. *Minutes of Proceedings of the Institution of Civil Engineers*, **81**, pp. 59–60.

4 Settlement of the ground over tunnels remains of perennial interest to tunnel engineers. A good modern review of existing knowledge of the problem is given by:
O'Reilly, M.P. & B.M. New (1982) Settlements above tunnels in the United Kingdom – their magnitude and prediction. *Tunnelling 82*. London (The Institution of Mining and Metallurgy) pp. 173–81.

5 Brunel, M.I. (1818) Forming tunnels or drifts under ground. *British Patent* No 2404. London (Great Seal Patent Office).

6 Greathead, J.H. (1896) The City and South London Railway; with some remarks on subaqueous tunnelling by shield and compressed air. *Minutes of Proceedings of the Institution of Civil Engineers*, **123**, Paper No 2873, pp. 39–73.

7 Moir, E.W. (1896) Discussion on Paper No 2873. *Minutes of Proceedings of the Institution of Civil Engineers*, **123**, p. 94.

8 Although largely self-educated, John Buddle the elder became noted for his knowledge of coal mining, so much so that in 1781 when a new colliery at Wallsend, Northumberland, was projected he was chosen for the post of colliery manager. He died in 1806 and was succeeded in the position by his son, John Buddle the younger (1773–1843), who also became a noted coal mining innovator and whose fame has overshadowed that of his father; details are given in:
Stephen, L. and S. Lee, Editors (1921–2) *The dictionary of national biography*. Volume 3. London (Humphrey Milford) pp. 222–3.

9 Galloway, R.L. (1898) *Annals of coal mining and the coal trade*. Volume 1. Newton Abbot (David and Charles Reprints, 1971) pp. 311–2.

10 Greenwell, G.C. (1869) *A practical treatise on mine engineering*. London (E. and F.N. Spon) 2nd edition, pp. 167–79 and Plates 37 and 39.

11 Greathead, J.H. (1874) Improvements in constructing tubular tunnels or subways, and apparatus for that purpose. *British Patent* No 1738. London (Great Seal Patent Office).

12 Watson, D.M. (1937) West Middlesex Main Drainage. *Journal of the Institution of Civil Engineers*, **5**, Paper No 5120, pp. 463–568.

13 Lance, G.A. (1981) Precast concrete tunnel linings – a review of current test procedures. *CIRIA Technical Note* 104. London (Construction Industry Research and Information Assn) p. 8 and Figure 3.

14 Groves, G.L. (1943) Tunnel linings, with special reference to a new form of reinforced concrete lining. *Journal of the Institution of Civil Engineers*, **20**, Paper No 5304, pp. 29–42.

15 Ward, W.H. (1967) Precast concrete tunnel linings. *Proceedings of the fifth International Congress of the Precast Concrete Industry*. London (Cement and Concrete Assn) pp. 140–51.

16 Skempton, A.W. (1943) Discussion on Paper No 5304. *Journal of the Institution of Civil Engineers*, **20**, pp. 53–6.

17 Terzaghi, K. (1943) Discussion on Paper No 5304. *Journal of the Institution of Civil Engineers*, **20**, pp. 349–61.

18 Ward, W.H. (1967) *op cit*, p. 141.

19 Scott, P.A. (1952) A 75–inch-diameter water main in tunnel: a new method of tunnelling in London Clay. *Proceedings of the Institution of Civil Engineers, Part 1*, **1**, Paper No 5844, pp. 302–17.

20 Tattersall, F., T.R.M. Wakeling & W.H. Ward (1955) Investigations into the design of pressure tunnels in London Clay. *Proceedings of the Institution of Civil Engineers, Part 1*, **4**, Paper No 6027, pp. 400–55.

21 Cuthbert, E.W. & F. Wood (1962) The Thames–Lee tunnel water main. *Proceedings of the Institution of Civil Engineers*, **21**, Paper No 6578, pp. 257–76.

22 Ward, W.H. (1967) *op cit*, pp. 140–51, and Discussion, pp. 152–73.

23 Craig, R.N. & A.M. Muir Wood (1978) A review of tunnel lining practice in the United Kingdom. *TRRL Report* SR 335. Crowthorne (Transport and Road Research Laboratory).

24 Donovan, H.J. (1974) Expanded tunnel linings. *Tunnels and Tunnelling*, **6**, (2), pp. 46–53.

25 Cuthbert, E.W., A.C. Lyons & B.L. Bubbers (1979) The Jubilee Line. *Proceedings of the Institution of Civil Engineers, Part 1*, **66**, Paper No 8250, pp. 359–406.

26 London Transport Executive (1984) Piccadilly Line: extension to Heathrow Terminal 4. (Unpublished note).

27 Bartlett, J.V., T.M. Noskiewicz & J.A. Ramsay (1971) Precast concrete tunnel lining for Toronto Subway. *Journal of the Construction Division, Proceedings of the American Society of Civil Engineers*, **97**, (CO2), November, pp. 241–56.

28 Goldsby, E.F. (1978) Improvements in the lining for tunnels. *British Patent* No. 1508040. London (Patent Office).

29 Smith, N. (1971) *A history of dams*. London (Peter Davies) pp. 225–6.

30 Peck, R.B. (1983) Karl Terzaghi, 1883–1963. *Géotechnique*, **33**, (3), pp. 349–50.

Pressurised face tunnelling shields

Bearing on shoulders immense,
Atlantean, the load.

We have seen how compressed air used in conjunction with the tunnelling shield enabled tunnels to be driven in soft ground beneath the water table. However, we have also seen that the use of compressed air brought with it a number of health hazards to the workforce. Even with compressed air, shield driving through gravel and sand below the water table was always potentially dangerous because of the risk of a blow followed by flooding of the tunnel. So if an alternative to compressed air tunnelling could be found, both the health hazards and the risk of a blow would be eliminated. These considerations provided a strong incentive to search for alternatives to the use of compressed air; the results of this search were the various pressurised face tunnelling shields which are the subject of this Chapter.

One problem which was to bedevil the designers of pressurised face tunnelling shields was that of how to cope with cobbles and boulders. Most of the pressurised face shields were designed for driving through sands and gravels and had spoil handling systems designed to cope with particles of these sizes. Such systems might become choked or might not be able to cope at all with larger sized particles. The designers attempted to deal with this problem in a number of ways but, as we shall see, cobbles and boulders were a contributory factor in the lack of success of one pressurised face system, and the success of the others lay to some extent in the choice of ground conditions substantially free from cobbles and boulders for their use. Another problem which had to be solved was a method of preventing whatever fluid was used for the pressurised face support from leaking back along the outside of the shield and entering the tunnel where the lining was being erected. A solution to the problem was necessary for two reasons: Firstly to prevent loss of pressure at the face, and secondly, to prevent dangerously slippery working conditions from developing at the spot where the miners would be handling heavy lining segments. The solution to this problem with the systems to be described was to fit a device called a 'tailseal' to the inside of the trailing edge of the tailskin of the shield (see Figure 9.2). The tailseal was intended to form a watertight seal between the tailskin and the lining so that if any fluid did leak down the outside of the shield it would not get into the tunnel. This idea was sound in principle, but, as we shall see, its practical realisation proved to be a difficult technical matter.[1]

In addition to the two problems discussed above, there was a third problem that faced the designers of pressurised face tunnelling shields, and this was the selection of an appropriate fluid to use in the system to balance the ground water pressure and support the face. The whole history of the development of the pressurised face tunnelling shield was the search for a fluid that would accomplish these objectives effectively, and as we shall see, many different substances were tried. As well as the removal of the health hazards, there was great advantage to be gained from using a liquid of some kind rather than compressed air at the face, because the pressure distribution from crown to invert would vary with depth in exactly the same way as that of the hydrostatic pressure in the ground water, so that the problem of unbalanced pressure at the face would be eliminated, thus greatly reducing the likelihood of a blow.

The development of the pressurised face tunnelling shield covers the period when the tunnelling shield itself was becoming mechanised, that is to say undergoing the change from a simple hand shield to a soft ground tunnelling machine. Therefore we will see that the early pressurised face tunnelling shields were hand shields but that the later ones were pressurised face tunnelling machines. Because of this, there is some overlap between this Chapter and Chapter 12 where soft ground tunnelling machines are considered in their own right.

9.1 The hydraulic shield

In a patent specification of 1874, Greathead described a tunnelling shield designed to drive tunnels in sands below the water table in which excavation was accomplished by water jets operating in front of a closed diaphragm at the front of the shield, the water jets being assisted by men operating long chisels which were worked through watertight spherical stuffing boxes in the diaphragm.[2] The loosened spoil mixed with water was pumped out from the face through a large-diameter return pipe in the centre of the diaphragm to settling tanks further back in the completed section of tunnel. In this patent specification Greathead described the mechanical and hydraulic arrangements in some detail but omitted to deal with the important matters of the operating pressure of the water and how it should be regulated. Later, he must have either realised this, or had it drawn to his attention, because in a patent specification of 1884 the hydraulic shield is again described, but this time in a more developed form and with a consideration of the hydrostatic pressure required. For these reasons, the hydraulic shield examined here will be from the 1884 specification.

Greathead's hydraulic shield is shown in Figure 9.1. The water for excavating the face was supplied by an outflow pipe DD to a set of nozzles NN which projected through the diaphragm at the front of the shield. The spoil mixed with water was removed through the inflow pipe EE to a large settling tank situated behind the shield. After the debris had settled out, the water was returned to the outflow pipe DD by means of the pump G which maintained the flow in the circuit. Pipe P was a cross connection for scouring the inflow pipe EE if it became choked with spoil, and L was a pocket behind the diaphragm in which cobbles and boulders could be

Figure 9.1 Greathead's hydraulic shield (1884)

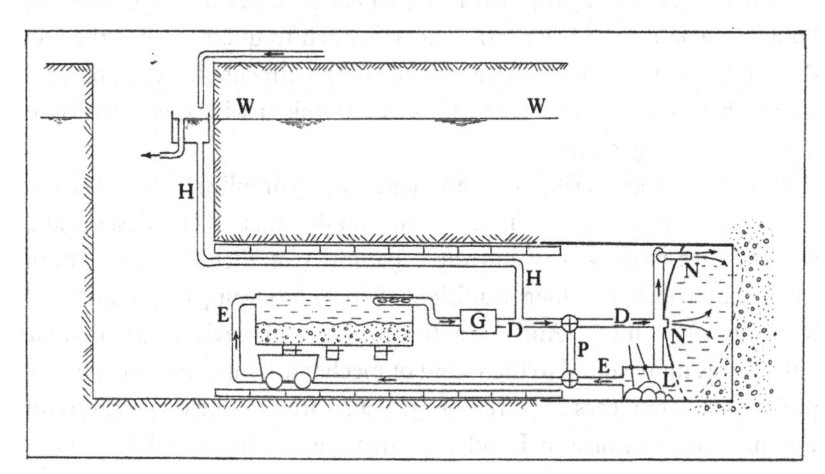

broken up by means of chisels worked through watertight spherical stuffing boxes. Periodically the debris was emptied from the settling tank into rail-borne muck skips via gate valves in the bottom of the tank. Greathead then considered the question of the water pressure, and referring to the hydraulic circuit that has just been described he said:

> I also put this closed circuit in communication with a column of water arranged in the shaft or elsewhere, of head equivalent to that of the external water, so that, in advancing the shield, there is free escape for the water displaced by its advance.[3]

Greathead did not show how this principle was to be put into practice, but in the redrawn diagram of the hydraulic shield (Figure 9.1) the present author has shown how pipe HH could have terminated in a constant-head header tank fixed at the level of the water table WW which would have regulated the pressure in the shield to be the same as the pressure due to the head of the ground water and so have accomplished Greathead's objective.

So far as is known, the hydraulic shield was never built although Greathead made a very brief reference to 'experiments upon a small scale, to remove the material from the path of the shield by a current of water'[4] which may relate to a model. However, in his patents of 1874 and 1884 he had designed a tunnelling system for subaqueous work in sands which contained the two essential elements of a pressurised face tunnelling shield – a method of balancing the hydrostatic pressure of the ground water at the face and a method of removing the spoil from the face. Whether Greathead's proposed arrangement for breaking up cobbles and boulders would have been satisfactory is open to question, but the fact that he had made some provision for dealing with cobbles and boulders shows that he recognised that this was a problem with the pressurised shield tunnelling method.

It is worth considering for a moment why hydraulic shields were not constructed and used. There is no doubt that their design and manufacture were well within the capabilities of Victorian mechanical engineers who had a long tradition of making steam, pneumatic and hydraulic machinery. And even the problem of devising a workable tailseal was more suited to their kind of mechanical ingenuity than it is to present-day engineers. The reason why a hydraulic shield was not built was probably because in London at that time there would have been limited scope for its use. The main underground railway lines were concentrated north of the Thames where they were sited so as to exploit tunnelling in the London Clay to which the ordinary shield was so well suited.[5] For short lengths of tunnel in water-bearing sands and gravels or for tunnels under the Thames, the use of compressed air together with the shield was found adequate, especially as bone necrosis was unknown at that time.

9.2 Shields with compressed air in the face

Although Greathead had described a pressurised face tunnelling shield that used water to balance the ground water pressure at the tunnel face, one of the first pressurised face tunnel shields that was constructed actually used compressed air as the support fluid. This was a pressurised face shield built to drive the 1440 ft long River Spree Tunnel in Berlin in 1899.[6] The tunnel was 13 ft outside diameter and it was driven 40 ft below mean water level with 10 ft of mud and sand over the top of the shield. The working area of the shield was divided into two compartments by a double bulkhead fitted with an air lock passing through both. The front of the shield had a hood inclined at 45°, the face of which was closed by a series of doors. The material was excavated by hand by miners digging it out through the doors using spades; the spoil was thrown to the bottom of the hood where it was removed by a sand suction pipe passing through the double bulkhead and out down the tunnel. The completed tunnel was lined with cast-steel flanged rings, off which the shield was shoved by hydraulic jacks. This system allowed only the miners who were excavating the face to be exposed to the full compressed air pressure required to balance the ground water pressure. The miners erecting the rings were on the other side of the double bulkhead also in compressed air, but at a lower pressure. There was another air lock fitted further back along the tunnel since this was a tunnel driven using double-stage compressed air working (see Section 7.1). This shield had all the elements needed for successful operation: compressed air to prevent ingress of water, doors to support the face when it was not being excavated, an inclined hood to the shield for additional roof support and a sand pump to remove the spoil from the fully pressurised section.

The next pressurised face shield considered also used compressed air as the support fluid but attempted to overcome the health hazards by restricting the compressed air to the face of the tunnel alone, the rest of the tunnel being in air at normal atmospheric pressure. This was the shield developed by Campenon Bernard and Grand Travaux de Marseille to drive the underground railway tunnel between La Folie and Etoile in Paris in 1962.[7] Ground conditions at tunnel level on this site consisted of limestone overlying a stratum of fine sand in which the ground water was under artesian pressure; the drive was intended to be mainly in the sand.

The 10 m diameter shield was divided into a forward compartment and a rear compartment by a pressure-resisting watertight bulkhead. There were three horizontal platforms across the shield which allowed the face to be excavated from six working positions, two on each platform. The miners stood on the platforms in the rear compartment and excavated the face by operating servo-assisted hydraulic tools which worked through the bulkhead into the front compartment. The excavation was controlled by the miners observing the action of their tools through windows in the bulkhead. Only the front compartment was in compressed air, the rear compartment where the miners stood was at normal atmospheric pressure. As well as operating the excavating tools, the miners could operate breastboards to support the face if required. Behind the working platforms in the rear compartment of the shield a precast concrete segmental lining was erected within the protection of the tailskin using an erector arm. The shield was jacked forward off the lining in the usual manner. The debris from the excavation process fell to the bottom of the front compartment where it was gathered in by a scraper. The muck was then hoisted by a clamshell grab to the centre of the shield where it was emptied through an evacuation air lock into the free air section of the shield. In order to remove the spoil in this way it was necessary to have two men in the forward chamber of the shield and, of course, this meant that they had to be in the compressed air. This vitiated the whole principle of the pressurised face shield which was to eliminate the need for men to work in compressed air. And the design of the shield was complicated by the fact that a special air lock was necessary so that they could enter and leave the forward chamber. The compressed air pressure used was 2.5 atm.

The La Folie–Etoile section passes under the River Seine at Pont du Neuilly and for the intended short length of drive under the river it was proposed to use the shield with the whole of the forward compartment filled with water,[8] like Greathead's hydraulic shield. However, had this been done the men who controlled the mucking out would have been in a communicating subcompartment in which compressed air would, of course, still have had to be maintained. This would have been highly dangerous because any sudden loss of compressed air would have led to a rise in water level in the subcompartment which could have drowned these men. In the event, this idea was never put into practice and the river crossing was made by another means.

The shield with compressed air in the face alone thus provided only a partial solution to the problems of compressed air tunnelling. Two of the eight miners still had to work in compressed air so that the hazards were reduced but not eliminated. There were also other serious problems with the operation of this shield; it was not a success in Paris and as far as is known it has not been used again. The solution lay with the use of other fluids to support the face and these will now be discussed.

9.3 The Universal soft ground tunnelling machine

In 1967 engineers of the Mitchell Construction Kinnear Moodie Group Ltd designed a soft ground tunnelling machine that embodied the pressurised face principle utilising a slurry made from spoil and water as the working fluid.[9] The machine was basically a shield, designed so that the digging compartment was a pressure chamber sealed off by a bulkhead capable of withstanding the full hydrostatic head of the ground water conditions of the site. A small air lock was provided in the bulkhead to permit access to the chamber for maintenance. Excavation at the face was accomplished by an oscillating cutter head fitted with 6 arms each carrying 15 picks; the cutter head had a V-shaped central cutter and oscillated in an arc sweeping through 60–70°. The shield was fitted with 22 shove rams capable of a total thrust of 1300 tons; in addition the cutter head itself could be thrust forward independently with a force of 500 tons. Spoil was removed via twin 6 in diameter outlet pipes located at the bottom of the bulkhead, through which the material was piped to slurrifier units prior to being pumped away from the shield. In the tailskin of the shield there was a hydraulic erector arm for the erection of a precast concrete segmental lining. Power for the entire equipment was supplied from hydraulic units located on sledges pulled along by the shield as it advanced.

Turning now to the pressurised face system, the whole of the space between the bulkhead and the tunnel face, including the cutter head, was filled with slurry under pressure. The slurry was obtained by mixing the spoil with water in slurrifiers and then piping a proportion of it back to the face via a hydraulically operated pinch-type pressure-control valve which maintained the pressure at the face between limits set by high and low pressure switches irrespective of the flow rate through the valve. If desired the machine could be operated as a non-pressurised face shield in which case only sufficient water was supplied to the face to remove the spoil; because of this facility of operating the machine as either a pressurised face shield or a non-pressurised shield as ground conditions required, it was called a 'Universal' soft ground tunnelling machine. It is worth noting that in spite of this name the machine would have been unable to cope with very stony ground. Boulders and cobbles would have accumulated in the pressure chamber and eventually blocked the spoil outlet pipes even though the openings of these were swept by agitators. It was claimed that the concept of using hydraulic pressure to maintain stable conditions at the face was new, but we have seen that this was not so, the idea having been thought of before by Greathead in his hydraulic shield. What was new was the realisation of the idea in a practical working machine.

Three machines were manufactured by Markham and Co Ltd and sent to Mexico City to drive sewer interceptor tunnels of some 20 ft in diameter through wet soft clay at a depth of 100 ft below the surface. Although no mention was made of the maximum chamber pressure that was designed to be used in the machines, because we know they were intended to take the full hydrostatic pressure of 100 ft of water at Mexico City the chamber must have been capable of withstanding at least 3 atm. On arrival in Mexico the machines passed out of the hands of the manufacturer and of the contractor to whom they had been originally supplied. Their subsequent history was something of a mystery until D.H. Scotney's discovery that two had been converted to conventional shields and one had been abandoned.[10] Only one machine had ever operated as a pressurised face shield and this one, in 1970, drove only 40 m of tunnel – the face pressure used was 3 atm and the average rate of advance was a reasonably good 3 m per day. However, overall slow progress together with other problems which were not associated with the pressurised face system caused the Mexican engineers to abandon the plan for a deep level sewer scheme in favour of one at a higher level, and the three Universal machines suffered the fate described above. As far as is known, no further Universal soft ground tunnelling machines were ever built.

The Universal soft ground tunnelling machine had, however, demonstrated that the principle of a pressurised face machine using a liquid as the support fluid could be made to work on a real tunnel drive. The tragedy was that this novel type of machine received its first practical trial in circumstances where its operation had passed out of the control of its designer, manufacturer and even the contractor for whom it was originally supplied. There is no doubt that in the early days of an innovation, a dedicated commitment to its success is needed on the site and it is now considered that the Universal soft ground tunnelling machine might have had a brighter future if its first application had been nearer home and under the control of its originators. This was not to be the sole occasion in the development of pressurised face shields that a British innovation had got off to a false start and failed because of it, as is shown in the next Section.

9.4 The bentonite tunnelling machine

An early practical pressurised face tunnelling machine was the bentonite tunnelling machine invented by John Vernon Bartlett of the firm of consulting engineers Mott, Hay and Anderson in 1964 and patented in 1967.[11] Bartlett first considered Greathead's system as described in the 1874 patent but concluded that water was unsuitable for a pressurised face system because, while it can balance the ground water, it offers no support to the particles of sand or gravel in the working face so that there was nothing to prevent a progressive collapse. This thought led to the realisation that bentonite slurry might be the ideal fluid to use.

The circumstances of the origin of Bartlett's idea are very interesting. His firm had been asked to advise on the construction of a Metro system in Milan, Italy. The subsoil of the city consists mainly of gravels in which the shield tunnelling methods developed for London Clay would not, of course, be suitable. Cut-and-cover tunnelling, already in use for constructing the first section of the Metro, would be immensely disruptive, and ground treatment from a small pilot tunnel very expensive and possibly of doubtful efficacy. These considerations led Bartlett to seek an alternative method which he found in the idea of a pressurised face system. Rejecting Greathead's idea of water as a support fluid, and being aware of the lack of success of the French shield with compressed air in the face alone, Bartlett chose bentonite slurry as a support fluid – an idea partly suggested by the fact that the trenches for the cut-and-cover section of the Milan Metro were being constructed by the diaphragm walling technique using bentonite slurry.[12] An additional factor that weighed in Bartlett's mind was the desire to replace compressed air working following the realisation that a proportion of workers in compressed air suffer permanent damage to their health from bone necrosis (see Section 7.6).

Bentonite is a naturally occurring montmorillonite clay which has been activated by suitable chemical treatment so as to enhance its thixotropic behaviour. (Thixotropy is the property of a slurry to be a liquid when it is agitated or in motion, but a gel when it is at rest – the familiar property of non-drip household paints.) A slurry of bentonite in water has marked thixotropy, and because of this it has long been used as a drilling mud in oil well drilling where its abilities to support the unlined borehole wall and to carry the drill cuttings back to the surface have been of crucial value to the petroleum exploration industry. Bentonite slurry has also been used in the civil engineering industry[13] for supporting the walls of deep trenches in the foundation construction process of diaphragm walling, as well as for other purposes, and this made it a natural choice for use with a pressurised face tunnel shield.

A tunnelling system was therefore designed which was based on the use of bentonite slurry maintained at a selected pressure in a working chamber in front of a diaphragm in the shield, the bentonite slurry having two functions: firstly, to support the face and control the ground water and secondly, to act as a carrier for the hydraulic removal of the spoil. A slurry containing 3.5–5 per cent of bentonite in water was found in later trials to be a suitable consistency to achieve these purposes. Although bentonite was a logical choice for a slurry shield system its choice has been questioned; it being suggested that a slurry made from a mixture of an ordinary clay such as London Clay and water might be equally effective and less costly.[14] As we have seen this idea had already been used in the Universal soft ground tunnelling machine, and it was later taken up and used by the Japanese very effectively (see Section 9.7).

The bentonite tunnelling machine is shown in Figure 9.2. It consisted of a tunnelling shield with a mechanical cutter head which operated in a pressurised chamber that could be filled with bentonite slurry. Fresh bentonite slurry was supplied from the surface via a pump (labelled Mono pump in the Figure) to the bottom of the pressurised chamber in which the cutter head operated. The slurry together with the spoil produced by the rotating cutter passed out of the chamber at the top where it was allowed to escape through an automatic pressure-control valve. In the pressure bulkhead of the machine chamber a feedwheel was fitted which

Figure 9.2 Bentonite tunnelling machine (1976)

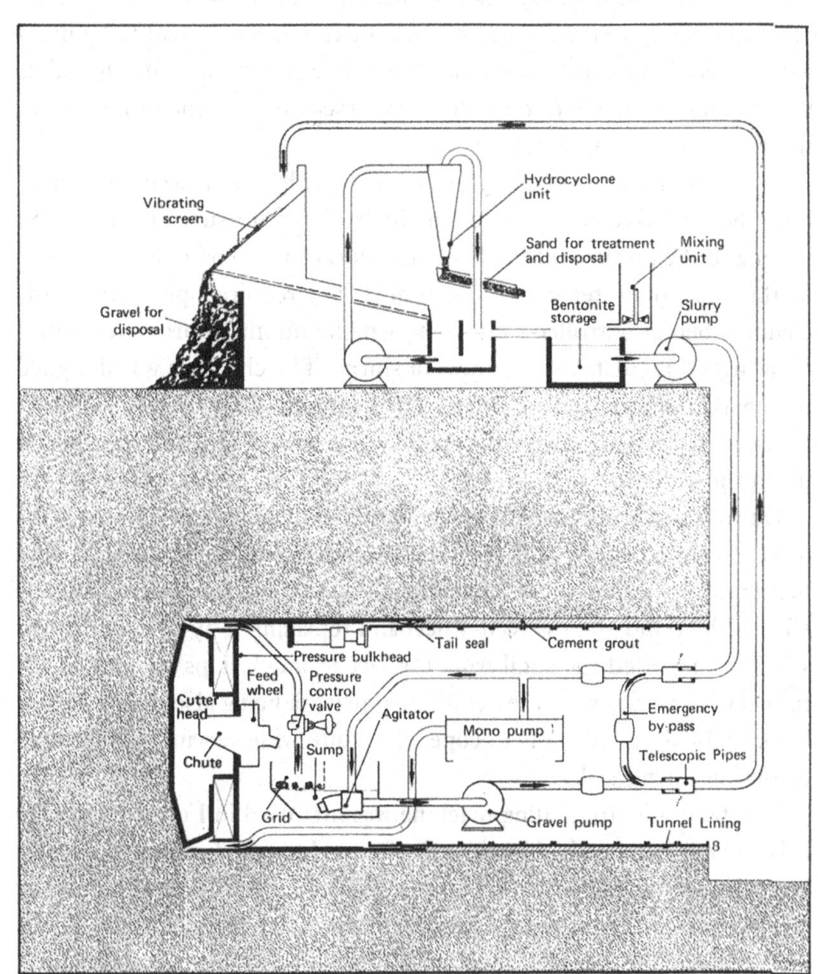

allowed any cobbles present to be passed out, as through a lock, without significant loss of slurry – this coarse debris collected in a grid above the sump and was periodically cleared away by hand. The mixture of spoil and slurry was then returned via a gravel pump to the surface where it was first screened to remove gravel and then passed through a hydrocyclone to remove sand. After this the cleaned slurry was returned to the bentonite storage tank for reuse. Both the supply and return pipelines were fitted with telescopic sections which allowed the pipework to extend while the machine was advancing. The bolted segmental tunnel lining was erected within a tailskin, in the usual manner, when the machine had advanced for the full stroke of the rams. After the machine shoved forward, the newly built ring was grouted in the usual manner. The shield had to be fitted with a special tailseal to prevent any bentonite slurry which had found its way down the outside of the shield from escaping into the tunnel where the rings were being erected.[15]

In the bentonite tunnelling machine, as with compressed air tunnelling, the objective is to balance the hydrostatic pressure of the ground water and the pressure of the bentonite slurry in the chamber is matched to the head of water above the tunnel, the required pressure in the chamber being controlled by the setting of the automatic pressure-control valve which regulates the outflow of slurry. The chamber was designed for a maximum pressure of 2 atm. The bentonite slurry also had another function, namely to convey the spoil away from the face and transport it up to the surface.

One of the features of the bentonite tunnelling machine was that if the drive ran into clay or soft rock the machine could be converted from bentonite operation to a conventional tunnelling machine. This was effected by removing the feed-wheel and replacing it by a belt conveyor which transported the spoil from the face to muck skips running in the tunnel in the usual way, the bentonite pipework being also disconnected. This enabled the machine to cope with quite wide variations in ground conditions along a drive.

The bentonite tunnelling machine was first used in February 1972 to drive an experimental 144 m length of tunnel for the London Transport Executive at New Cross.[16] The tunnel was driven at a depth of 8 m below ground surface in gravel and sand with the water table between tunnel axis level and the crown of the tunnel. The outside diameter of the shield was 4.1 m because the short length of tunnel was to be of standard running tunnel size: it was lined with cast-iron bolted segmental lining. The experimental drive was financed jointly by the National Research Development Corporation and the London Transport Executive, the latter body foreseeing the need for a machine of this kind if the underground railway network were to be extended as planned south of the Thames where water-bearing gravels and sands were likely to be encountered on the tunnel drives as well as London Clay. The experiment showed that the basic concept of the bentonite tunnelling machine was practicable, a maximum advance rate of 4 m per ten hour shift being achieved, but there were a considerable number of details in the machine that called for redesign, particularly the feedwheel, tailseal and pressure-control valve. The surface treatment plant also required some modification.

The second job on which the bentonite tunnelling machine was used was in 1976 on the construction of a trunk sewer tunnel at Warrington, Cheshire.[17] For this, the first commercial contract on which the bentonite tunnelling method was used, a new machine was built with a shield of 2.87 m outside diameter: it incorporated the lessons learnt from the New Cross experiment. For example, a new tailseal was developed that was made from an abrasion resistant multi-filament urethane material. Its design was based on the principle of the 'lip seal' in which the higher the pressure acting on it the better it seals onto the tunnel lining. Two tailseals of this type were fitted to the tailskin of the machine. The tunnel was 1.37 km long of which the bentonite system was used to drive 1.04 km. The ground conditions at tunnel level were sand overlying sandstone with a layer of bouldery gravel at the interface; the water table was at tunnel axis level. The tunnel drive was fraught with difficulties[18] and took three years to complete, giving an average rate of advance of about 1 m per day. This progress rate was much less than that achieved by Greathead in sand and gravel using a shield and compressed air in the 1880s (see Section 7.5) and was a great disappointment to the British tunnelling industry which had had high expectations of the bentonite tunnelling machine. With hindsight it can be seen that, for the following reasons, the site was an unfortunate choice for a new method that had had only a short experimental trial: the tunnel had to be fitted into a narrow strip of ground between the Manchester Ship Canal and a frontage of terraced houses with basements, the depth of cover was shallow, only about 4–5 m, the sand was of low density and the layer of gravel contained hard granite and dolerite cobbles and boulders of up to 500 mm diameter which were a problem because the feedwheel could only handle particles less than 150 mm in size.

The experience of the bentonite tunnelling machine at Warrington effectively put a stop to its further use in Great Britain and so far it has not been used again. For the further development of the pressurised shield concept we have to go abroad – to Germany and Japan. Nevertheless it is important to realise that even though overall the Warrington job was not a success, when the system was working satisfactorily there were short periods when advance rates of 3.9–5.6 m per day were achieved[19] and this showed that the principle was not at fault and that the bentonite tunnelling system could work satisfactorily if ground conditions were right for it.

9.5 The Hydroshield

The Hydroshield is a bentonite tunnelling machine invented in Germany by Erich Jacob of the firm of Wayss and Freytag AG of Frankfurt. The arrangement of the Hydroshield system is shown in Figure 9.3. The basic system is very similar to the British bentonite tunnelling method, the machine consisting of a mechanical cutter head operating in a shield having a pressurised compartment at the face which can be filled with bentonite slurry. Pipelines lead to the surface where there is a slurry cleaning plant. It can be noted that the cutting wheel is slightly tilted forward which aids the stability of the face. As with the British system, the bentonite slurry controls the soil and ground water, and the spoil loosened by the cutter goes into suspension in the slurry and is pumped away to the surface.

However, there is one important difference in the German machine. It will be recalled that on the British machine the pressure of the slurry in the working chamber was controlled by the setting of the automatic valve which controls the outflow of slurry. Now with this arrangement any sudden loss of slurry from the face would result in a sudden loss of pressure followed by a lack of support with consequent risk of a cave-in at the face. With the Hydroshield an 'air accumulator' is provided. This is a chamber partially filled with compressed air and fed by an air compressor at the surface via a pipeline having a pressure-control valve. The air pressure acts on a free surface of slurry in communication with that in the main chamber and has the effect of maintaining a constant fluid pressure at the face even if there are abrupt variations in the rate of advance of the shield or sudden losses of slurry into the surrounding ground. Since the air pressure is maintained constant, the level of the interface between air and slurry fluctuates within limits controlled by the feed and discharge of slurry to and from the main chamber; because of this the air accumulator is sometimes called the 'fluctuation chamber'. The air accumulator on the Hydroshield, by safeguarding against ground loss at the face, is probably the biggest single improvement over the previous machine.

Behind the air accumulator, the Hydroshield has an air lock; through this the crew can enter the front chamber of the machine for maintenance purposes after the slurry has been displaced with compressed air. It is interesting to note that this idea of an air lock fitted close up to the diaphragm of the shield was first thought of by Greathead in 1874 and

Figure 9.3 Hydroshield (1978)

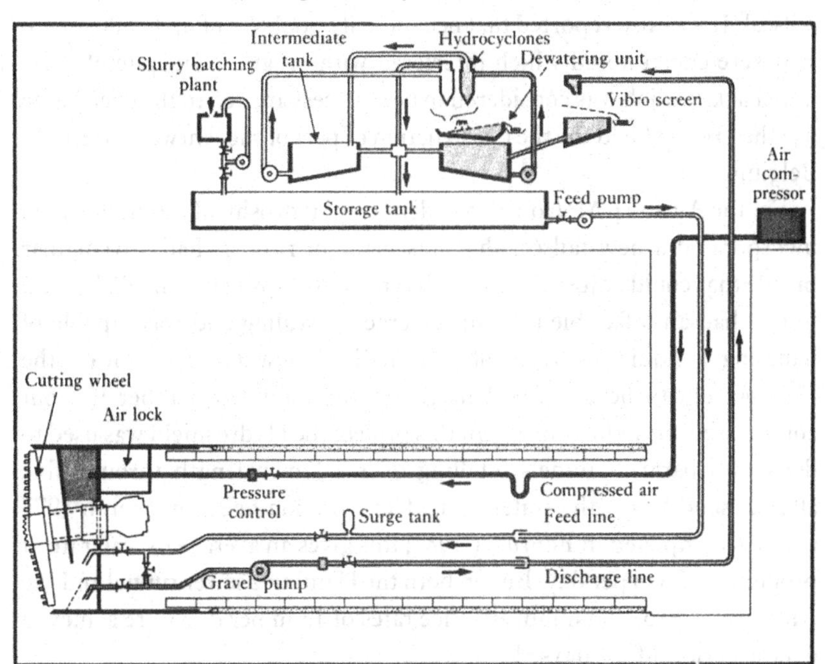

incorporated into a shield that was actually built to drive the proposed North and South Woolwich Subway tunnel, even though the project was later abandoned for non-engineering reasons and the shield was never used.[20] Indeed, W.C. Copperthwaite considered this idea of such potential importance that he described Greathead's Woolwich shield in his textbook.[21]

The Hydroshield was first used to drive the 4.6 km long, 4.32 m outside diameter, Sammler Wilhelmsburg sewer tunnel through coarse and fine sands below the water table in Hamburg in 1974; on this drive an average advance rate of about 3 m per day was achieved. However, for the Hamburg sewer tunnel drive the tailseal of the Hydroshield was, from the outset, considered to be so unreliable that the tunnel was driven under compressed air as well, an air lock being provided at the top of the shaft for this purpose.[22] It would seem that because of this the whole point of the pressurised face system was vitiated and the slurry system was merely used as a means of removing the spoil. In justifying the use of compressed air as well, Wayss and Freytag have said that the job was used to demonstrate the feasibility of the bentonite slurry circuit and gain experience in its use whilst experiments were in progress on designing a tailseal. It was also reported that occasionally boulders of up to 800 mm in size were encountered which interfered with progress. In spite of these drawbacks the job was considered to be a success and led to the decision to use the Hydroshield for the construction of part of the Antwerp Metro in Belgium.

For the Antwerp Metro a 6.55 m diameter Hydroshield was built which incorporated a new tailseal that was made of rubber, had a triangular profile that could adjust itself to different sized gaps between tailskin and lining, had an inflatable tube for emergency sealing and was capable of handling a liquid pressure of 2 atm. This new tailseal allowed the Hydroshield to be used at Antwerp in the intended manner without compressed air in the tunnel. On this project the Hydroshield was used to drive two parallel tunnels totalling over 4 km in length through fine alluvial sand below the water table. Construction began in March 1977 and was completed in February 1981; this gives an average overall rate of progress of 2.8 m per day. But on both the Hamburg sewer tunnel and the Antwerp Metro, maximum advance rates of 13 m per day were achieved over a period of 20 days.[23]

It is interesting to consider the reasons given for the choice of the Hydroshield to drive the tunnels for the Antwerp Metro. These were that the early experience of Metro construction using cut-and-cover methods in the narrow, historic city streets caused too much disturbance, ground water lowering was not permitted because of the fear of concomitant settlement, and compressed air working was ruled out because of the incidence of the bends on previous projects.

The tunnels of the Antwerp Metro were lined with precast concrete bolted segments. Each ring consisted of seven 1.1 m long segments plus a key piece, and a novel feature was that a watertight sealing of the lining was achieved by use of a neoprene gasket set in a side groove running around all four sides of each segment.

It can be noted that just as a compressed air tunnel can suffer from a blow if the air pressure exceeds the ground water pressure (see Section 7.6), so too can a pressurised face shield if the fluid pressure inadvertently rises above the preset pressure or if the preset pressure is too high for the site conditions. This can result in a loss of bentonite slurry, or even in extreme cases 'bentonite fountains' at the surface. In 1985 on the drive beneath the river Rhône for the new Metro being constructed in Lyon, France, a blow occurred from the bentonite tunnelling machine being used. When this happened there was just 4.5 m of cover between the machine and the river bed and the bentonite pressure was 1.6 atm. The blow resulted in a 0.8 m diameter chimney being pushed out through the alluvium over the top of the shield at tunnel face. The drive was recovered by grouting the void, coupled with the time-honoured measure of dumping ballast on the river bed. The bentonite tunnelling machine used for this drive was a 6.5 m diameter version of the Hydroshield that embodies the Hochtief system for producing a continuously extruded concrete lining reinforced with steel fibres, and which will be referred to again in Section 13.4.

A blow with an incompressible fluid like bentonite slurry, because there is no potential for expansion on release of pressure, is likely to be much less dramatic than a blow with compressed air.

9.6 The earth pressure balanced shield

In the pressurised face shields previously examined, water, compressed air, spoil-and-water slurry or bentonite slurry was used to support the face. In the earth pressure balanced shield, the idea of using the excavated spoil itself to support the face is employed. Like many good ideas, once it has been thought of it seems both obvious and simple, but in fact it was not applied until after the other methods.

The earth pressure balanced shield is shown in Figure 9.4. The ground at the tunnel face is excavated by a large disc which carries cutters to loosen the ground and has slots through which the spoil can enter a chamber that is integral with the cutter frame. The excavated spoil is transported from the chamber by a screw conveyor which discharges it into a large tank called the mucking adjuster. Between the screw conveyor and the mucking adjuster is a gate controlled by a hydraulic jack: by adjusting the setting of the gate, the pressure of the spoil in the cutter frame chamber can be regulated to support the face. The machine operator, situated in a control booth about 10 m behind the chamber, controls the shield advance jacks, the rotary speeds of the cutting head and the screw conveyor and the setting of the gate between the screw conveyor and the mucking adjuster. Water is supplied to the mucking adjuster to convert the spoil into a slurry which, after cobbles have been removed, is conveyed via the muck discharge pipe to a treatment plant on the surface.

Figure 9.4 Earth pressure balanced shield (1977)

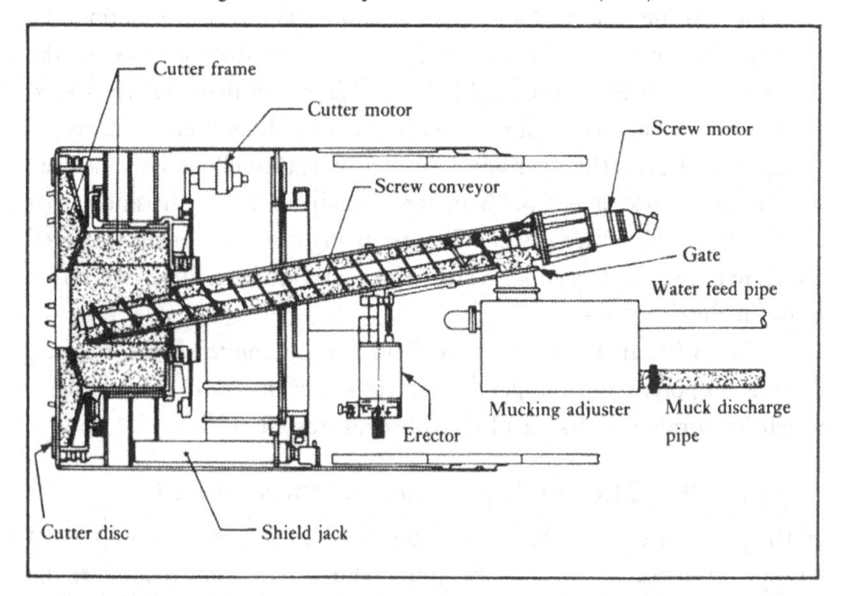

If the machine is working in clay, or in sand *above* the water table, the mucking adjuster can be removed and the screw conveyor can discharge the spoil directly into rail-borne muck skips. The earth pressure balanced shield, therefore, has a similar versatility to the British bentonite tunnelling machine in being adaptable to different ground conditions.

Earth pressure balanced shields originated in Japan in the 1970s; the first report of the use of one was in 1974 in that country where a 3.35 m diameter tunnel 1.9 km long was driven for an aqueduct. Since then, between 1976 and 1980, the Japanese have used earth pressure balanced shields to drive a number of tunnels and have claimed that these have been very successful. It is therefore worth while examining them more closely. Table 9.1 shows the construction details of the tunnels. All the tunnels were in sandy or gravelly soils and all were below the water table. Tunnels 1 and 5 were over 1 km in length but the others were shorter. The information above the dashed line in Table 9.1 comes from a paper by Hisakazu Matsushita,[24] but the entry below the dashed line has been derived by dividing the tunnel length by the construction time – it is, therefore, the average progress rate. By modern tunnelling standards these rates are low; little better, in fact, than the progress rate for the Warrington tunnel which ended the hopes of the British bentonite tunnelling machine. The only project for which a maximum progress rate was quoted was Tunnel 1 for which an advance of 45 m per week was reported over a short specially selected section of the site. The low overall progress rates for what are claimed to be successful jobs suggests that they may have been done with earth pressure balanced shields primarily to gain experience with the new method; this is a point that will be taken up again in the Conclusion.

The shields listed in Table 9.1 are from about 2–5 m in diameter, a typical range for most tunnels, but in 1978 the Japanese built and operated a larger earth pressure balanced shield of 8.48 m diameter to drive a metropolitan main sewer in Tokyo.[25] Part of the drive was in water-bearing sand with an overburden of only 13 m. To have driven a tunnel of this large diameter with shallow cover under compressed air would have been very hazardous because of the large unbalanced pressure at the face and the consequent danger of a blow. This application illustrates the advantage that has been gained by developing alternatives to compressed air tunnelling. Even with this large shield, an overall

Table 9.1. *Details of tunnels in Japan constructed using earth pressure balanced shields*

	Tunnel				
Details	1	2	3	4	5
Location	Tokyo	Tokyo	Osaka	Kanagawa	Tokyo
Type	Sewerage	Electricity	Electricity	Sewerage	Water
Length, m	1634	150	374	678	1045
Diameter, m	5.24	3.49	1.98	4.94	3.74
Depth, m	10–12	10–12	7–8	7	16–25
Ground conditions	Fine sand	Silt and gravel	Gravel	Gravel	Fine sand
Depth to water table, m	1	5	2	3	3
Construction time, months	31	31	11	20	25
Progress rate, m per day	1.8	0.2	1.1	1.1	1.4

average progress rate of 2.9 m per day was achieved on this job. The development of earth pressure balanced shields in Japan has been reviewed by K. Naitoh.[26]

9.7 Japanese pressurised face tunnelling shields

The foregoing account of the Japanese earth pressure balanced shield provides an appropriate introduction to the subject of Japanese pressurised face tunnelling shields in general. For as well as the earth pressure balanced shield, the Japanese have been very active in developing other types. These are: firstly, the slurry shield, which is similar in principle to the bentonite tunnelling machine; secondly, the pressurised mud shield in which the chamber between the rotary cutter and the diaphragm is kept filled with a plastic mixture of spoil and slurry – a sort of cross between the slurry shield and the earth pressure balanced shield; and thirdly, the partial pressure shield, in which compressed air is applied to the head of the shield and the spoil is mucked out to atmospheric pressure through a rotary feeder incorporating a special airtight seal. Descriptions of these shields and details of some of the tunnels driven with them have been given by the International Tunnelling Association[27] and by Piers Harding.[28]

Some of these shields have been of large size. For example, the 10 m diameter slurry shield shown in Figure 9.5 was in 1980 used to drive a 844 m long section of double-track railway tunnel in northwest Tokyo.[29] The drive has been through gravel and sand containing cobbles and occasional boulders of up to 400 mm size; the whole drive being well below the ground water level. Right from the outset of designing the shield attention was given to ensuring the stability of the face by providing methods for careful control of both the properties of the slurry and its pressure in the chamber at the face, and to equipping the machine with a crusher at the bottom of the chamber to crush any cobbles and boulders before they could block the discharge pipe. Provision was also made to displace the slurry by compressed air and an air lock was provided just behind the chamber to enable miners to enter the chamber for repair and maintenance purposes and to replace damaged cutter bits. The shield was fitted with triple wire-brush-type tailseals which were found to be effective in preventing slurry leakage. The weight of muck being discharged from the face in the slurry was measured by a combination of electromagnetic and nuclear methods and the result continuously displayed on television screens at a control console. In this way a balance could be maintained between shield advance and the amount of ground excavated, thus preventing a collapse of the face. The average advance rate was 3.7 m per day.

Not only have the Japanese used these various types of pressurised face shields for driving tunnels in their homeland, but in 1981 they won contracts for their use abroad outside Asia. The first of these was for the use of a slurry-type shield of advanced design which utilised a slurry made from a mixture of the clay in the spoil and water rather than using bentonite. The 1.4 m diameter, 240 m long tunnel was at Bordeaux in France and advance rates of 10 m per day have been reported through ground consisting of a sandy clay.[30] The second contract was in the United States and was for the use of an earth pressure balanced shield to drive the N-2 tunnel contract on San Francisco's huge sewerage project.[31] The shield was 3.7 m in diameter and the drive was 930 m long at a depth of 12 m. The ground consisted of bay fill which included old wooden piles in one area which the machine was able to cut through without difficulty. The average rate of advance for the job was 9 m per day with a maximum

Figure 9.5 The 10 m diameter slurry shield (1979)

advance of 30 m being achieved on the best day.[32] The success of a Japanese firm both in winning the contract and completing the tunnel was something of a surprise to the American tunnelling industry.

The Japanese have even been trying to find a solution to the problem of cobbles and boulders. In one of their latest slurry shield machines, the firm of Iseki Poly-Tech Inc have provided a crusher chamber immediately behind the cutter head but separated from it by hydraulic doors. Here, any boulders and cobbles are broken down to fragments of less than 100 mm in size which can be transported away by the slurry system. The machine can cope with boulders of up to 500 mm in diameter, and rocks as hard as dolerite have been successfully crushed. The machine has been termed the 'crunching mole' because of this facility. The first Japanese shield to be used to drive a tunnel in Great Britain was one of these boulder-crushing slurry shields having a diameter of 1.8 m. It was used to excavate a 600 m long sewer tunnel near Gateshead and had completed two-thirds of the drive by December 1983. During January 1984 considerable difficulties arose due to problems with the operation of the hydraulic doors and the machine had to be removed from the tunnel for repairs.[33] The machine's hydraulic doors in part utilise the principle of the earth pressure balanced shield, and it is this aspect of their operation that gave rise to the trouble and not the boulder-crushing part of the mechanism. The slit plates of the hydraulic doors were replaced, parts of the mechanism were strengthened and the machine was reinstated in the tunnel: the drive was completed in May 1984 without further damage to the doors. If the machine eventually proves successful, it would seem that one of the most serious problems facing the pressurised face tunnelling shield – that of coping with boulders and cobbles – has been at last solved. It can be noted that the crusher chamber of the Japanese 'crunching mole' is similar in concept to the pocket behind the diaphragm L in which cobbles and boulders were to be broken up in Greathead's hydraulic shield (Figure 9.1) of 1884.

The Japanese pressurised face shields of the various types, earth pressure balanced shield, slurry shield and mud shield, are all intended to utilise the same principle of operation, namely that the pressure in the machine chamber balances the total pressure in the ground. This is shown diagrammatically in Figure 9.6 where the total pressure in the ground is shown represented by its two components – the earth pressure acting from the surface and the water pressure acting from the water table. The pressure in the machine chamber is said to be adjusted so that the total ground pressure is balanced, that is $C = A_I + A_2$ and $D = B_I + B_2$ (see Figure). In practice it is suspected that this is not in fact done since to do so with the slurry and mud shields would be to risk producing a blow; from the limited data available of chamber pressures on actual drives it seems that the chamber pressure is kept higher than the water pressure ($C > A_I$ and $D > B_I$) but somewhat lower than the sum of the earth pressure and the water pressure ($C < A_I + A_2$ and $D < B_I + B_2$). This supposition is corroborated by published diagrams showing these pressurised face shields with the pressure in the machine chamber set to balance the water pressure plus a height of 'loose soil' that is considerably less than the total height of ground above the tunnel.

A feature of all the Japanese pressurised face tunnelling shields is the use of sophisticated electronic and computerised monitoring and control systems for the mechanical and hydraulic operations of the shield, cutter and spoil handling mechanisms. This, no doubt, is due to the influence of Japan's now well-established electronics industry. The flowering of innovation in the development of pressurised face tunnelling shields in Japan, hitherto a country not noted for innovation, is a subject that will be returned to in Chapter 13.

Figure 9.6 Principle of earth, mud and slurry shields

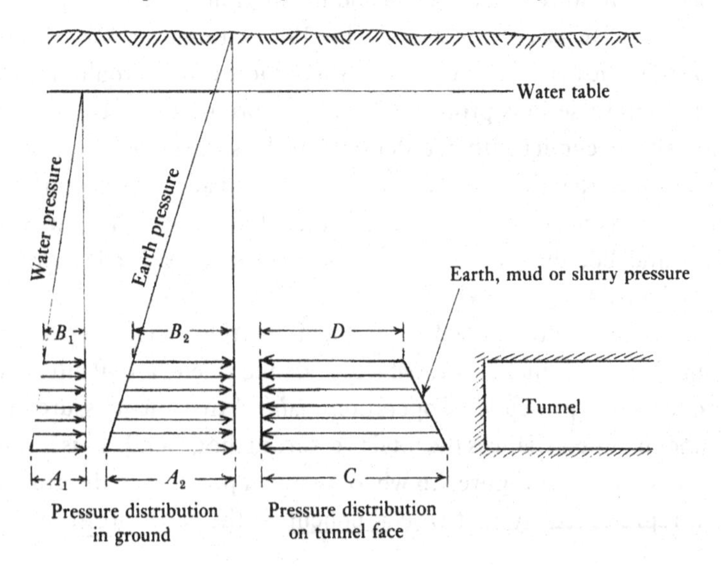

9.8 Summary

Growing realisation of the disadvantages of compressed air tunnelling led to the search for other methods and we have seen that the outcome was a succession of pressurised face tunnelling shields. The concept of a pressurised face shield had occurred to Greathead in the nineteenth century and he described his ideas in two patents of 1874 and 1884. Greathead envisaged using water as the working fluid, but in fact the first working pressurised face shields used compressed air at the face which only partially solved the problems of compressed air working. Next came the Universal soft ground tunnelling machine which utilised a slurry made from spoil and water as the working fluid. This worked successfully but did not develop further because of the completely unsuitable operational arrangements on site during its first trial. Next followed the bentonite tunnelling machine, designed by Bartlett in 1964. Bartlett was aware of Greathead's first patent but not his second, and also knew of the unsuccessful French shield built in 1961 which used compressed air in the face alone. The bentonite tunnelling machine made a successful trial drive, but because of the choice of an unsuitable site, was not a success on its first commercial application. The concept was taken up in Germany, and the Hydroshield – really an advanced form of the bentonite tunnelling machine – was developed. At about the same time the Japanese devised the earth pressure balanced shield, a pressurised face tunnelling shield in which the excavated spoil itself supports the face. The Japanese also developed slurry shields which worked on the same principle as the bentonite tunnelling machine, together with other pressurised face systems. Pressurised face tunnelling shields, therefore, were both invented and developed within the tunnelling industry.

Lastly, a point can be made concerning the use of bentonite slurry in the bentonite tunnelling machine. We have seen that the choice of bentonite slurry as the fluid to use in a pressurised face system was suggested partly by its use as the support fluid for trench walls in the diaphragm walling technique used in civil engineering, and partly by its use as a drilling mud in petroleum drilling. The use of bentonite slurry in the bentonite tunnelling machine was, therefore, borrowed from these two industries.

Notes and references for Chapter 9

1 The problem was not so much one of making an effective seal but of making a seal that would withstand mechanical damage when the shield was shoved forward, particularly when the shield had run a little off true so that it was not lying parallel to the lining.

2 Greathead, J.H. (1874) Improvements in constructing tubular tunnels or subways, and apparatus for that purpose. *British Patent* No 1738. London (Great Seal Patent Office).

3 Greathead, J.H. (1884) Tunnelling apparatus. *British Patent* No 5665. London (Patent Office).

4 Greathead, J.H. (1896) The City and South London Railway; with some remarks on subaqueous tunnelling by shield and compressed air. *Minutes of Proceedings of the Institution of Civil Engineers*, **123**, Paper No 2873, p. 66.

5 If underground railways had been constructed south of the Thames there might have been more scope for the hydraulic shield. As it was, the area was served by a surface network of suburban railways (now part of the Southern Region of British Railways) which precluded the development of an underground system.

6 Stauffer, D.M. (1906) *Modern tunnel practice*. London (Archibald Constable and Co Ltd) pp. 118–20.

7 Anon (1963) Travaux de construction de la ligne régionale est-ouest de la RATP: Section La Folie–Etoile. *Annales de l'Institute Technique du Batiment et des Travaux Publics*, Supplément, 187–8, p. 2–16.

8 Anon (1963) Travaux de construction de la ligne est-ouest du réseau express régionale: La traversée sous-fluviale de Pont de Neuilly. *Construction*, **18**, (4), pp. 199–202.

9 Mitchell Construction Kinnear Moodie Group Ltd (undated) *Universal soft ground tunnelling machine*. Peterborough (Mitchell Construction Kinnear Moodie Group Ltd).

10 Stack, B. (1982) *Handbook of mining and tunnelling machinery*. Chichester (John Wiley and Sons) p. 396.

11 Bartlett, J.V. (1967) Improvements in or relating to tunnelling apparatus and methods of tunnelling. *British Patent* No 1083322. London (Patent Office).

12 This account is based on a discussion the author had with Mr Bartlett on the origin of the bentonite tunnelling machine.

13 Boyes, R.G.H. (1972) Uses of bentonite in civil engineering. *Proceedings of the Institution of Civil Engineers, Part 1*, **52**, Paper No 7461, pp. 25–37.

14 Muir Wood, A.M. (1974) Discussion on Paper No 7670. *Tunnels and Tunnelling*, **6**, (5), pp. 21–3.

15 This description is based in part on that given in the brochure:
Edmund Nuttall Ltd (1977) The bentonite tunnelling machine. London (Edmund Nuttall Ltd).

16 Bartlett, J.V., A.R. Biggart & R.L. Triggs (1973) The bentonite tunnelling machine. *Proceedings of the Institution of Civil Engineers, Part 1*, **54**, Paper No 7670, pp. 605–24.
During the drive at New Cross measurements were made of ground movements:
Boden, J.B. & C. McCaul (1974) Measurement of ground movements during a bentonite tunnelling experiment. *TRRL Report* LR 653, Crowthorne (Transport and Road Research Laboratory).

These measurements showed that subsidence at the ground surface was no more than caused by existing tunnelling methods in London Clay, thus giving confidence in the new method.

17 Walsh, T. & A.R. Biggart (1976) The bentonite tunnelling machine at Warrington. *Tunnelling 76*. London (The Institution of Mining and Metallurgy) pp. 209–18.

18 These difficulties were reported at the time in the technical press as, for example, in the following two articles:
Hayward, D. (1977) Bentonite mole on shaky ground. *New Civil Engineer*, 1 December, pp. 16–17.
Hayward, D. (1978) Bentonite tunneller struggles on. *New Civil Engineer*, 26 October, pp. 23–4.

19 Anon (1976) Nuttall explains the problems of bentonite shield. *Construction News*, 18 November, p. 19.

20 Greathead, J.H. (1896) *op cit*, p. 66, Figure 18.

21 Copperthwaite, W.C. (1906) *Tunnel shields and the use of compressed air in subaqueous works*. London (Archibald Constable and Co Ltd) pp. 16–17, Figure 11.

22 Jacob, E. (1976) The bentonite shield: technology and initial application in Germany. *Tunnelling 76*. London (The Institution of Mining and Metallurgy) pp. 201–7.

23 Jacob, E.J. & V.O. Meldner (1979) Contractors' experience with the Hydroshield tunnelling system. *Proceedings of the Rapid Excavation and Tunnelling Conference*, 1. New York (The American Institute of Mining, Metallurgical and Petroleum Engineers Inc) pp. 467–77.

24 Matsushita, H. (1979) Earth pressure balanced shield method. *Proceedings of the Rapid Excavation and Tunnelling Conference*, 1. New York (The American Institute of Mining, Metallurgical and Petroleum Engineers Inc) pp. 521–9.

25 Saito, T. & T. Kobayashi (1979) Driving an 8.48 m diameter sewer tunnel by use of an earth-pressure balancing shield. *Tunnelling 79*. London (The Institution of Mining and Metallurgy) pp. 295–304.

26 Naitoh, K. (1985) The development of earth pressure balanced shields in Japan. *Tunnels and Tunnelling*, **17**, (5), pp. 15–18.

27 International Tunnelling Association's Working Group on Research (1981) Development and future trends of shield tunnelling. *Advances in Tunnelling Technology and Subsurface Use*, **1**, (3), pp. 199–365.

28 Harding, P. (1981) Japan takes the lead. *Civil Engineering* (UK). February, pp. 20–32 and 58.

29 Watanabe, T. & H. Yamazaki (1981) Giant size slurry shield is a success in Tokyo. *Tunnels and Tunnelling*, **13**, (1), pp. 13–17.

30 Harding, P. (1981) Japanese TBM moves into Europe. *Civil Engineering* (UK), July, p. 23.

31 Paulson, B.C. (1981) Japanese tunnel design: lessons for the US. *Civil Engineering* (USA), **51**, (3), pp. 51–3.

32 Anon (1981) Kudos for Japanese on US bore. *Engineering News Record*, 16 July, pp. 56–61.

33 Hayward, D. (1984) Japan beats UK at slurry game. *New Civil Engineer*, 16 February, pp. 30–1.

Immersed tube tunnels

And the children of Israel shall go on dry ground through the midst of the sea.

So far, in our treatment of subaqueous tunnelling we have considered only tunnels driven through the ground by what might be called conventional tunnelling methods. However, there is another method of constructing a subaqueous tunnel which consists of excavating a trench in the river, estuary or sea bed, laying a tube in the trench and then backfilling it: this is termed an immersed tube tunnel. Construction of immersed tube tunnels is, of course, feasible only where the bottom is suitable for excavation using marine dredging plant but this has usually been the case in the lower reaches of rivers, in estuaries and in harbours where most immersed tube tunnels have been built. These locations are usually busy navigation channels and, as we shall see, there has sometimes been a conflict between immersed tube tunnel construction and river traffic.

It has been shown in Chapters 6 and 7 that the pressing need for tunnels as transport links across rivers, particularly in cities, gave rise to the means of driving them using the shield in combination with compressed air. But already, before this solution had been found, the problems of constructing a subaqueous tunnel had given rise to the idea of the immersed tube tunnel, and it is the purpose of this Chapter to discuss this innovation and the problems that had to be overcome before it could be successfully realised. First we must return to the Thames driftway.

10.1 Trevithick's idea for a subaqueous cut-and-cover tunnel

It will be recalled (Section 6.3) that when the Thames driftway was flooded in February 1808, Trevithick proposed to recover the works by sinking a caisson in the river, but that the Thames Archway Co did not accept this plan. In July Trevithick put forward an alternative scheme for constructing a tunnel under the Thames[I] which will now be described. His new idea is shown in Figure 10.1. A watertight wooden caisson, 50 ft long by 30 ft wide and reaching above high water level, was to be placed on the river bed. Wooden piles were then to be driven into the river bed just inside the caisson and to a depth just below the level of the bottom of the proposed tunnel, to form a close-boarded watertight cofferdam inside the caisson and projecting beneath it. The water in the caisson was to be pumped out and the earth inside the cofferdam was then to be excavated down to the depth of the bottom of the proposed tunnel. Any surplus water which had leaked into the excavation was to be drained away through a pipe into the driftway underneath, which was to be used as a drainage gallery for the new works. In the excavation it was then intended to construct a 50 ft length of twin-tube brick tunnel, which when complete was to be covered with earth back up to river bed level. The piles forming the cofferdam were then to be withdrawn and the caisson moved 50 ft further on across the river. The whole process was to be repeated as many times as was necessary to cross the river, 50 ft of tunnel being added at each stage. The caisson was to be fitted with a platform at the top

Figure 10.1 Trevithick's subaqueous cut-and-cover tunnel (1808)

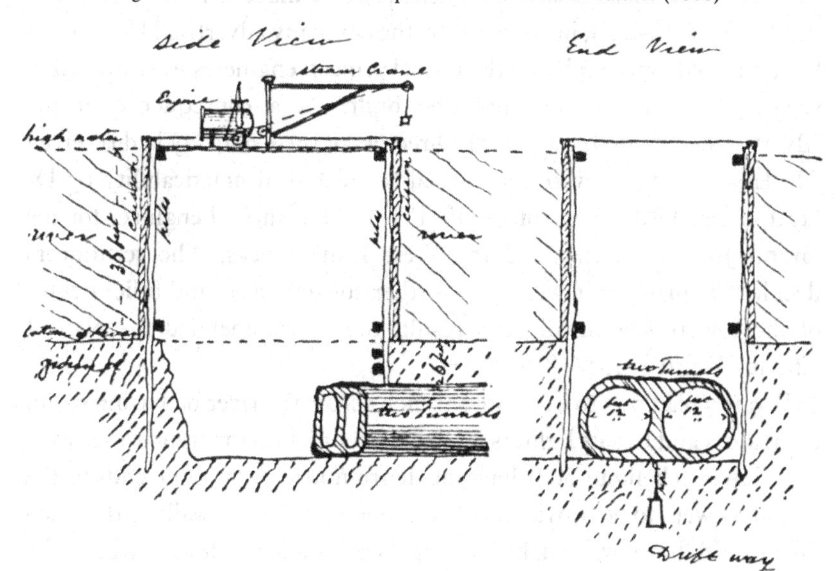

carrying a steam crane – invented by Trevithick in 1805 – which was to be used for drawing the piles and for hoisting the spoil and lowering materials for constructing the tunnel. Each tube was to be 12 ft in diameter and was to accommodate an 8 ft wide waggon road and a 4 ft wide footpath. Only 50 ft of river would be occupied at any one time during construction, which Trevithick said was less than a 400 ton ship lying at anchor.

Trevithick's plan, had it been put into practice, would have amounted to a cut-and-cover tunnel with the novel feature of being constructed beneath a river bed instead of on dry land. However, in September 1808 Trevithick proposed that instead of brickwork, the section of tunnel should be made of cast-iron. Here, we have then, an early proposal for a method of tunnel construction that approaches closely the concept of an immersed tube tunnel. If it had been proposed that the cast-iron cylinders could have been sunk into position in a trench in the river bed rather than placed in a cofferdam excavation, then the idea would have been a true immersed tube tunnel. As it was, this concept quickly followed, but it sprang from the mind of another engineer.

10.2 Wyatt and Hawkins' immersed tube tunnel

The Thames Archway Co rejected Trevithick's plan for a cut-and-cover subaqueous tunnel just as it had earlier rejected his scheme for recovering the driftway. Instead, in March 1809 the company advertised for plans to be submitted for completing the tunnel project: engineers were invited to suggest how the tunnel could be built. As a consequence of the advertisement, by May 1810 the Directors of the Company had received 54 plans, and these were scrutinised for technical practicability by Dr Hutton and William Jessop (1745–1814), the principal engineer for the Grand Junction Canal and the West India Docks. The scrutineers decided six plans were worthy of further consideration and fuller details of these were examined. As a result, they recommended the plan of Charles Wyatt (1751–1819).

Wyatt's plan was to excavate a trench across the river bed using steam excavators and ballast lighters, and then to sink into the trench a series of cylinders, each about 50 ft long, made from brickwork. The ends of the cylinders were to be provided with temporary spherical walls so that each one would be a watertight floating vessel. Each cylinder was to be provided with a cock to admit water so that it could be sunk, and a pump to empty it of water after it was in position. The trench was to be deep enough so that the cylinders would be covered with 6 ft of earth without raising the level of the river bed, this thickness of cover being considered sufficient protection against damage from ships' anchors. It can be seen that Wyatt's plan made provision for excavating the trench, floating out the cylinders, sinking them into their final position, covering them, and finally pumping out the water from them. There remained, however, four vital questions upon which the success or failure of the plan would depend. These were: (i) Could cylinders be made from brickwork that would be strong enough to be towed into position and sunk? (ii) Could the cylinders be placed in the trench on the river bed with sufficient accuracy to be joined together to form a tunnel? (iii) When the cylinders were joined together and the temporary ends removed, would the joints be watertight? (iv) Could the whole operation be carried out in a busy waterway without being imperilled by collision from passing ships? These then were the problems that Wyatt's plan faced and the Thames Archway Co decided to see if solutions could be found by carrying out a preliminary trial before going ahead with the main project.

Accordingly, the Company authorised John Isaac Hawkins (1772–1855) to build two cylinders of brick and Roman cement, each 25 ft long, 9 ft internal diameter, 11 ft 3 in external diameter and to sink one of them in a horizontal position lying across the bed of the Thames at about low water mark near Horse Ferry Stairs.[2] The second cylinder was to be laid against the first the two meeting end to end. It was intended to sink the cylinders at high water in such a situation that the tops of the cylinders should be visible at low water so that the Directors of the Company could inspect the position of the cylinders and the fit of the junction between them. The watertightness of the join was also to be assessed by entering one of the cylinders via a manhole to be provided in the top of it.

In September 1810 Hawkins carried out some preliminary experiments to determine the strength of brickwork made from 'malm paviour' bricks and Roman cement made to Parker's patent, and concluded from these that brickwork cylinders of sufficient strength could be made if the walls of the cylinders were 13½ in thick. The cylinders were built remarkably quickly: construction of the first cylinder commenced on 2 October and it was launched on 8 November; construction of the second

Figure 10.2 Wyatt and Hawkins' immersed tube trial (1811)

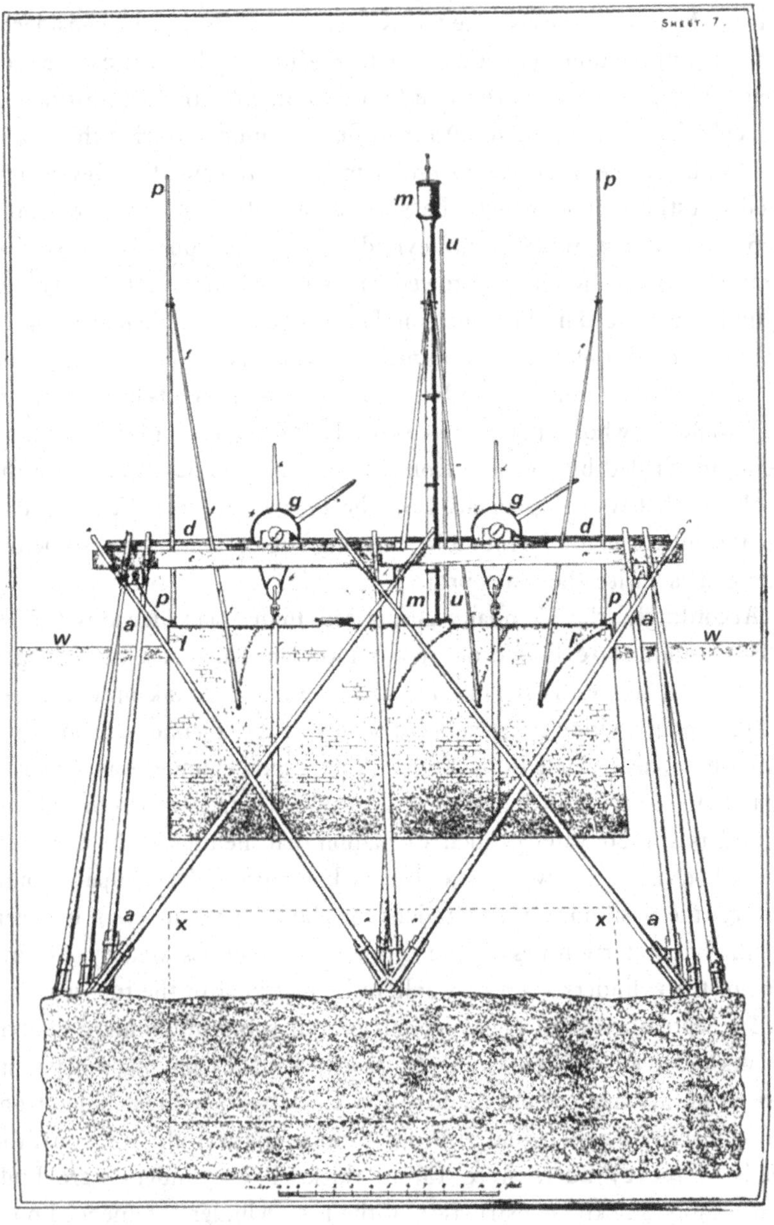

cylinder began on 14 November and it was launched on 19 December. Each cylinder weighed 52 tons and required 8–10 tons of ballast to sink it. The cylinders were built inside a semicircular wooden cradle constructed in the bottom of a 40 ton barge moored in a basin of the Grand Surrey Canal. The inside of the cylinder was supported with wooden centring and ribs and the outside with two iron hoops having eyes at the top. After construction, one-sixth of the cradle was removed from each side leaving one-third of the cylinder supported on the bottom. All the cocks on the cylinder were closed and a cock on the bottom of the barge was opened; the barge then sank leaving the cylinder afloat, the top being 2 ft out of the water. The barge was then refloated and the second cylinder built. The cylinders were kept floating in the basin for four months – this was because of bad weather, there being ice in the river.

The excavation in the river bed was made from a ballast barge equipped with winches and steel-mouthed leather bags, each holding about 12 cwt and directed by means of a handle 50 ft long. Excavation was done at high water to give a realistic idea of what an actual job might be like. The Conservators of the River allowed a scaffold to be erected in the river from which to lower the cylinders into the excavation. The scaffold had to be surrounded by defences to protect it from collision by vessels using the river. Moored vessels nearby restricted the navigable channel to only 200 ft wide at this point; Hawkins reported that on average 200 vessels per hour passed through this gap and on one occasion as many as 400 colliers came up on a single tide. Vessels ran upon the scaffold as a daily occurrence and three-quarters of their time was spent by the workmen in repairing the damage done in this manner. This was the first real problem thrown up by the trial and showed that a solution to question (iv) would have to be found before the method could be used for a real tunnel.

The scaffold was erected at the end of March 1811 and on 4 April the first of the cylinders was brought out of the basin and down to the scaffold, being lashed to the side of a barge for the journey. Figure 10.2 shows the arrangements of the scaffold. On arrival at the scaffold *aa*, the cylinder *ll* was moored under the platform *dd*, hooks from the windlasses *gg* being attached to the eyes on the iron hoops of the cylinder to secure it in place. After some lowering and raising trials, masts *pp* were attached to the cylinder and it was finally lowered to the bed of the excavation, position *xx*, by admitting water. This operation was carried out at high water, level *ww*, the masts being used to indicate whether the cylinder was in the correct position. Two barges full of gravel were then positioned above and the gravel was thrown down and rammed in under and alongside the cylinder by men in boats. The whole operation took just over an hour.

The scaffold was then moved shorewards into the position required for lowering the second cylinder. The second cylinder was lowered in the same manner as the first, the masts being used to line it up with the first cylinder. On inspection at low water the brickwork was in contact at the top of the junction, but $1\frac{3}{4}$ in apart at the bottom. At the next high water, two barge loads of mud and gravel were thrown down over the junction until a mound 6 in thick covered the top of the junction. Water was then pumped out of the cylinders using the pump *mm*, air being admitted through the pipe *uu*. On entry of the cylinders it was found that water was leaking in the join but it was considered that this could easily be sealed with puddled clay.

The trials were now brought to a close by the Company, and presumably the cylinders were raised and removed from the site and the scaffold dismantled although Hawkins does not mention this. Hawkins had shown that questions (i)–(iii) could all be answered affirmatively and that Wyatt's plan for an immersed tube tunnel was completely feasible technically. However, as we have seen, question (iv), the problem of posing an obstruction to navigation had proved troublesome. Indeed the Company blamed the 'accidental and unforeseen circumstances arising from the crowded state of the River' for the fact that the cost of the trials had far exceeded the estimate.

In November 1811 the Directors of the Thames Archway Co declared that they considered it proper to suspend operations, being deterred from proceeding with the main project by the cost of the trials. They were still in debt from the preceding works and recommended selling off the property of the company to pay these debts. So ended the Thames Archway Co's plan for a tunnel under the Thames, and as we have seen, it was left to Brunel to be the first to complete a Thames tunnel some 30 years later.

The Wyatt and Hawkins immersed tube trials have been described in some detail, because although they did not lead to a completed immersed tube tunnel, they demonstrated successfully on a practical scale all the features that construction of an immersed tube tunnel would need to possess.

10.3 Waterloo and Whitehall Pneumatic Railway

The next immersed tube project was for a pneumatic underground railway to run between College Street near Waterloo on the south bank of the Thames and Scotland Yard near Whitehall on the north bank.[3] Work on the project commenced in October 1865. For the river crossing, an immersed tube tunnel was proposed. This was to be constructed from four lengths of tube of 10 ft internal diameter, each 235 ft long. The end of each length was to be supported on a pier formed from a 21 ft diameter cylinder sunk in the bed of the river which was to be dredged sufficiently deep between the piers to allow the tube to lie in a channel between them and have its top well below the river bed. The two piers nearest the south bank were sunk and the river bed between them dredged ready to receive the first length of tube, but although this had been constructed it was never laid. The project was brought to a halt in 1866 due to financial collapse and the cylinders sunk as piers were removed. The lengths of tube were to have been made from $\frac{3}{4}$ in thick iron boiler-plate, lined inside and outside with brickwork. The tubes were to have been fitted with bulkheads at both ends and then floated up river from Messrs Samuda's yard at Poplar, where they were constructed, to their destination just above Hungerford Bridge. Here just enough water was to be admitted to sink each tube upon its piers.[4] As mentioned, one of these had been constructed and two more had been partially constructed when the scheme was brought to a halt.[5] The company was finally wound up in 1882, and so the second plan for an immersed tube tunnel beneath the Thames was brought to nought. It is interesting to note that both the Wyatt scheme and the Waterloo and Whitehall scheme failed for financial reasons and not because of their engineering impracticability, although for both tunnels, had they gone ahead, a solution would have to have been found to the problem of coping with river traffic while the tubes were being laid.

10.4 The Detroit River Tunnel

Although there are brief references to earlier small schemes[6] and several patents, the first large completed immersed tube tunnel of which we have a record was the Detroit River Tunnel,[7] constructed to carry the Michigan Central Railroad across the Detroit River between the cities of Detroit, USA and Windsor, Canada. The situation of this tunnel was very similar to that of the St Clair River Tunnel described in Section 6.10. Before construction of the tunnel, the rail traffic was carried across the river by large self-propelled railcar ferries for which the trains had to be uncoupled into sections and then reassembled on the other side of the river, a process which took between three and eight hours. The large volume of traffic across the river at this point gave rise to the name 'Strait of Detroit' for this half-mile wide crossing. From December to April ice in the river frequently interrupted the ferry service and matters were brought to a head during the winter of 1887–8 when the most powerful ferry was stuck fast in the ice for four days and rendered unfit for service for a month. The United States Board of Engineers looked into the problem and recommended crossing the river by tunnel, a bridge being considered to be an unacceptable interference with river navigation. In fact, a small tunnel had been started in 1872 by the driving of drifts from both sides of the river. They were abandoned, however, within a few hundred feet of meeting because of the many obstacles encountered and the high cost. The two greatest obstacles were the presence of sulphur, present as a gas[8] and in the ground water, and the high artesian pressure of the ground water which was higher than the hydrostatic head of the water in the river.

Following these deliberations the Detroit River Tunnel Co was formed to construct the tunnel and one of their first tasks was to consider different designs for the crossing. The four alternative methods proposed for constructing the subaqueous portion of the tunnel were:

Design A. Dredging a trench across the river. Sinking and connecting cores or forms. Surrounding these with concrete. Dewatering interior of cores. Lining with reinforced concrete.

Design B. Dredging a trench across the river. Lowering of completed sections of reinforced concrete tunnel, the sections to be built in a dry dock, towed to position, sunk and joined under water.

Design C. Similar to Design A but the cores or forms were specified to be made of steel.

Design D. A shield driven tunnel lined with cast-iron segments. A final lining of concrete to be placed afterwards.

The design that was accepted, submitted by the contractor Butler Brothers – Hoff Co, was a modification of Design A which embodied some of the elements of Design C.[9] The reason why a conventional shield driven tunnel (Design D) was rejected was the problems that had been encountered in the 1872 drifts. The design accepted consisted of dredging a trench of the required depth and width and sinking in it watertight steel tubes, constructed in pairs, and surrounding the tubes with concrete deposited under water. An inner lining was to be provided after the tubes had been dewatered. Construction commenced in October 1906.

The tubes were each 23 ft 4 in in diameter and were built from $\frac{3}{8}$ in steel plates rivetted together with lap joints and caulked as for boiler construction. Each tube was to accommodate a single railway track and was 262 ft 6 in long. The tubes were built by the Great Lakes Engineering Works of Detroit using similar methods of construction to those they used for building lake freighters. All the tube sections were built within a period of 20 months. Thus the problem of fabricating the huge steel tubes required for the immersed tube tunnel was solved by drawing on the existing technology of the indigenous shipbuilding industry. The sections of tube were in pairs and to ensure a good fit at the twin joints between sections when they should be in place, adjacent sections were fitted together in the shipyard and tested for watertightness under pressure. Each section of tubes was fitted with external steel diaphragms supporting wooden sheathing to form a box, open at the top and bottom, which served as shuttering for the concrete which was to be placed around the tubes after they had been sunk in position. The ends of the tubes were closed with heavy wooden temporary bulkheads designed to take the full hydrostatic pressure of the river.

The trench in the river bed to receive the tubes was excavated by a 3 yd^3 clamshell bucket operated from a large barge. The ground below the river bed consisted of blue clay with occasional seams of sand and gravel. The river was between 18 and 48 ft deep at the tunnel crossing. After excavation, the trench was swept for obstructions and inspected by divers; following this, grillage was placed on the trench bottom to receive the tubes.

Each of the twin tubes in a section was fitted with a gate valve and an air escape tube, and the section was sunk by admitting water to the tubes. Air cylinders were fitted over the tubes to control the buoyancy during sinking; these were detached after the tubes were sunk and used over again. The position of each tube section, both line and level, was determined by steel masts fitted one on each end of a tube section. These masts were carefully set up at right angles to the horizontal diameter of a tube, on axis, and were graduated in feet. The tube sections were then adjusted by manipulating holding lines whilst sighting on the masts with instruments sited on shore. The tubes were finally positioned by divers using shims between the tube base and the grillage. The air cylinders allowed perfect control, it being found possible to raise or lower the tubes by a fraction of an inch. Large barges fitted with derricks and hoisting engines were used to lower the tubes. The barges were moored to concrete anchors buried in the bottom of the river upstream and opposite the ends of the section to be lowered. The time required for taking a tube section from its moorings, placing it over the trench and sinking it into position was 12–13 hours. The lining up of the section with the previous one and the bolting of the flange connections took a further two days, this bolting being carried out by the divers who were an essential component of the workforce. It will be recalled that Wyatt and Hawkins' plan utilised a scaffold which consisted of a fixed platform supported on legs resting on the river bottom as the means of lowering the immersed tube sections. At the Detroit River, instead of a fixed platform, we see that a floating platform moored to the river bed was used as the solution to the problem of lowering the sections, the river being between 18 and 48 ft deep at the crossing location. This point will be returned to later.

For depositing the concrete around the sunken sections, a large barge was specially built by the contractor. It was fitted out with concrete mixers, derricks, hoisting engines and hoppers and screens for the gravel. Along one side of the barge three tremie pipes were fitted and the concrete was loaded into hoppers at the tops of the tremie pipes down which it was channelled to the shuttering box around the tube sections. After the concrete had been placed, the installation was inspected by a diver who made sure that the concrete was carried up to at least 6 in above the top of the steel diaphragm in the shuttering at all points. The remainder of the trench was then backfilled. Gravel was used to a depth of 11 ft and the remainder was filled with clay that had been excavated from the trench.

The gravel was placed by means of wrought-iron pipes and the clay was dumped from drop-bottom barges. For a distance of several hundred feet near the middle of the river the top of the tunnel was 3–7 ft above the bed of the river leaving a 41 ft deep main channel, but under the remainder of the river the top of the tunnel was below the original bed of the river.

The water valves on all the tube sections were closed by divers and the sections were dewatered one at a time, the water being pumped out by a centrifugal pump placed in the first section prior to sinking. The placing of the internal reinforced concrete lining now took place using steel shutters, the invert, side walls and arch being formed in that order. The internal lining was completed in twelve months and the construction contract was completed in July 1910. The subaqueous section of the tunnel was 2668 ft long and its successful completion was a convincing demonstration of the practicability of immersed tube tunnelling and showed that the technical problems had been solved.

10.5 Subsequent development

Following the successful construction of the Detroit River Tunnel, immersed tube tunnelling became an established technique for river and estuary crossings, particularly in the United States where the method had been pioneered, and in the Netherlands where the geography of the country presented many suitable sites for its application. However, in Great Britain no immersed tube tunnels were built until as late as the 1970s[10] and it is worth considering briefly why this was so. The reason is undoubtedly the early establishment of a tradition of shield tunnelling combined with compressed air for subaqueous conditions which was based on the early successes that these methods had in Great Britain (see Chapter 7 and particularly Table 7.1). There is no doubt that some of the Thames tunnels could have been constructed by the immersed tube method but were not because a proven tunnelling system was already available.

Two types of construction of immersed tubes soon established themselves. The first was the steel shell type, of which the Detroit River Tunnel is an example, and which became the standard type in the United States. These could be readily fabricated in the many shipyards that already existed in America; even if the shipyard is not near at hand the cost of transport is so low that it has been found economic to tow the units considerable distances. The ready availability of good shipbuilding facilities also explains the popularity of this form of construction in Japan (see Table 10.1). The iron tube constructed by Samuda for the abortive Waterloo and Whitehall Pneumatic Railway in 1865 can be thought of as the precursor of the steel shell type of immersed tube. The other type of immersed tube construction is the reinforced concrete immersed tube, of which the tunnel for the Hong Kong mass transit railway, shortly to be described, is an example. This type was pioneered in the Netherlands and is now popular in Europe and is finding application elsewhere. The steel shell immersed tubes are usually circular in cross section but the reinforced concrete immersed tubes are particularly suitable for rectangular cross sections which lend themselves well to the design of multi-lane traffic tunnels. Table 10.1 lists the numbers of each type of construction that have been used for the world's major immersed tube tunnels up to 1981.[11]

The development of immersed tube tunnelling in the Netherlands was a direct response to the problem of conflicting communications. Because of the Netherlands' situation on the North Sea coast and at the deltas of the Rhine, Maas and Scheldt rivers, seaborne and inland navigation are important in the economy of the country. The rivers and estuaries are, therefore, important communications channels which it is clearly desirable to keep unobstructed with bridges. However, roads, railways and other services have to cross the waterways and the solution to this conflict of interests is a tunnel. The low lying delta area is characterised by soft alluvial subsoils with a very high water table and if tunnels were to be constructed by shield tunnelling, compressed air with all its disadvantages would have to be used to keep the workings safe from flooding. The alluvium, however, was ideally suited to dredging and the Dutch had plenty of experience and plant for this sort of operation. They therefore decided to use the immersed tube tunnelling method and pioneered its use in Europe by constructing the first immersed tube tunnel outside the United States,[12] the Maas Road Tunnel, built between 1937 and 1942. Since then they have constructed eleven further major immersed tube tunnels[13] and Dutch engineers are now among the world's leading authorities on this method of tunnel construction.[14] As previously mentioned, the reinforced concrete tube method of immersed tube tunnel construction was introduced in the Netherlands and the reason for this is

Table 10.1. *Location and type of construction of major immersed tube tunnels*

Location	Type of construction		Total
	Concrete	Steel shell	
United States	2	17	19
Netherlands	12	0	12
Japan	3	11	14
Others	14	1	15

not because shipbuilding facilities were not available for the construction of steel shell units, but because of the high cost of steel in Europe compared with the United States. Reinforced concrete tubes were not only cheaper than steel shell ones, but were more easily made in rectangular cross sections which were more suitable for multi-lane road tunnels, it being the case that ten of the twelve Dutch tunnels listed in Table 10.1 are road tunnels. The reinforced concrete tube unit is now firmly established and seems likely to become the standard method for immersed tube tunnels world wide.

As well as two types of construction of immersed tube tunnels, two different ways of placing the immersed tubes became established. The first of these utilised a fixed platform with legs resting on the river bed for lowering the tubes and the precursor of this type is clearly the scaffold that Hawkins constructed for his trials in the Thames in 1811. The fixed platform, however, is best suited to relatively shallow water conditions and for deeper water a laying barge or pontoons became the accepted method. It will be recalled that the Detroit River tube sections were sunk from a laying barge. In fact for both deep and shallow water conditions a laying barge or pontoons are often preferred by contractors to a fixed laying platform. The reason for this is that a laying barge or pontoons can be easily adapted or improvised from an existing vessel or marine plant and then can be returned to their original use after the tunnel tubes have been laid. On the other hand a fixed platform is a purpose-made item of plant which is costly to manufacture and has no further use after the job is completed unless another immersed tube tunnel is planned.

Modern versions of these two ways of placing an immersed tube tunnel are shown in Figures 10.3 and 10.4. In Figure 10.3 a rectangular cross

Figure 10.3 Laying immersed tube from fixed-leg platform (1978)

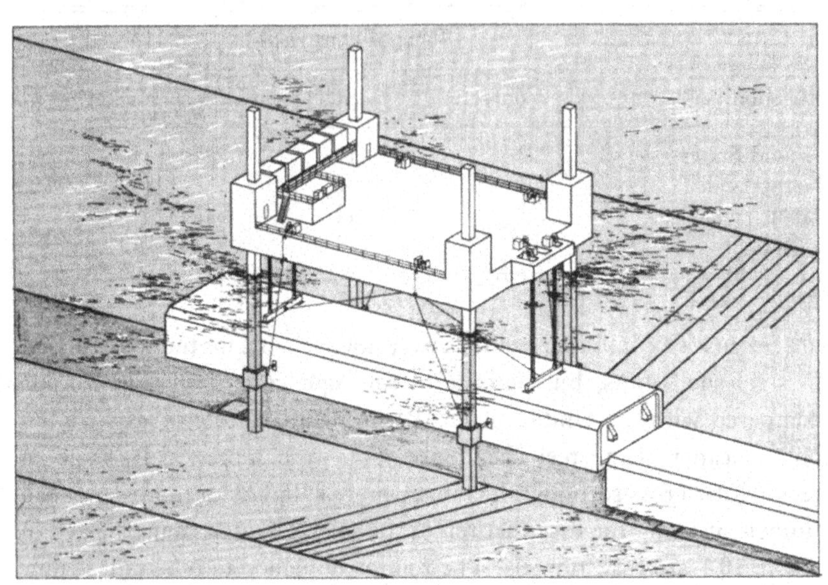

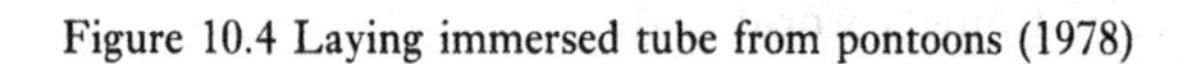

Figure 10.4 Laying immersed tube from pontoons (1978)

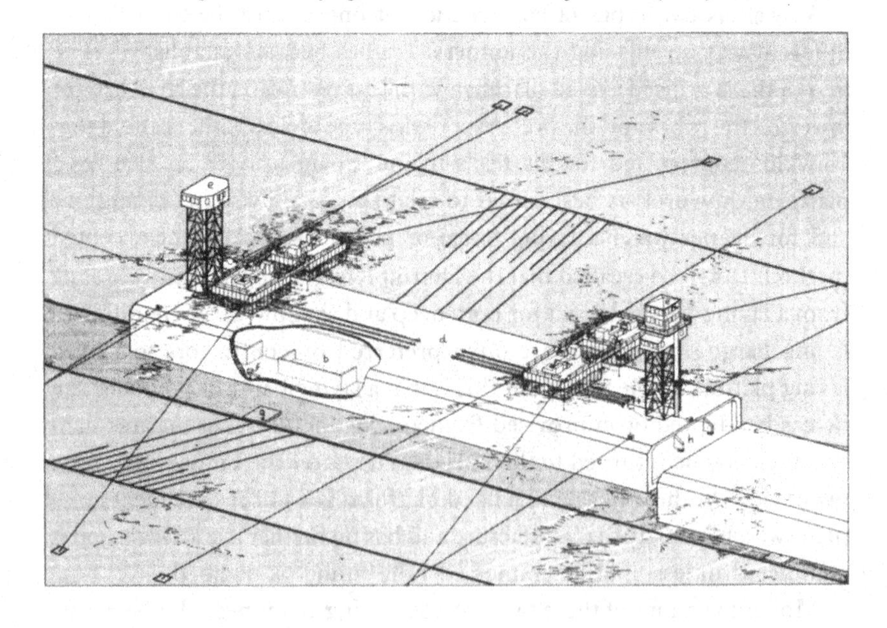

section reinforced concrete immersed tube tunnel unit is shown being lowered into position from a fixed-leg platform whilst in Figure 10.4 the same kind of unit is shown being lowered from four pontoons which are kept in position by holding lines attached to achors on the river bed.[15] An example of the construction of a modern immersed tube tunnel is that for the Hong Kong mass transit trailway.

10.6 Immersed tube tunnel for Hong Kong railway

Where the railway tunnels for the Hong Kong mass transit railway cross Victoria Harbour from Hong Kong Island to Kowloon on the mainland they do so via an immersed tube tunnel constructed between 1976 and 1979.[16] The tunnel is formed from 14 twin-track precast reinforced concrete immersed tube units each 100 m long and weighing 7800 tonnes. The two tracks in the immersed tube are in separate 5.3 m diameter compartments. A dry dock for the prefabrication of the immersed tube units was constructed at the north-east corner of Hong Kong Island. Dredging of the trench across the harbour was carried out at the same time as construction of the units. The dredging of the trench, and the subsequent backfilling afterwards, was carried out with clamshell dredgers. After construction, the immersed tube units were fitted with watertight temporary bulkheads to enable them to float and were then towed to a special sheltered mooring area nearby. From here they were in turn towed to their final position in Victoria Harbour when required. A specially built laying vessel consisting of a huge pair of pontoons which straddled the tunnel unit, then lowered the unit by lifting cables on to a gravel screed which had been previously placed on the trench bottom. During the lowering of the units they were flooded to a negative buoyancy of 300 tonnes to allow them to sink. Once on the trench bottom, the newly placed unit was joined to the previous one by being brought into contact with it using a rubber gasket; the hydrostatic pressure on the further end of the new unit was then utilised to press the units tightly together. Units were placed and joined at the rate of one per month. After the units had been laid in position and joined the trench was backfilled, part of it being covered with a protective cover of rockfill. The water was then drained out of the tubes and secondary water seals fitted to the joints.

Some comments will be made on the problems encountered at Hong Kong and how they were overcome. The trench dredging and tube laying operations had to be carried out in a busy international harbour amidst sea-going shipping and therefore had to comply at all times with the requirements of the local Marine Department. The clamshell dredger operators were accustomed to manoeuvring in the harbour traffic. The special laying vessel used to lay the tube sections was actually a cross between the two types shown in Figures 10.3 and 10.4. When it was being used to place the gravel screed it worked as a fixed platform standing on four legs with the pontoons jacked up above the level of wave action, but for the sinking of the tubes the legs were retracted and the tunnel sections were lowered from the floating pontoons. The reason that Kumagi Gumi, the contractors for the job, had the special laying vessel built was that they considered that a floating mode of operation was best for positioning and laying the immersed tube units, but that a fixed-leg mode of operation was necessary for placing the gravel bed for the units which had to be screeded to an accuracy of ± 10 mm. It will be recalled that immersed tube tunnelling requires a bottom that can be readily excavated. In certain parts of the trench large boulders were encountered which could not be handled by the clamshell dredgers. This problem was overcome by blasting the boulders so that they were broken up into fragments small enough to be lifted with the clamshell grabs.[17]

10.7 A use for hydrostatic pressure

So far this account has dealt mainly with the technical problems of immersed tube tunnelling and how they were overcome, but there is one aspect of the subject that is worth considering briefly because it is an interesting example of a technical advantage that was exploited. One problem that faced the engineer in immersed tube tunnelling was that of making a satisfactory tight joint where the end of one unit joined the end of another. It will be recalled that in Hawkins' trial in 1811 this aspect of the job had not been altogether satisfactory, there being a gap of $1\frac{3}{4}$ in at the bottom of the joint. In the Detroit River Tunnel the tube sections had bolted flanges and the joints seem to have been satisfactorily made with these. However, in the account of the Hong Kong tunnel it was mentioned that hydrostatic pressure acting on the end of a unit was used to force the unit into tight contact with the previously laid unit. The basis of this method of effecting a tight joint will now be described.[18] The joint between the immersed tube sections was fitted with a thick soft-rubber gasket and the temporary bulkheads at the end of the units were set back a little from the ends as shown in Figure 10.5. The newly placed unit had then only to be positioned so that the soft-rubber gasket was slightly compressed. The water in the small chamber between the two temporary bulkheads was then pumped out and the newly placed section was forced against the previously placed section by the hydrostatic pressure of the water acting on its further end (see Figure 10.5); this pressure caused the gasket to compress completely forming a perfectly watertight joint.[19] In this way the hydrostatic pressure which is always present in immersed tube tunnelling has been turned by the engineer to his advantage and used to assist him in overcoming the technical problem of joining the sections together closely. It will be recalled that in constructing subaqueous shield driven tunnels, hydrostatic pressure was an enemy that had to be overcome by the use of compressed air (Chapter 7), but in immersed tube tunnelling the engineer has turned hydrostatic pressure to good advantage.

The idea of using hydrostatic pressure in this way can, in fact, be traced back well over a century, for in 1861 James Chalmers described a scheme[20] that would have made use of the principle. Chalmers' plan was

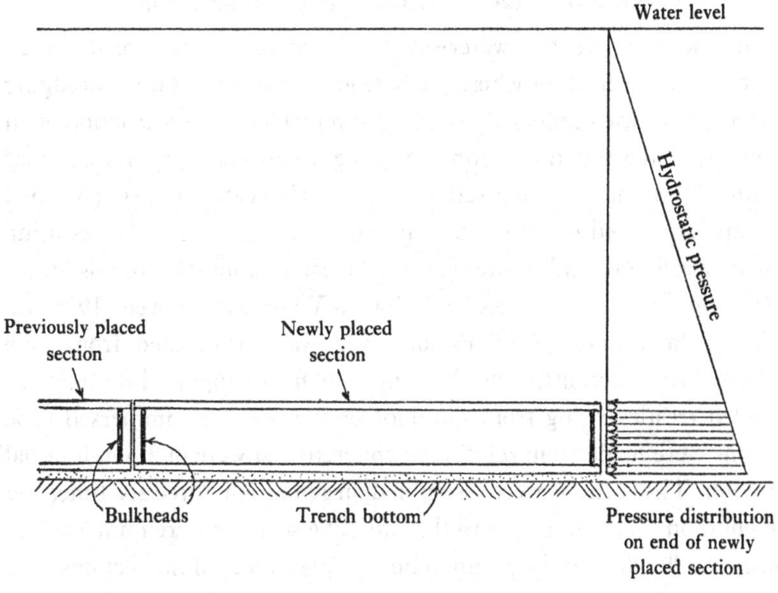

Figure 10.5 Using hydrostatic pressure to join immersed tube sections

to construct a Channel Tunnel by laying an iron tube on the sea floor and covering it with an embankment of stone. The tunnel tube was to have been 30 ft in diameter and made from iron plates, double riveted and caulked. The tube sections, each 400 ft long, were to have been fabricated in a harbour or river, fitted with temporary bulkheads, launched and towed out to sea to their final positions. Water was then to have been admitted and the sections were to have been drawn down to the sea bed by chains anchored on the bottom. In Chalmers' design for the tube section the temporary bulkheads were set back (just as shown in Figure 10.5) so that when the flanges were brought together, water could be drained out of the space between the bulkheads into the previously placed section, allowing hydrostatic pressure to push the tubes tightly together. The flanges were to be provided with an elastic seal to make a watertight joint when this was done. The water was then to be pumped out enabling workmen to pass into the space between the bulkheads and make a permanent joint. Chalmers calculated that the force available to close the tubes together in this way was 2000 tons, given the depth of water in the Channel. Each section of tube was to be covered with stone as soon as it was placed, in order to keep it safely down on the sea bed when the water was pumped out.

10.8 Unburied immersed tube tunnels

In situations where the waterway to be crossed is too deep for an immersed tube tunnel or where the bottom is of rock so that the dredging of a trench for the immersed tubes is not possible, the construction of an unburied immersed tube tunnel may be a solution. The idea is that instead of burying the immersed tube sections beneath the sea or river bed they are supported on piers built up from the bed so that the resulting structure is like an underwater bridge. An early application of this idea is the tunnel constructed across Bakar Bay in Yugoslavia between 1974 and 1978.[21] The tunnel is 400 m long and was constructed from nine immersed tube sections, each 40 m long, which are supported at their ends by piled trestles rising from the floor of the bay. The immersed tube sections were made from reinforced concrete and were of 3.5 m internal diameter. The sections were assembled on a pontoon, three at a time, and the pontoon was then towed to the site. The sections were immersed by controlled flooding of the pontoon but the placement of the sections onto the underwater piers was done using a floating crane. Sealing of the units to each other was done by compressing a rubber gasket running around the end face of each section using hydrostatic pressure as described in Section 10.7. On completion of installation of all the immersed tube sections the tunnel was dewatered and a belt conveyor system was installed inside it, the purpose of the tunnel being to transport coal to a coking plant.

In 1982, the technique was used again, this time for a 590 m long tunnel at Karmoey Island in Norway.[22] The support piers were constructed on six natural undersea peaks on the bed of the fjord and the five immersed tube sections vary in length from 90 to 150 m, being tailor-made to fit the individual peak-to-peak spans. The immersed tube sections were of prestressed concrete, 6.5 m wide by 7.5 m high, and they were towed to the site from the dry docks where they were made at Stavangar and Kristiansand some distance away. When in its correct position, each section was kept there by two tugs on either side, while cables running down to the sea bed piers and then to the shore were used to winch the section down after controlled flooding. Underwater video cameras were used to observe the operation and a floating command centre co-ordinated the movement of the tugs and the operation of the winches. The tunnel was successfully completed, followed by the installation of the two gas pipelines that it was built to carry.

One very important difference between the unburied immersed tube tunnel and the normal immersed tube tunnel buried in a trench in the sea or river bed is that the unburied tunnel is vulnerable to damage from being struck by a sinking ship or collision with a submarine. It is significant that the two that have been constructed to date are both for the transport of materials, namely coal and natural gas, and not for people.

Both the tunnels that have just been discussed have negative buoyancy in their final position, that is to say their own weight is sufficient to keep them resting securely on their piers even though they are totally immersed in water. However, the idea of an unburied immersed tube tunnel having positive buoyancy has been proposed as a means of crossing the Eidfjord on the main route between Bergen and Oslo in Norway.[23] Instead of being supported on piers, the immersed tube sections for this kind of tunnel would be suspended below the surface of the sea or river at sufficient depth not to obstruct navigation, by means of cables anchored to the sea or river bottom. The tunnel sections would have to be made of sufficiently lightweight final construction so that, when dewatered, they would have a tendency to float rather than sink. The idea, it is claimed, would find application for tunnels across waterways where the depth of water was too great for the construction of piers from the sea bed. Tunnels of this sort would be even more vulnerable than those previously described and it remains to be seen whether this idea is ever put into practice. It is interesting to note that this idea is not new, the idea of a buoyant tube tunnel moored to the sea bed having been proposed as a means of constructing a Channel Tunnel by James Wylson as long ago as 1855.[24] Wylson's plan was to construct a cylindrical tube from wrought-iron staves held in place with hoops, the whole to be lined throughout on the inside with brickwork and made watertight on the outside by a flexible covering. The internal diameter of the tube was to be over 20 ft, large enough to accommodate a single-track railway with a walkway on either side. Some of the hoops were to be fitted with eyes to which chains were to be attached, these chains leading to mooring weights for permanently restraining the tube at a uniform depth from the surface of the Channel. Although nothing came of the scheme, Wylson's idea is the same in principle as that proposed for the Eidfjord crossing. We can note, finally, that an idea for an immersed tube tunnel consisting of a wrought-iron or cast-iron tube resting on the sea bed was proposed by H. Horeau as a method of constructing a Channel Tunnel several years before Wylson's scheme, in 1851.[25]

In 1984 it was announced that the new North Wales Coast Road (A55) will cross the estuary of the Conway River by means of a 700 m long immersed tube tunnel consisting of six prefabricated twin-tube units of rectangular cross section, made either of steel or conrete. Construction of the tunnel commenced at the end of 1986 and is expected to take about $4\frac{1}{2}$ years to complete. Apart from the outfall tunnels mentioned in Note 10, this will be the first immersed tube tunnel in Great Britain, and comes some 175 years after Hawkins had demonstrated the practicability of Wyatt's idea.

10.9 Summary

The immersed tube tunnel has its origin in the trials carried out by Hawkins in 1810–11, although it was not until 1906–10, a century later, that the first large completed immersed tunnel, the Detroit River Tunnel, was built. From then on immersed tube tunnels, either of the concrete or the steel tube type became an established method for constructing river crossings, although the technique did not find application in Great Britain until the 1970s because of the fact that the use of the shield combined with compressed air had become so firmly entrenched. We have seen that almost the whole of the development of immersed tube tunnelling, from the early trials by Hawkins right up to the present day, took place entirely within the tunnelling industry, although, very significantly, the development of the steel tube immersed tunnel took place in the United States and Japan where many good shipyards were available for the fabrication of the steel tube units. The concrete immersed tube tunnel was developed in the Netherlands, firstly because of the high cost of steel in Europe and secondly because of the suitability of concrete for making the rectangular cross sections which are appropriate for road tunnels, most of the Dutch immersed tube tunnels being for this purpose.

Notes and references for Chapter 10

1 Trevithick, F. (1872) *Life of Richard Trevithick, with an account of his inventions.* Volume 1. London (E. and F.N. Spon) pp. 273–5.
2 Hawkins, J.I. (1840) Report upon two experimental brick cylinders laid down in the River Thames at Rotherhithe, in the year 1811, for the purpose of ascertaining the practicability of forming a tunnel under the river, by laying a continuous line of brick cylinders in a trench, excavated across from shore to shore in the bed of the river, and covering them with earth. (Unpublished manuscript in Institution of Civil Engineers' Library).
3 Wragge, E. (1902) Correspondence on subaqueous tunnelling. *Minutes of Proceedings of the Institution of Civil Engineers*, **150**, pp. 83–4.
4 Anon (1866) The pneumatic railway. *Engineering*, **2**, 5 October, p. 257.
5 Hadfield, C. (1967) *Atmospheric railways: a Victorian venture in silent speed.* Newton Abbot (David and Charles) pp. 100–1.
6 A list of early proposals for immersed tube tunnels together with brief references to two actual schemes which seem to have predated the Detroit River Tunnel are given by:
Carson, H.A. (1911) Discussion on the Detroit River Tunnel. *Proceedings of the American Society of Civil Engineers*, **37**, pp. 1165–7.
7 Kinnear, W.S. (1911) The Detroit River Tunnel. *Proceedings of the American Society of Civil Engineers,* **37**, (6), pp. 897–965.
8 Probably hydrogen sulphide – which is poisonous even at low concentrations.
9 The account given here has been slightly simplified, concentrating on essential principles, but a full description of the different designs, including comments on the one finally adopted, is given by:

Hoff, O. (1911) Discussion on the Detroit River Tunnel. *Proceedings of the American Society of Civil Engineers*, **37**, pp. 1313–27.

10 The first immersed tube tunnel in Great Britain seems to be that constructed in 1970 for the outfall of Hunterston 'B' nuclear power station in Ayrshire, Scotland. The tunnel units were made of reinforced concrete and were 3.5 m inside diameter and 6.5 m long. The immersed tube tunnelling method must be particularly well suited to the construction of outfall tunnels because all the subsequent immersed tube tunnels in the United Kingdom have been built for this purpose; they are the following:
   (i) Kilroot power station, Northern Ireland. Twin-compartment reinforced concrete outfall tunnel with immersed tube sections 94 m long. Constructed 1977–8.
   (ii) Peel Common outfall, the Solent. Twin-compartment reinforced concrete outfall tunnel with immersed tube sections 15 m long. Completed 1980.
   (iii) Anchorsholme outfall, Blackpool. Single-tube welded steel pipe outfall with sections 6 m long. Constructed 1981. This immersed tube tunnel was constructed by a novel method; instead of sinking the individual sections and joining them in the trench, the sections were joined together on a slipway on land and the whole 930 m long tube was then launched and towed 17 miles to the trench position and sunk.

11 A description of the development of immersed tube tunnelling and lists of the world's major immersed tube tunnels have been given by D.R. Culverwell in a series of three articles:
Culverwell, D.R. (1976) Immersed-tubes and the Tees. *Tunnels and Tunnelling*, **8**, (1), pp. 27–33.
Culverwell, D.R. (1976) Immersed-tube tunnels. *Tunnels and Tunnelling*, **8**, (3), pp. 91–8.
Culverwell, D.R. (1981) World list of immersed tubes. *Tunnels and Tunnelling*, **13**, (8), pp. 17–23.
American immersed tube tunnels are described in:
Jeffery, A.H.G. (1962) *Immersed tube tunnels – a study in achievement*. London (Kemp's International Publications Ltd).

12 Omitting a small pedestrian footway tunnel constructed in 1927 at Friedrichshafen, Berlin.

13 These are listed with full details of their construction in:
Royal Institution of Engineers in the Netherlands (1978) *Immersed Tunnels: Proceedings of the Delta Tunnelling Symposium, Amsterdam*. Amsterdam (Royal Institution of Engineers in the Netherlands) pp. 62–70.

14 Modern techniques of immersed tube tunnelling have been described in the following series of three articles:
Glerum, A. (1979) Designing immersed tunnels. *Tunnels and Tunnelling*, **11**, (2), pp. 29–32.
Vos, J. (1979) Constructing immersed tunnels. *Tunnels and Tunnelling*, **11**, (3), pp. 42–4.
Janssen, W. (1979) Efficient waterproofing of immersed tunnels. *Tunnels and Tunnelling*, **11**, (4), pp. 25–9.

15 Molenaar, V.L. (1978) The immersing of the tunnel units. *Immersed Tunnels: Proceedings of the Delta Tunnelling Symposium, Amsterdam*, pp. 44–7.

16 Haswell, C.K. *et al* (1980) Hong Kong Mass Transit Railway Modified Initial System: design and construction of the driven tunnels and the immersed tube. *Proceedings of the Institution of Civil Engineers, Part 1*, 68, Paper No 8391, pp. 627–55.

17 Edwards, J.T. *et al* (1982) Discussion on Papers Nos 8389–8392. *Proceedings of the Institution of Civil Engineers, Part 1*, **72**, p. 97.
18 Brakel, J. (1971) Some considerations of submerged tunnelling. *Proceedings of the Institution of Civil Engineers*, **48**, Paper No 7384, pp. 599–620.
19 Depending on the depth of water and the size of the units, the force available can be immense. For example, for a rectangular cross section unit 25 m wide by 8 m high sunk in 20 m depth of water, the force on the further end will be 4000 tonnes.
20 Chalmers, J. (1861) *The Channel railway, connecting England and France.* London (E. and F.N. Spon) *passim.*
21 Polz, K., H. Kreuder-Sonnen, G. Benrath & D. Delać (1980) Immersed tube carries coal conveyor across bay. *Tunnels and Tunnelling*, **12**, (2), pp. 57–60.
22 Anon (1982) Norwegian tunnel spans subsea peaks. *New Civil Engineer*, 18 November, pp. 30–3.
23 Martin, D. (1981) First suspended immersed tube tunnel may be built in Norway. *Tunnels and Tunnelling*, **13**, (8), p. 38.
24 Wylson, J. (1855) Proposed Anglo-Gallic submarine railway. *The Illustrated London News*, **27**, (769), p. 570.
25 Anon (1851) Submarine railway between England and France. *The Illustrated London News*, **19**, (530), pp. 612–3.

Hard rock tunnelling machines

More than any other people, the Americans are mechanically-minded.

We have already seen how the introduction of compressed air rock drilling machines transformed the operation of hard rock tunnelling by the drill-and-blast method, freeing the miners from gruelling hand drilling. And in soft ground tunnelling as well, the tunnelling shield is a simple machine for supporting the ground although for some considerable time after its introduction the ground had still to be excavated by hand tools; in both hard rock and soft ground tunnelling the spoil had to be loaded into skips by hand before it was carried away. These two innovations, then, are both examples of partial mechanisation, but a tunnelling machine is a device in which the whole process of excavation of the ground at the face and the transporting away of the spoil is done by machinery. Tunnelling machines developed along two separate lines, those for hard rock and those for soft ground and this will be reflected in their treatment here. Hard rock tunnelling machines will be dealt with in this Chapter whilst soft ground tunnelling machines will be dealt with in Chapter 12. It can be remarked, however, that in spite of their different ancestries, some modern hard rock and soft ground tunnelling machines have many features in common and often look alike. As J.H. Stephens[1] points out perceptively, the two machines made for the ill-fated 1975 Channel Tunnel project exemplify this: one was a Priestley soft ground tunnelling machine and the other was a Robbins hard rock tunnelling machine but both machines had been adapted to cut chalk, a material having many properties intermediate between soft ground and hard rock. Another distinction that is sometimes made is to distinguish between tunnelling machines that excavate the whole of the face at the same time – these being termed full-face tunnelling machines,[2] and those that excavate only part of the face at any given time – these being termed partial-face tunnelling machines.[3]

Once the compressed air rock drilling machine had been perfected and nitroglycerine explosive had come into general use, drill-and-blast tunnelling in hard rock was so successful that it is, at first sight, difficult to see what the incentive was to develop tunnelling machines. However, for all its strengths, drill-and-blast tunnelling was essentially a sequential operation, with the inevitable consequence that the utilisation of the plant was low: when the shot holes were being drilled the rock drills were in use but the muck trains were not, and when the rock spoil was being removed the muck trains were in use but the drilling machines were idle. The concept of a tunnelling machine was essentially one where the two operations of excavating the rock and moving it away down the tunnel could take place simultaneously and in a continuous operation – the utilisation of the plant would then be high because the only time it need be stopped would be for maintenance and repair. A further incentive to develop tunnelling machines, which did not matter so much in the early days when labour was cheap but became increasingly important as labour costs rose, was that a tunnelling machine needed far fewer miners at the face to operate it than did drill-and-blast tunnelling. We shall see, however, that the development of a successful hard rock tunnelling machine posed problems to the early designers, and took many years before it was perfected.

The designers of the first hard rock tunnelling machines were faced with a number of problems. These were (i) to devise a mechanical means of excavating the rock at the tunnel face, (ii) to devise a mechanical means of continuously removing the spoil produced by the excavation, (iii) to devise a suitable power supply for operating the machine and (iv) to design machines robust enough for the job. We will now look at some early hard rock tunnelling machines, bearing these problems in mind.

11.1 Mont Cenis and Hoosac tunnelling machines

For the first attempts to make and use tunnelling machines for driving hard rock tunnels we must return again to the Mont Cenis and Hoosac Tunnels which featured so prominently in the development of compressed air rock drills as described in Chapter 2.

The inventor of the first hard rock tunnelling machine was Henri-

Joseph Maus. Maus was born in Namur (now in Belgium, then in France) on 22 October 1808. After an early career in mining, he became involved with railway construction and in 1845 was asked by the Sardinian Minister for Public Works to review the proposed future railway system for Piedmont which was then a province of the Kingdom of Sardinia. This involved the proposed main line between Chambéry and Turin (see Figure 2.5) with its important tunnel between Modane and Bardonecchia - the Mont Cenis Tunnel through the Alps. To drive the tunnel, Maus designed, constructed and tested a full-face tunnelling machine which will shortly be described. In 1854 Maus left Piedmont and returned to Brussels; his contributions to the development of Piedmont railways were recognised by the Sardinian Government who conferred several honours upon him. On his return to Belgium, Maus was appointed Inspector-General of Highways and later, Director of Mines. He died on 11 July 1893. Details of Maus' tunnelling machine were scanty until the elucidation made by Stack who came upon an original report made by Maus to the Sardinian Government in 1849; the description given here is based upon her account.[4]

Maus' tunnelling machine consisted of a metal framework of square cross section, 7 ft by 7 ft, which carried at the front an array of long chisels. The chisels were arranged in five horizontal rows of 16 chisels per row, together with two columns of 18 chisels down each side, giving a total number of 116. The action of the machine was, therefore, to cut the rock into four blocks lying one on top of the other, but still attached to the rock mass at their further end. These blocks were then removed by driving wedges into the grooves between them, causing them to break off at the back. The chisels were actuated by a system of cams which drew them back against the action of powerful co-axial springs which then released them suddenly against the rock at the rate of 150 blows per minute.[5] The cams also caused the chisels to rotate slightly with each impact to obviate jamming in the hole. A water jet kept each chisel cool and flushed the hole free from debris. The whole frame was moved forward by a manually-operated crank, the speed of advance being governed by the hardness of the rock. Maus' idea was that half the tunnel face would be cut by the machine at any one time. When the chisels had cut to their full depth, they would be retracted and the whole machine shifted sideways and put to work on the other half of the face. While this was going on, miners would be breaking out the blocks from the cut half of the face. The machine obtained its power from a main pulley-driven shaft; had it come into use, the power was to be transmitted to the machine via a series of endless cables and pulleys running down the tunnel, driven by water turbines at the portals. In 1847 a prototype machine was constructed at the Valdocco arms factory with funds provided by the Sardinian Government and over the next two years practical tests involving cutting numerous blocks of rock were carried out. As a result of these trials, Maus estimated that the machine should be able to advance at the rate of 16–24 ft per day. Working from both portals and allowing for breakdowns, Maus estimated that the Mont Cenis tunnel could be driven in five years. However, because of political troubles during 1848–9 the tunnel project was shelved and when it came to be revived in 1857, the tunnel was constructed using the drill-and-blast method utilising the new compressed air drilling machine as described in Chapter 2. Maus' machine, therefore, was never used to drive a tunnel.

We shall now examine how this, the first hard rock tunnelling machine, sought to overcome the problems outlined at the end of the introduction to this Chapter. Maus' solution to problem (i) was to use an array of 116 chisels to cut grooves in the rock. This was perfectly sound in principle and obviously based on the old method of hand drilling. However, we can see that the problems of keeping the machine supplied with sharp chisels would have been extremely difficult; this was already a serious problem with the much smaller number of tools required for drill-and-blast tunnelling (see introduction to Chapter 4). His solution to problem (ii) was to move the whole machine sideways and allow miners to wedge the blocks out of the face. The blocks would then have been loaded on to rail skips and transported out of the tunnel. Here we see another flaw in the operation – the delay while the machine was moved to one side and set up again – but otherwise there was the good feature that breaking out the blocks could take place while the other side was being cut which meant both operations could take place simultaneously. Regarding problem (iii), Maus himself estimated that a considerable amount of power would be lost down the cables: he said that of every 100 hp transmitted by the turbine only 22 hp would reach the tunnelling machine, the rest being lost in the friction of the pulleys. And this was a problem that would become worse as the tunnel got longer. There is little doubt that this was a most serious shortcoming of the machine and would probably have been its downfall had it ever been used. Also the cable system of power transmission down the tunnel would have been an appalling potential hazard to the workforce due to the ever-present possibility of a cable break. Lastly, problem (iv), was the machine sufficiently robust? The original model of the prototype was not, and Maus made modifications to the machine to make it more robust during the two years of trials. This showed the advantage of making trials before committing a machine to a tunnel, because modifications can be made much more readily in the factory than underground.

The second hard rock tunnelling machine to be built was for the Hoosac tunnel in 1851. The machine was designed by Charles Wilson of Springfield, Massachusetts, and constructed by Richard Munn and Co of South Boston.[6] It was tested at the east portal of the tunnel in 1853. The body of the machine consisted of a horizontal frame which slid longitudinally upon a wheeled carriage running on rails. At the front of the machine was mounted a 24 ft diameter short hollow cylinder, around the leading edge of which were fixed a series of rolling disc cutters made of steel and 15 in in diameter. As the cylinder rotated, a 24 ft diameter annular groove 13 in wide was cut into the tunnel face by pressing the discs into the rock and feeding them forward. Additional rolling disc cutters mounted on the axis of the cylinder cut out a small diameter central borehole. After the machine had advanced 2 ft it was intended to withdraw the machine and remove the core of rock that was left in place by blasting or by wedging. Wilson provided simple mechanical arrangements for disposing of the spoil produced when the machine was cutting. During the test at the east portal the machine actually cut rock at the rate of 14–24 in per hour, but the total distance that it cut was only some 25–30 ft. The machine was then abandoned and sold for scrap.[7] As we saw in Chapter 2, the Hoosac Tunnel was finally driven using the drill-and-blast method with compressed air rock drills and nitroglycerine explosive.

We can now consider how this machine coped with the problems previously outlined. Wilson's solution to problems (i) and (ii) was to use the machine to cut a circular groove and central borehole, but to rely on blasting or wedging to remove the bulk of the rock. The machine had to be withdrawn from the face for the miners to be able to do this so that there was the serious disadvantage that continuous operation was not possible.

The Chief Engineer of the Hoosac Tunnel at the time, A.F. Edwards, tried to make a virtue out of necessity by suggesting that the time the core of rock was being broken out could be used for generally maintaining the machine. For the power supply for the machine, problem (iii), steam was used, but as we have already pointed out in Section 2.2 a steam engine could not be operated inside a tunnel nor could steam be piped very far from the portals. Therefore the Wilson machine was impracticable as a tunnelling machine on this score alone. On robustness, problem (iv), the machine's main flaw during the trials were the cutter bearings which could not take the pressure exerted on them.

The Mont Cenis and Hoosac tunnelling machines are of great interest because they are the first two tunnelling machines actually built and tried out. They are different from each other, one being based on the percussive action of chisels and the other on the rotary action of disc cutters, but they both worked on the principle of removing only a slot of rock, leaving the main bulk of the material to be removed by wedging or blasting. In this sense they were not fully developed tunnelling machines because only part of the excavation was mechanised. They would be better described as tunnelling assisting machines. The other factor common to them both is that neither had any provision for mechanised removal of the spoil – this had to be done by men loading the rock into skips as in existing methods of tunnelling. Both were completely impracticable as tunnelling machines because of the unsuitability of their power supply systems – cable and steam – for tunnel operation.

Following the trials at the Mont Cenis and Hoosac tunnels, a great many tunnelling machines were designed and patented, but few were built.[8] Two main problems awaited solution, firstly to find a suitable power supply for a tunnelling machine, and secondly – a problem that had not arisen with the Mont Cenis or Hoosac machines because they had not been used for driving any real length of tunnel – to equip them with tools or cutters suitable for excavating hard rock without suffering excessive wear. We now turn to the Channel Tunnel scheme of 1881 where the first of these problems, but not the second, was solved. But before examining the Channel Tunnel tunnelling machines it is appropriate to say a few words about the Channel Tunnel project itself.[9]

11.2 The Channel Tunnel

The idea for a tunnel under the Dover Strait between France and England can be traced back to Napoleonic times, but the first serious attempts to assess the practicability of such a tunnel can be dated from 1833 when Thomé de Gamond (1807–76), a French engineer began his lifelong examination and promotion of a fixed cross-Channel link.[10] Interest gradually increased, and with the realisation that a continuous stratum of Lower Chalk – which would be ideal for tunnelling – probably existed beneath the Channel from Calais to Dover, cautious plans to build a Channel Tunnel were made. By the 1870s these had reached the stage where a joint Anglo–French venture was envisaged, each country to drive a tunnel from its coast to meet half way across the Dover Strait. In 1872 a British Channel Tunnel company[11] was formed and in 1875 a Parliamentary Bill (The Channel Tunnel Co (Ltd) Act, 1875) was passed allowing investigations and preliminary work to go ahead. The French Assembly passed a similar bill at the same time and a French Channel Tunnel company[12] was also formed. In 1876 a Protocol was signed between Great Britain and France, valid for 20 years, which decided questions of international law posed by the tunnel and laid down conditions under which it should be built and operated.[13]

In 1881 tunnelling work commenced with the sinking of three shafts beside the railway line along the coast between Folkestone and Dover. Tunnels were driven from two of these shafts in a direction more or less parallel with the cliffs, through the Lower Chalk and for some distance under the sea;[14] details of these shafts and tunnels are given in Table 11.1, and the general location of the scheme is shown in Figure 11.1. The two 7 ft diameter tunnels constructed from Shafts Nos 1 and 2 were driven by means of a tunnelling machine that will shortly be described (Section 11.5). The intention was eventually to build two 14 ft diameter tunnels, and another machine was to be constructed to enlarge the 7 ft pilot out to the full 14 ft finished diameter, this machine to follow the first at a suitable interval. When completed, the Channel Tunnel trains were to be worked with compressed air locomotives, the exhaust air from which would be more than adequate to ventilate the tunnels. Mains carrying compressed air at a pressure of $6\frac{1}{2}$ atm were to be provided in both tunnels from which the locomotives could replenish their supply if they inadvertently ran out

Table 11.1. *Location of main English works for 1881 Channel Tunnel scheme*

Shaft No	Shaft depth ft	Shaft location	Tunnel details
1	74	Western end of Abbot's Cliff Tunnel[a]	7 ft diameter 2600 ft length
2	160	Western end of Shakespeare Cliff Tunnel[a]	7 ft diameter 6078 ft length
3	?	East of Shakespeare Cliff	No tunnel driven from this shaft

[a]These were existing tunnels on the Dover to Folkestone railway line.

of air.[15] The whole plan, both for constructing the tunnels and for operating them, seems to be thoroughly workmanlike and it seems certain that the Victorian engineers could have made a success of the Channel Tunnel. Work also commenced in France on the other side of the Dover Strait, for at Sangatte French engineers sank a 280 ft deep shaft from which they drove a pilot tunnel for a distance of over 6000 ft. As part of these works they excavated a working chamber by the conventional drill-and-blast method, but found it was not well suited to tunnelling in chalk because it fractured the rock, making it prone to leaks. However, in May 1882 the British Board of Trade reported that the tunnel could not be made secure from military seizure[16] and all work was stopped. Nothing further happened until 1922–3 when a 12 ft diameter, 480 ft long adit was driven from Abbot's Cliff to join the tunnel that had been driven from Shaft No 1; the adit was driven to regain access to the tunnel, the shaft having been filled in. For this adit drive another tunnelling machine was used which will also be described later (Section 11.6).

During the period from 1882 until 1970 enthusiasm for a Channel Tunnel waxed and waned in proportion to the cordiality of Anglo-French relations.[17] During the years 1957–60 an Anglo–French Channel Tunnel Study Group[18] made a comprehensive site investigation of the proposed Dover to Sangatte crossing, which once again concluded that the scheme was technically feasible, but it was not until 1973 that Great Britain and France signed another treaty to build a Channel Tunnel. For the new scheme three tunnels were planned, two running tunnels of 7.1 m internal diameter, each carrying a single railway track, and a third service tunnel of 4.5 m internal diameter situated beneath and between the two running tunnels.[19] Work commenced on construction of the access area, but in January 1975 when the driving of the service tunnel was just about to commence the British Government again cancelled the scheme, their ostensible reason being the high estimated cost of the associated new rail link from Dover to London that was considered necessary as part of the project.[20] Because the tunnelling machine and all its ancillary equipment had just been installed when the scheme was cancelled, it was decided to drive 260 m of the service tunnel on an experimental basis.[21] As in 1882 the French were dismayed by the British cancellation; they had again started work on their side of the Channel. There can be no doubt that this twice cancelling of a scheme once it was underway by the British Government has given the French a mistrust of the seriousness of our intentions for a Channel Tunnel and they have indicated that they would require guarantees against cancellation in any future scheme.[22] In 1982 interest in a fixed cross-Channel link was again in the ascendant, and plans for a bored tunnel, an immersed tube tunnel, a bridge and all combinations of these were put forward. At the same time the ferry operators and harbour authorities claimed there was no need for a fixed

Figure 11.1 Location of the 1881 Channel Tunnel scheme

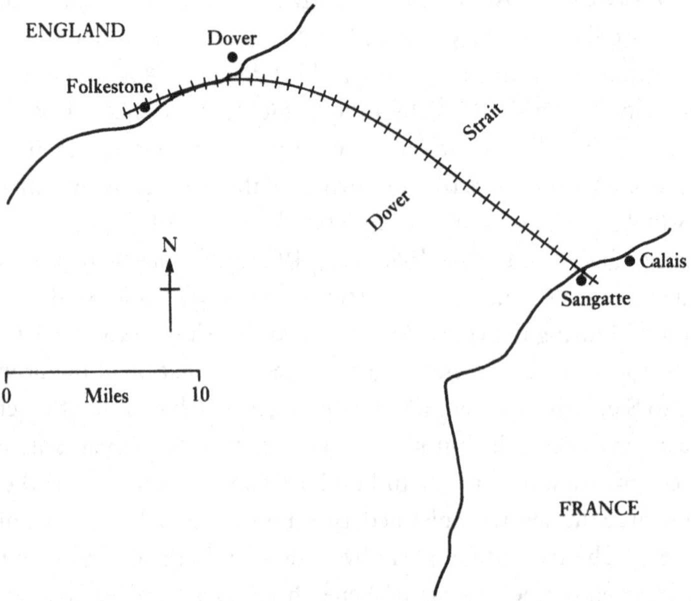

link, all anticipated future traffic being capable of being handled by expansion of the ferry services. The Government considered that a review of these schemes was necessary before a decision could be made and a UK/French Study Group was set up to do this. The Study Group reported that they considered that a twin-tube 7 m tunnel offers economic returns which exceed those normally required for public sector investments, and that of all the alternative schemes the bored tunnel is the only one on which work could be put in hand with the minimum delay.[23] After deliberation, the British Government's response to the Study Group's report was to state that the Government's firm position is that any project would have to be financed entirely without the assistance of public funds and without commercial guarantees by the Government.[24]

In April 1985 the British and French governments produced guidelines for potential Channel link builders and in response to this four different schemes were proposed by four different consortia; in October these schemes were submitted for governmental scrutiny by independent assessors. In December the House of Commons Select Committee on Transport decided in favour of the Channel Tunnel Group's scheme and as the French also favoured this scheme it was announced as the preferred one by the British and French governments jointly in January 1986. During the rest of 1986 plans were made to pass a Bill through Parliament and the technical planning for the works commenced. Concurrently, efforts were made to raise the financial backing for the scheme which, as remarked before, has to come entirely from the private sector. The Channel Tunnel Group's scheme comprises twin running tunnels each of 7.5 m diameter plus a service tunnel of 4.5 m diameter located between them. The tunnels will be 50 km in length and will be bored in the Lower Chalk. The twin running tunnels will each carry a single railway track and will be large enough in diameter to accommodate rail shuttles carrying cars, coaches and lorries in enclosed special wagons as well as the normal British Rail and French Railways mainline trains. At intervals there will be cross passages connecting the running tunnels to the service tunnel. The estimated construction time is 6 years. If this latest scheme comes to fruition, it will do so over a century after the first practical tunnelling trials for a Channel Tunnel had demonstrated its technical feasibility.

The designer of a tunnelling machine for the Channel Tunnel possessed one enormous advantage that it is impossible to overestimate, namely the nature of the rock on site. As we have said, the tunnel was planned to be driven entirely within the Lower Chalk stratum and chalk is an almost perfect tunnelling material. It is soft enough to be cut easily with a penknife (and can even be scratched with a fingernail) while at the same time it is strong enough for an excavation to stand unsupported. The designer was, therefore, freed from two of the greatest headaches of the tunnelling machine designer right up to the present day – that of providing tools suitable for excavating the rock without excessive wear, and that of supporting the tunnel from collapse as soon as it is excavated. There were, of course, the problems enumerated earlier to be overcome and we shall see how this was done, but it is important to realise that the nature of the rock at the Channel Tunnel site presented the tunnelling machine designer with an ideal opportunity to make a tunnelling machine that would work successfully.

11.3 Brunton and Trier's stone dressing machine

Before examining the first Channel Tunnel tunnelling machine it is necessary to look at a stone dressing machine invented by J.D. Brunton and F. Trier. The reason for this is that the first Channel Tunnel machine was designed by Brunton and a study of the stone dressing machine, although it is slightly later, sheds considerable light on the design of his tunnelling machine. Not only this, but a study of the stone dressing machine also shows that Brunton and Trier had grasped some extremely important points about the basic principles of cutting rock with metal tools.

Brunton and Trier's stone dressing machine, they claimed, was based upon an entirely new principle of rock cutting. Most hard rocks, they argued, consisted of particles that were hard enough to cut and wear away even the hardest steel, but these particles were held together by cohesive forces that were far less than the molecular forces in the steel. Because of this, a rock cutting machine should be designed so that attrition of the tools was avoided and the rock was cut by pressure alone. Their method of achieving this was to design a circular rotating cutter which operated by rolling its edge along the surface of the rock and, by pressure alone, detached pieces from it. Noting that previous machines had used chisels to chip or scrape away irregularities of the stone surface they remarked:

> The authors' machines diverge from the beaten path, and take

> hold of a new principle of action; namely of rotating cutters, operating by rolling to chip off from the stone the inequalities of its surface.[25]

So anxious were they that there should be no attrition involved in the action of their machine that they provided a mechanism to turn the cutters, not being content to let them free-wheel on the surface of the stone.

Brunton and Trier's stone dressing machine is shown in Figure 11.2. The lower drawing shows the complete machine and the upper drawing, details of one of the cutters. The cutters were mounted on a circular cast-iron chuck A attached to a horizontal hollow shaft C rotated by the pulley K which was driven by belting. The cutter spindles D were rotated via bevel pinions E by a solid shaft G turned by the pulley H, also belt driven. In this way the chuck was rotated and the individual cutters were rotated independently as well, the rates of cutter rotation and chuck rotation being so adjusted that the cutter edge rolled in its track. For a 2 ft diameter chuck track and 8 in diameter cutters, for every rotation of the chuck, the cutters made three rotations. Ordinarily, the speed of the chuck was 300–350 revolutions per minute and that of the cutters 900–1050. It was found that the best results were obtained when the angle between the edges of the cutter was 70° and the angle of inclination of the cutter to the surface of the stone was 45°. For cutting sandstones, gritstones and moderately hard limestones the cutters were made from chilled cast-iron. The cast-iron cutters were chilled on their outer conical face so that as the cutter wears the cutting edge is always formed against the hard surface produced by chilling the metal. For cutting hard limestones and granites steel cutters were used. Cutters were changed after about 7–8 hours use and were resharpened on a grindstone and used again, taking about 20 grindings before they were finally worn out. This showed that Brunton and Trier were aware that although they had designed a machine which attempted to minimise attrition by the exact roll of the cutters, this ideal was not being realised in practice and some wear of the cutters had to be provided for.

Another feature of the stone dressing machine's design is important. Brunton and Trier remark that the desirable feature of its construction is not so much a question of strength but of massiveness: 'Absolute solidity is required to prevent anything approaching spring in the tool when at

Figure 11.2 Stone dressing machine (1881)

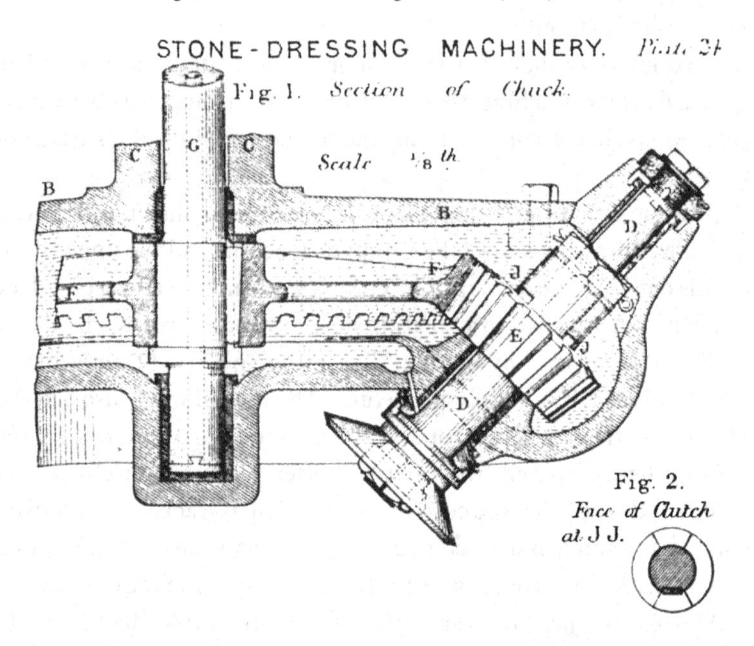

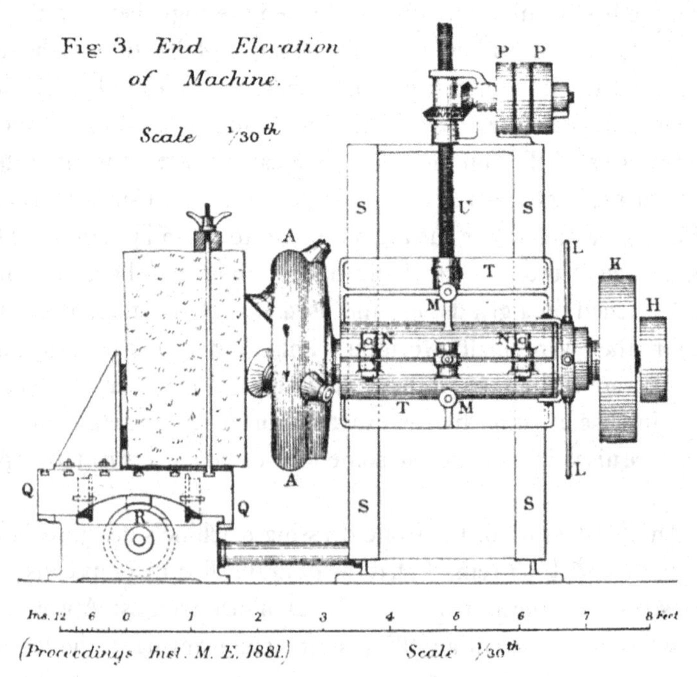

work.' The importance of this observation cannot be overestimated and the point will be taken up again when discussing Brunton's tunnelling machine, and when discussing modern tunnelling machines. What Brunton and Trier had discovered was that a rock cutting machine not only had to be strong, but it had to be *stiff* as well. The reason for this is that if the machine is not stiff, no matter how good the design of all other aspects, the tool is not kept pressed hard against the rock all the time but can spring back and forth to an extent directly proportional to the lack of stiffness of the machine.

Finally it must be remembered that the machine shown in Figure 11.2 was not a tunnelling machine but a stone dressing machine. Thus there was no mechanism for driving the machine forward into the rock; instead the block of stone was moved slowly sideways on the holder Q so that the action of the cutters was to produce a dressed surface. We will now see how the two rock cutting principles embodied in the stone dressing machine's design had also been used in the design of the first Channel Tunnel tunnelling machine, these principles being those of a rolling disc cutter and of a stiff machine.

11.4 Brunton's tunnelling machine

The first tunnelling machine intended to excavate the Channel Tunnel was designed by Brunton and is shown in Figure 11.3. The view is from the rear looking forward. It consisted of a massive iron frame which was held in position in the tunnel by six radial arms fitted with wheels, and was braced against the tunnel walls by two screw jacks.[26] The lower sets of wheels ran on rails placed on the tunnel floor. The cutter head at the front of the machine was made up of two face plates, each of which carried six steel disc cutters 10–20 in in diameter. The face plates were mounted axially on pivots which were attached to either end of a crosshead which was turned by the central drive shaft of the machine. The face plates could be moved a small distance outwards on the crosshead which allowed the correct tunnel diameter to be maintained even if the discs became worn. The cutter head was advanced by means of a massive screw thread cut on the outside surface of a hollow shaft inside of which the central drive shaft ran right through from the rear to the front of the machine. The two screw jacks which braced against the tunnel were attached to a huge cast-iron nut in which the advance screw turned; this position was therefore

Figure 11.3 Brunton's tunnelling machine (1874)

stationary, the machine slowly advancing past it on its wheels as the advance screw rotated. When the screw had reached the end of its travel the jacks were withdrawn and the cast-iron nut was run up to the front. The jacks were then rebraced against the tunnel walls and a fresh length of advance was commenced. The central shaft was rotated via bevel gears, by a large belt-driven pulley which can be seen at the extreme left in Figure 11.3. The hollow outer shaft was also rotated by the large pulley, via the pair of small pulleys and belting that can be seen on the right of the machine – these in turn caused the outer shaft to rotate by means of a worm acting on a large worm wheel attached to the outer shaft. From the compound motions of the cutters and of the advancing screw, the tunnel face was progressively cut, the relative speeds of the outer and inner shafts regulating the rate of advance which was governed by the hardness of the rock to be cut. Power was transmitted to the machine by belting running from a prime mover to the large pulley. No mention was made of what the prime mover was but in the preliminary trials in a quarry this was probably a steam engine; for use in a tunnel it would have been a compressed air engine. The tunnelling machine was also fitted with a mechanism for periodically advancing the rails upon which it ran. In the Figure the debris from the excavation at the face is seen lying in the invert of the tunnel, but Brunton envisaged that in a fully working machine an arrangement of scoops would collect the debris from the invert and dump it into a shute from where it would be removed to the rear of the machine by a belt conveyor. This is exactly how modern tunnelling machines remove the debris from the face.

The first time the Brunton machine was tried seems to be in the period September 1870 to February 1871 when one was tested in a chalk quarry at Snodland, Kent. This was a 7 ft diameter machine and it behaved very satisfactorily, giving advance rates of between 44 and 49 in per hour. In the event, the Brunton machine was never used on the English Channel Tunnel drive, but it was used for a short experimental drive on the French side of the works as will be described in the next Section. Brunton machines, including one of 9 ft diameter, were later tried in North Wales, Westmorland and South Wales but in all these cases the ground was hard rock, with the result that the cutter wear was so great that the machines were withdrawn. This was the end of the Brunton machine, which was, nevertheless, the first practical full-face tunnelling machine that actually drove a section of tunnel.[27]

We can see that Brunton's tunnelling machine employed the same kind of rotating steel disc cutters that were used on the stone dressing machine described in the previous Section. The only difference is that on the tunnelling machine it seems that the individual disc cutters were not turned but allowed to rotate freely. The 'face plates' of the tunnelling machine correspond to the 'chuck' of the stone dressing machine. The tunnelling machine, of course, had to have a mechanism for driving forward and we can note that this was achieved by advancing the cutter head by means of a screw thread – a mechanism that has a high stiffness. In contrast, modern full-face tunnelling machines have a relatively low stiffness because the cutter head is pressed against the face by hydraulic rams and hydraulic systems are not very stiff. Research in the early 1980s has confirmed Brunton and Trier's findings that a high degree of stiffness is a desirable property for a tunnelling machine to possess because this leads to more efficient rock cutting.[28] If future tunnelling machines are designed to have a high stiffness, then we may see a reversion to the screw thread method of advancement that this old machine possessed.

We will now consider how the Brunton machine solved the problems facing the tunnelling machine designer. Problem (i), that of excavating the face mechanically, we have seen was solved by means of steel disc cutters mounted on rotating face plates. Brunton even provided a means of keeping the tunnel cut to gauge when the discs became worn. Problem (ii), that of mechanically removing the spoil was solved by the intended belt conveyor system which remains the modern way of doing this part of the job. Regarding problem (iii), that of providing a power supply, we have noted that Brunton's machine did not have its own engine, but was belt-driven from a prime mover, which would probably have been a compressed air engine, situated a little further back in the tunnel. Problem (iv), to design a machine robust enough for the job is a matter that will be referred to later on when the Sangatte trials are described, but it would seem that in spite of the massive appearance of the Brunton machine there was weakness in several of the main components. It will be recalled that although the Brunton machine was very successful in cutting chalk it was not able to cut hard rocks. This was a shortcoming of the cutters rather than of the machine as a whole and would probably have happened with any tunnelling machine built at that time. The successful way forward with full-face tunnelling machines was to get them to work satisfactorily with a soft rock such as chalk and then progress to harder rocks. We shall now look at the machine that was used to drive the lengths of Channel Tunnel listed in Table 11.1.

11.5 The Beaumont–English tunnelling machine

The tunnelling machine that was actually used for driving the English section of the Channel Tunnel in 1881–2 was designed by two Army officers, Frederick Edward Blackett Beaumont and Thomas English. Beaumont was born on 22 October 1833 and was commissioned as a Second Lieutenant in the Royal Engineers in 1853. After service in the Crimea, India and Canada, Beaumont was employed on the defences of Dover where his interest in tunnelling through chalk may have been initiated. He retired from the Army in 1877 with the rank of Colonel and died on 20 August 1899. English was born in 1843 and was commissioned in the Royal Engineers in 1862. From 1867 to 1884 he was involved in the design and construction of armour plate for defensive works. He retired from the Army in 1889 as Lieutenant-Colonel and then went on to hold a number of posts in industry. During the Great War he was Assistant Inspector of Steel for the Navy. He died on 20 June 1935.

There is some confusion as to which of the two men should be credited with the design of the machine, though at the time it was referred to as the 'Beaumont machine' as it still is in many present-day accounts. The facts are as follows. In 1875 Beaumont was granted a patent[29] for tunnelling machinery. Beaumont's specification described a full-face tunnelling machine specifically 'for driving tunnels or galleries in soft rock, shales or strata such as chalk'. The machine consisted of a rail-mounted carriage carrying a horizontal shaft at the front end of which was a conical cutter head comprising two or more radial arms bearing chisel-shaped steel cutters. At the centre of the cutter head was a conventional drill bit but the cutters were arranged in steps so that by the rotation of the head the rock would be cut away in a series of circular ledges or steps. The main cutter head shaft was rotated by a large driving pulley acting through gearing; no prime mover was specified. The machine was to be advanced by means of a screw spindle, bearing against the carriage and reacting on a fixed abutment placed behind it, but this arrangement is not shown in the

Figure 11.4 The Beaumont–English tunnelling machine (1881)

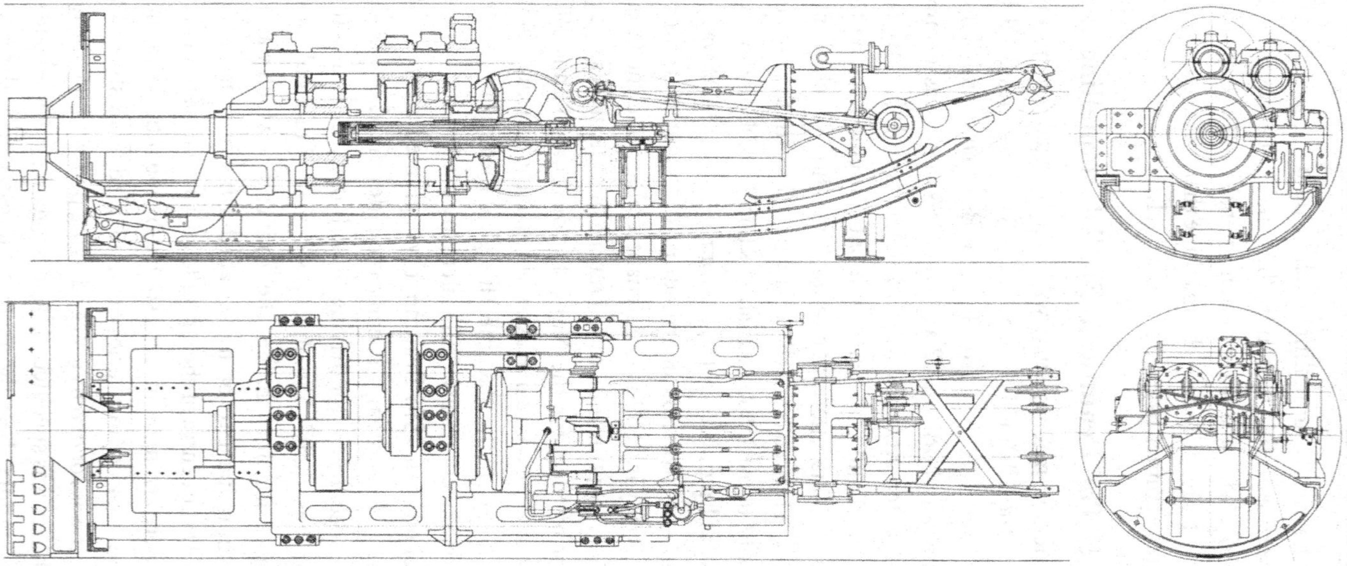

drawings accompanying the patent specification. Debris falling from the cutter head was to be collected on a belt conveyor delivering into trucks further back in the tunnel. It can be seen that the specification is a mixture of the detailed and the general: the cutter head and carriage are specified in some detail, but other important items like the source of power and the method of advancing the machine are only touched upon in general terms. A lot more work would have to be done before such a machine could be built and operated, and as far as is known the original Beaumont tunnelling machine was never manufactured. In 1880 English was granted a patent[30] for improvements in tunnelling machinery. At the outset English refers to Beaumont's patent of 1875 and says 'my invention relates to improvements in machinery of this class'. English's tunnelling machine is shown in Figure 11.4 (redrawn from the original for clarity) and it is apparent that it represents a great advance over that of Beaumont. Furthermore it is a practicable machine in which all the details have been worked out. The tunnelling machine consisted of a massive under frame or bed of trough-like form, curved underneath to fit the bottom of the tunnel. Fitting on this was an upper frame which carried all the machinery and which slid on the under frame. A twin-cylinder compressed air engine was carried on the upper frame and this was supplied with compressed air piped in from outside the tunnel. By means of gearing the engine gave a slow rotation ($1\frac{1}{2}$ revolutions per minute) to the central shaft which carried the boring head at its front end. The boring head consisted of two flat arms carrying a series of chisel-shaped steel cutting tools. The tools are not shown in Figure 11.4, but their sockets are. The debris from the face was collected by a bucket chain which was powered by the engine. The bucket chain discharged into rail-borne trucks further back in the tunnel. A sloping mouth in the cutter arms directed the spoil into the buckets but platforms were provided on either side for workmen to stand on to shovel debris into the buckets as well. The engine also drove a hydraulic pump, the stroke of which could be varied. The main shaft of the machine was tubular and its interior formed a hydraulic cylinder fitted with a double-acting piston. This hydraulic cylinder was supplied by the hydraulic pump and the piston rod was connected to the under frame by means of a projecting bracket. There were two cycles of operation. In the first, during normal cutting, the piston reacted on the under frame which was stationary, and the main shaft and boring head together with the whole of the upper frame was carried slowly forward while the head rotated. When the full advance of the piston had taken place the machine was stopped and the second mode of operation took place. The whole of the upper frame was raised on three jacks and at the same time the under frame was also lifted just clear of the tunnel floor. The piston was now made to reverse direction; it now reacted on the upper frame and moved the under frame forward. When this had been completed the jacks were retracted and the upper frame was lowered on to the under frame and the latter allowed to descend onto the tunnel floor once more. Boring was then resumed as described above. An iron ring was provided up near the boring head to brace the under frame firmly against the tunnel during the cutting cycle. The machine was operated by a crew of seven. It is English's machine that was used for the Channel Tunnel drives in 1881.

From the foregoing it can be seen that what English did in his patent specification of 1880 was to take Beaumont's idea and turn it into a practicable tunnelling machine. He also provided a simpler and probably more effective cutting head. So, whilst Beaumont proposed the original idea, English turned it into a practical proposition. Because of this, it will be referred to here as the Beaumont–English tunnelling machine.

The Beaumont–English machine solved problem (i) with a very simple double-arm cutting head equipped with steel chisel-shaped tools. Problem (ii) was solved by sloping mouths in the arms which channelled the debris onto a bucket conveyor which transported it back to trucks at the rear of the machine. A compressed air engine was the solution to problem (iii) and it can be noted that on this tunnelling machine the single engine provides power for all the operations of the machine: it turns the boring head, drives the hydraulic pump for thrusting forward, and drives the bucket conveyor. The solution to robustness, problem (iv), was to provide a massive under frame which rested directly on the tunnel floor and to brace the under frame by means of an iron ring close up by the cutting head. We may note that men had to assist in loading the bucket conveyor indicating improvements could be made in the arrangements for transporting the spoil away from the face. Whilst Brunton's machine was the first practical full-face tunnelling machine, the importance of the Beaumont–English machine is that it is the first practical full-face tunnelling machine that was used to drive significantly long lengths of tunnel, totalling over $1\frac{1}{2}$ miles (see Table 11.1). Because of this it has a place in tunnelling history, and we can note there is a fine one-tenth scale model of the Beaumont–English tunnelling machine in Gallery 9A of the Science Museum, South Kensington, London.

After construction of their shaft at Sangatte, the French wished to see how well a tunnelling machine would work and obtained permission to try both the Brunton and the Beaumont–English machines. One of each was acquired and they were both installed down in the Sangatte shaft and commenced driving test headings from it. The Beaumont–English tunnelling machine used for the Sangatte trials, built by the French firm of locomotive manufacturers Société de Construction des Batignolles, is shown in Figure 11.5. The trials took place between February 1882 and March 1883.[31] The Brunton machine was found unsatisfactory because of the weakness of several of its main components, and failure of these led to the machine being dismantled and withdrawn from the trials. The Beaumont–English machine, however, achieved an average advance rate of 50 ft per day with a maximum of 81 ft per day being recorded, and in all the machine excavated 6036 ft of tunnel. Apart from a sudden inrush of water, which was nothing to do with the machine, no problems were encountered during the drive. The Sangatte trials confirmed that the Beaumont–English tunnelling machine was a thoroughly practical proposition, over 1 mile of tunnel having been driven without any difficulties coming to light. In fact in all, over 2¾ miles of tunnel were driven with the machine – some nine per cent of the total length of the proposed tunnel.

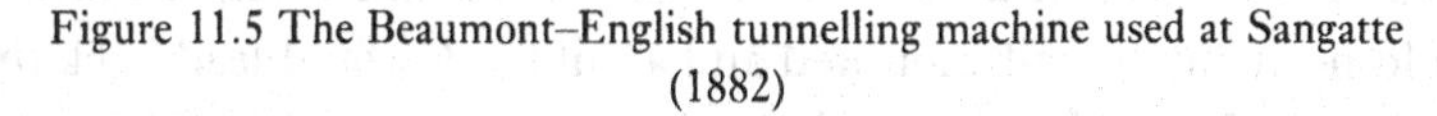

Figure 11.5 The Beaumont–English tunnelling machine used at Sangatte (1882)

It is of interest to note that the Beaumont–English machine went on to further success when in 1883 one was installed at Birkenhead and commenced driving a 7 ft diameter ventilation tunnel for the Mersey Railway Co's tunnel under the River Mersey. In driving this tunnel from Birkenhead to Liverpool, a distance of 6750 ft, the machine achieved advance rates of 192 ft per week. The success of this second use of the Beaumont–English machine was again due to a suitable rock being chosen for the machine to cut. On the Mersey Railway Tunnel the rock was Bunter Sandstone, which although abrasive, is of only moderate strength so that the chisel-shaped tools of the machine were able to cut into it and excavate it satisfactorily although there must have been frequent tool changes due to abrasive wear with this rock. Although the drive was a success, sadly nearly all the miners who operated the machine on this job died not many years later from silicosis resulting from inhaling the quartz-rich rock dust produced by the machine.[32] (On later machines this problem was quite simply solved by fitting water sprays which suppressed the dust.) Like the Lower Chalk, the Bunter Sandstone was able to stand unsupported in the heading while the tunnelling machine was driving through it.

Before leaving the 1881 Channel Tunnel machines there is one matter that requires further discussion. It will be recalled that problem (ii) which faced the tunnelling machine designer was to devise a mechanical means of removing the spoil from the excavation and we have seen that the designers of all the machines we have reviewed used a belt conveyor of some sort which carried the spoil away from the face and tipped it into muck cars which ran up to the rear of the machine. These muck cars would normally be rail-mounted and a rail track would lead back to the tunnel portal or to the nearest shaft where the spoil would be disposed of. This arrangement works very well for tunnels of normal length but for the Channel Tunnel these muck trains would eventually have had to make a journey of 15 miles. This posed a problem that contemporary commentators on the Channel Tunnel scheme were quick to notice.[33] This difficulty led T.R. Crampton to suggest that water should be piped to the tunnel face, mixed with the chalk debris in the ratio of three parts water to one part chalk, and the resulting slurry could then be pumped out to the portal via another pipe.[34] The two pipes would occupy very little space in the tunnel and the whole process would be continuous so long as the tunnelling machine was cutting. This was a neat solution to the problem, and an early suggestion of the hydraulic transport of tunnel spoil which presages part of the systems later used in the pressurised face tunnelling machines described in Chapter 9. Crampton also suggested that the same water supply could be used to drive the tunnelling machine if it were fitted with a hydraulic motor – thereby also giving a neat solution to problem (iii), that of providing a power supply to the machine.

11.6 Whitaker's tunnelling machine

The third tunnelling machine associated with the Channel Tunnel was that designed by Douglas Whitaker and constructed by Sir William Arrol and Co Ltd, Dalmarnock Ironworks, Glasgow, Scotland. The machine had its origin in the Great War as part of a plan to overcome the stalemate on the Western Front. Once the more-or-less static system of trenches had been established and it was realised that even massive artillery barrages could not achieve a breakthrough, special companies of sappers were formed to tunnel beneath the enemy lines and explode huge mines under their positions which would destroy from below what could not be destroyed from above.[35] These tunnels were excavated by hand and the rectangular headings were supported with timbering. To speed up these operations the War Office commissioned the design of the Whitaker machine, but in the event the Armistice came before the tunnelling machine could be used in action, although one was built and was undergoing trials in Great Britain when hostilities ceased.

The Whitaker tunnelling machine is shown in Figure 11.6. Machines varied in size between 7 and 12 ft in diameter.[36] The following account refers to a 12 ft diameter machine which was the size of machine used at the Channel Tunnel site; the Figure is also of a 12 ft diameter machine. From the contemporary photographs it seems that the smaller-diameter machines could be fitted with a shield if required for soft ground tunnelling. The cutter head at the front of the machine consisted of a $6\frac{1}{2}$ ft diameter cast-steel centre piece carrying a central cutter, with steel buckets fitted to the periphery to make up the complete diameter of 12 ft. Each of the five arms was fitted with a renewable chisel-type cutter set at a different radius so that different parts of the face were covered. The five buckets were also fitted with cutters at different radii and with gauge cutters to trim the tunnel profile. Because of the original military purpose

Figure 11.6 Whitaker's tunnelling machine (1922)

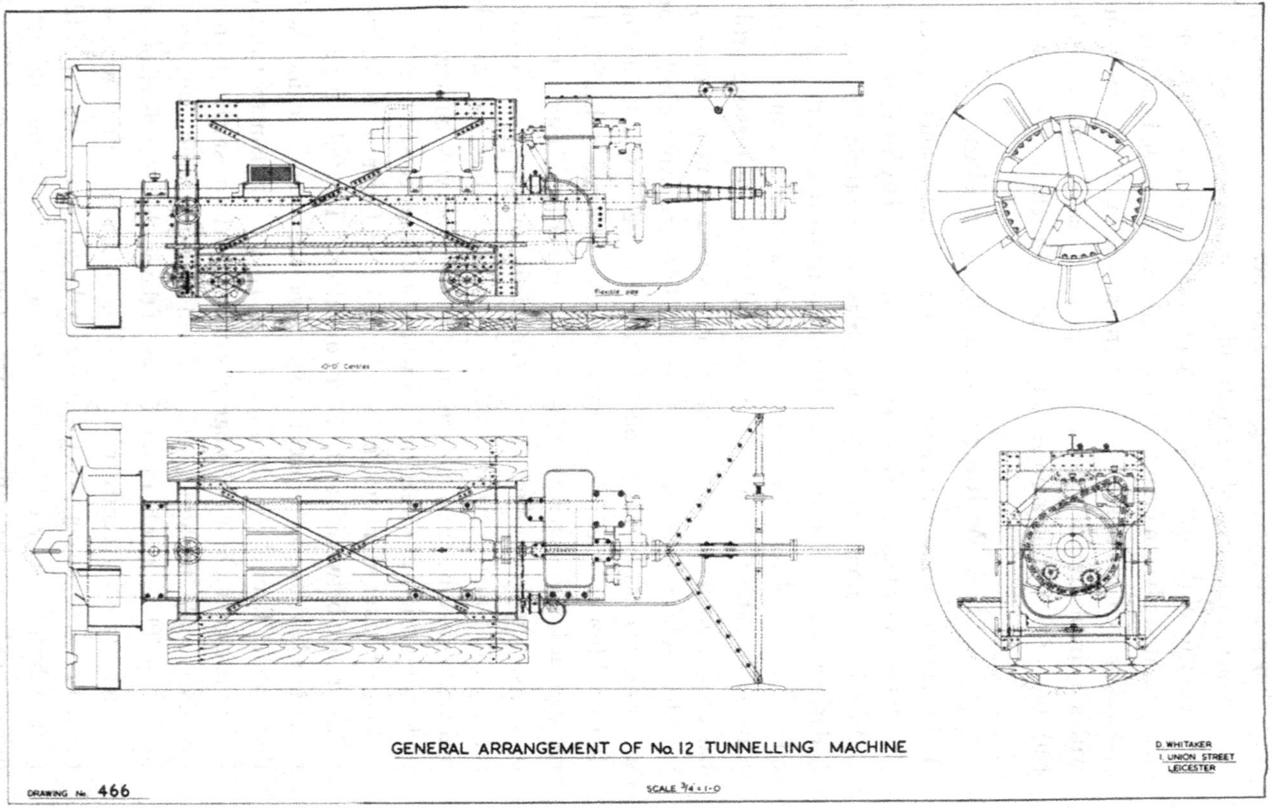

GENERAL ARRANGEMENT OF No. 12 TUNNELLING MACHINE

SCALE 3/4" = 1-0

DRAWING No. 466

D. WHITAKER
1 UNION STREET
LEICESTER

for which the tunnelling machine was designed, by taking off the buckets the machine could be withdrawn from the heading and taken back out of the tunnel without further dismantling. The whole machine was mounted in a rail-borne carriage which was moved forward by a hydraulic ram acting directly on the cutter shaft. The ram reacted against a large stretcher crosshead fitted with feet which was jacked against the tunnel walls. When the machine had advanced a distance equal to the full stroke of the ram the stretcher was collapsed and repositioned further forward ready for another advance. A single electric motor of 120 hp rotated the cutter head and provided power for the whole machine. Electric current at 500 V DC was supplied by a steam driven generator outside the tunnel. The hydraulic pump that actuated the ram was driven from the motor shaft by a light roller chain; the pump itself was of variable stroke so that the forward travel could be regulated. The spoil was carried to the rear of the machine by a pair of spiral screw conveyors (one is shown by broken line in the Figure) which were turned by chain drive from the motor shaft. The motor rotated the main cutter shaft through a gearbox and by means of a heavy roller chain acting on a large toothed wheel mounted on the main cutter shaft. The spiral screw conveyors discharged the spoil onto a belt conveyor (not shown in the Figure) which in turn lifted it, carried it to the rear and dumped it into waggons running on a 2 ft gauge twin-track manual tramway. The belt conveyor was powered by its own electric motor, but moved forward on wheels with the tunnelling machine. An overhead joist projecting from the rear of the tunnelling machine carried a travelling hoist for supporting the hydraulic ram mechanism. The machine commenced driving the adit at Abbot's Cliff in June 1922 and an average driving rate of 9 ft per hour was achieved. Work stopped in September 1923 when the 480 ft long tunnel had been completed. The machine was withdrawn to the portal where a shed was built to protect it. The shed has long since been demolished but the rear of the machine and the joist can still be seen projecting from the half-collapsed tunnel portal.

Turning again to the problems facing the tunnelling machine designer, problem (i) was solved in this machine by a rotary cutter head consisting of renewable cutters and buckets. The cutter head of the Whitaker machine, with its small cutting chisels and large buckets, is more like the cutter head of a soft ground tunnelling machine than that of a hard rock machine. In fact it is doubtful if the Whitaker machine with this kind of head would be able to excavate rock harder than chalk. To solve part of problem (ii) Whitaker provided a spiral screw conveyor to carry the spoil from the front of the machine to the rear. This innovation does not seem to have been taken up by tunnelling machine designers in subsequent machines until we see it again in the Japanese earth pressure balanced shield in the 1970s (see Section 9.6). The solution to problem (iii) was an electric motor which represented a new power source, the previous machines having been powered by compressed air. Heavy industrial electric motors had been available since the end of the nineteenth century and their use to power a tunnelling machine was an obvious application. In fact, as we shall see in Chapter 12, the electric motor had been used much earlier to drive soft ground tunnelling machines. The advantages of electric motors in a tunnel are obvious – they are clean, have no exhaust or heat problems and are easily supplied with power by means of cables which occupy little space in the completed section of tunnel. Electric motors became the standard method of powering tunnelling machines henceforth, sometimes by direct drive as in the Whitaker machine and sometimes indirectly by driving hydraulic power supply systems which in turn powered hydraulic motors. As regards problem (iv), the machine was clearly robust enough for the Abbot's Cliff adit drive although to modern eyes the stretcher crosshead which provided the reaction for the hydraulic ram looks very flimsy and not nearly as robust as the reaction jacks on modern machines or even those on the much earlier Brunton machine.

Table 11.2 summarises the occasions of use and compares the performance in terms of advance rate, for the machines used for the Channel Tunnel. It can be seen that the Brunton and the Beaumont–English machines of 1870–80 had similar advance rates of 4 and 3 ft per hour. The Whitaker machine of 1922 had an advance rate of 9 ft per hour. Taking the total proposed tunnel length of 30 miles and assuming drives from the English and French sides to have taken place simultaneously, then a pilot tunnel could have been bored with Beaumont–English machines in three years of continuous working, and with Whitaker machines in one year of continous working. In practice, of course, machine utilisation would have been much less than 100 per cent and these times would have therefore been longer. But even so, they show that a pilot tunnel could have been driven in a reasonable period of operation. For interest, the advance rate of the Priestley machine used for the 1975

Table 11.2. *The Channel Tunnel tunnelling machines*

Date	Machine	Diameter ft	Average advance rate ft/hour	Location
1870–1	Brunton	7	4	Snodland
1882–3	Brunton	7	Not known	Sangatte
1882–3	Beaumont–English	7	3	Sangatte
1881–2	Beaumont–English	7	3	Shakespeare Cliff No 2 Shaft
1881–2	Beaumont–English	7	Not known	Abbot's Cliff No 1 Shaft
1922–3	Whitaker	12	9	Abbot's Cliff adit
1975	Priestley (picks)	17.8	13.5	Shakespeare Cliff[a]
1975	Robbins (discs)	16.2	—	Not used

[a]The 1975 Channel Tunnel works were on the same site as the 1881 No 2 Shaft.

works is shown below the dashed line in the Table; the Robbins machine for the project was delivered to the French working site and assembled, but unfortunately, was not used before the scheme was cancelled so that no performance figure can be given. A comparison between the two would have been instructive since one was equipped with picks and the other discs as shown in Figures 11.7 and 11.8.

Folllowing the 1922–3 use of the Whitaker tunnelling machine there was little or no further development of full-face tunnelling machines for use in hard rock. The reasons for this are undoubtedly the earlier failures of tunnelling machines and the realisation that those few that had been successful had been restricted to relatively soft or weak rocks, coupled with the fact that an extremely effective method of hard rock tunnelling already existed in the drill-and-blast system. This dormant state of affairs continued for the next 30 years until developments which took place in the United States after the Second World War. The United States had long enjoyed a well-deserved reputation for mechanical engineering, and American capability in this field was enormously enhanced under the impetus of war production during the years 1940–5 when the United States became the 'arsenal of democracy'. American industry emerged from the Second World War with much improvement having been made in mechanical engineering. In the 1950s resurgence of the need for tunnels in the United States together with the realisation that considerable progress had been made in mechanical engineering led to the belief that a successful full-face hard rock tunnelling machine was now a practicable proposition. It is to the United States, therefore, that we must go for the next episode in the story of the development of the hard rock tunnelling machine, and in particular to the Oahe Dam.

11.7 The Oahe Dam tunnelling machines

The Oahe Dam project,[37] constructed in South Dakota, USA, between 1950 and 1960 by the US Corps of Engineers, involved the construction of seven diversion tunnels and five power tunnels. To drive these tunnels a number of full-face tunnelling machines were designed by the firm of James S. Robbins and Associates of Seattle, Washington. Two of the machines were of large size. The first machine, Model 910, was of 25 ft 9 in diameter and had a cutting head which consisted of two contra-rotating units. The inner section of the head was a rotating circular plate on which were mounted three radial cutting arms, while the outer section of the

Figure 11.7 Priestley tunnelling machine used on the Channel Tunnel works (1975)

head was fitted with six short cutting arms and rotated in the opposite direction to the inner section. Each arm, on both inner and outer sections of the head, carried fixed tungsten carbide pick cutters. Set behind these cutters was a row of rolling steel disc cutters which were positioned so that each disc was midway between a pair of picks. The idea behind this complex arrangement was that the fixed pick cutters were intended to cut the rock, while the rolling disc cutters were merely to break off the upstanding ridges of material left after the picks had made their cuts. The head was driven by two 200 hp electric motors whilst a 25 hp motor moved the tunnelling machine forward hydraulically. The rock at Oahe was a shale and the machine achieved an average advance rate of 50 ft per day while actual advances of 140 ft in one day and 635 ft in one week were recorded. It drove four of the tunnels on the Oahe Dam project and must

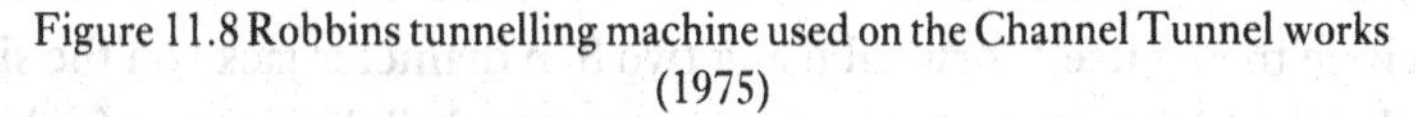

Figure 11.8 Robbins tunnelling machine used on the Channel Tunnel works (1975)

be considered the first modern hard rock full-face tunnelling machine. The only major problem encountered was the high replacement rate of the picks.

The second large-diameter full-face tunnelling machine designed by Robbins for the Oahe Dam tunnels was of 29 ft 6 in diameter and was built by the Goodman Manufacturing Co of Chicago, Illinois; it was designated Model 351.[38] Figure 11.9 shows the tunnelling machine in the erection shop. The head consisted of a single rotating unit fitted with pairs of fixed pick cutters behind which were mounted single rolling disc cutters, the cutters being set out in radial lines as shown in the Figure. Around the circumference of the head were fitted eight buckets which also carried cutters themselves. The spoil was picked up by the revolving buckets and deposited into a conveyor hopper at the top. The tunnel was supported by circular steel ribs set at intervals of 4 ft. The tunnelling machine thrust itself forward using two 8 in diameter jacks on the side at axis level which pressed against the last-erected rib by means of a flexible shoe. Rib stretchers were fixed at axis level to distribute the jacking force

Figure 11.9 Robbins tunnelling machine Model 351 (1960)

back along the previously erected ribs. The tunnelling machine was supplied with electricity at 2300 V, the total power requirement being 600 kW, and the head was rotated by eight electric motors positioned on a 12 ft diameter circle. The average advance rate of the machine was 42 ft per day with best advance rates of 93 ft per day and 503 ft per week recorded. The main problem encountered was rockfalls. When the bedding planes of the shale dipped down towards the face, pieces of rock fell into the buckets before being cut so that a cavity grew ahead of the face quicker than the machine was advancing. This showed up a shortcoming of the design of this particular machine, namely the large unsupported space between the buckets which is well seen in Figure 11.9.

Returning to the problem of the high rate of pick wear experienced on the first Oahe Dam tunnelling machine, on a later Robbins machine the same problem was encountered again and in desperation all the picks were removed and the machine was tried with discs alone. It was found that the machine worked better like this and, thereafter, disc cutters used by themselves have become the standard tools on the heads of hard rock full-face tunnelling machines. Later Robbins machines also abandoned the contra-rotating head that was used on the first Oahe Dam machine in favour of a single rotary head as used on the second Oahe machine and this too has become normal practice. Modern machines all have some kind of scoop or bucket arrangement around the circumference of the head to pick up the spoil, carry it to the top of the machine and deposit it onto a belt conveyor which removes it to muck wagons at the rear. This method still sometimes gives trouble as it did with the second Oahe machine, the problem being that if the opening between the buckets is too large there can be rockfalls but if the opening is made too small it can become jammed with blocks of rock, so that trying to overcome one difficulty creates another. Gripper pads which can be hydraulically expanded tight against the tunnel wall are now the usual method for providing both thrust and torque reaction for a tunnelling machine, and the method of jacking off the rib supports described above is no longer generally used. In this respect modern machines are very much like the old machines in principle.

It can be noted that on the Oahe machines, as on nearly all subsequent tunnelling machines fitted with rolling disc cutters, in order to increase the forward pressure on the rock surface the cutters were fitted normal to the rock face and not at an angle as on Brunton and Trier's stone dressing machine and on Brunton's tunnelling machine. It can also be remarked that Brunton and Trier's concern that a cutter should roll on the surface of the rock rather than scrape along it has been fully vindicated, because it has been often observed that if the bearing of a rolling disc cutter jams or seizes causing the disc to stall and cease to rotate, then the rate of wear of the disc becomes so great that it is quickly damaged beyond repair.

It is instructive to consider briefly the process of excavating hard rock with rolling disc cutters. As the cutter head rotates, each cutter crushes a narrow path of rock immediately beneath its rolling edge. With successive turns the penetration of the disc becomes great enough to cause a shear or tensile failure in the rock and a crack then propagates across to the next cutter path, dislodging a platy chip of rock from the face. The process is at its most efficient if the penetration for a single pass is great enough to do this, and as with percussive rock drilling discussed in the Introduction to Chapter 4, the higher the proportion of rock chips and the smaller the proportion of crushed rock dust, the more effective is the excavation process. In order to increase the depth of penetration, the edge angle of cutter should theoretically be very acute but if it is made too acute it is then unable to sustain the loads imposed upon it during cutting: an edge angle is therefore chosen which is a compromise between depth of penetration and strength of the steel. The process explains, among other things, why it is important that the tunnelling machine, particularly its cutter head, should be stiff, for if it is not, some of the thrust force will be absorbed by elastic deformation of the mechanism instead of being used to make the disc cutter penetrate into the rock.

11.8 Later developments

The tunnelling machines for the Oahe Dam tunnels and other machines built soon after by the Robbins Co demonstrated that successful hard rock full-face tunnelling machines could be made and established the basic lines on which later machines were to develop. Soon hard rock full-face tunnelling machines were in manufacture by a number of other American engineering firms and also by manufacturers in Europe, notably Germany; these developments have been described in detail elsewhere.[39] However, there are two aspects of machine design that have emerged that will be commented on.

The first is extremely important and concerns the trouble that full-face tunnelling machines have encountered when they run into broken rock. It will be recalled that hard rock full-face tunnelling machines do not have any means of supporting the ground and rely upon the rock in the heading being self-supporting over some distance until the machine has moved well forward and some support can be erected. If this is not so and the rock is in a broken condition, usually because of a fault zone, a shatter zone or a region of open jointing, then blocks of rock fall out of the tunnel roof and sides behind the head of the machine, jam it and bring it to a halt. This is exemplified by the experience with a tunnelling machine at Dawdon Colliery in County Durham. Here, a Thyssen FLP/35 full-face machine was used to drive two 3.65 m diameter exploratory roadways to establish the area of coal reserves offshore from the colliery. The intended length of the roadways was 3000 m but during drivage of the first, after 1040 m had been achieved the machine entered a major faulted area. The machine drove another 40 m through this highly disturbed ground but took 10 weeks to do so. After this the machine was removed.[40] On the second roadway, the machine drove only 183 m of tunnel before faulted ground was again encountered which stopped the machine. Once again the machine could not cope with this and was removed. With the Thyssen FLP/35 machine the first place that support for the tunnel in the form of wire mesh and steel rings could be installed was 5m back from the head. This was too great a length of heading to stand unsupported in broken rock and was the main cause of the machine being brought to a stop in both roadways. One of the main conclusions in a report[41] on the second roadway drive was that any future tunnelling machines should have provision for support to be put up within 1 m of the tunnel face. We shall see later how this advice was followed, at least in principle, for the next full-face tunnelling machine made to operate in a colliery had provision for erecting support close behind the cutter head, though not as close as 1 m which proved to be too small a distance to be practicable.

The second aspect of machine design that has received attention concerns the tools. It will be recalled that the effectiveness of disc cutters used alone was soon established and that these are now the standard tools for hard rock full-face tunnelling machines. For very strong or abrasive rocks, however, the rapid wear of the disc cutters is a problem since the tools themselves are expensive and there is also the cost of stopping the machine while they are replaced. The cutter discs themselves are wheels of about 350 mm diameter with a tapered sharp edge and are manufactured from hard steel. Hard steel has proved very suitable for making the disc cutters used on tunnelling machines engaged on excavating tunnels in rocks of medium to hard strength and abrasiveness. However, an attempt was made to drive a tunnel in quartzite using a full-face machine.[42] Quartzite is probably one of the hardest rocks to tunnel through, being of very high strength *and* very high abrasiveness. Cutter wear was extremely high and to try to combat this, special disc cutters were fitted which had tungsten carbide buttons fitted all round the periphery of the discs. These buttons were similar to those fitted to the drill steel bit shown in Figure 4.2. These button discs were an improvement over plain steel discs but not as successful as was hoped. The problem has not been solved and it would seem likely that future tunnels in quartzite will be driven by the drill-and-blast method. However, it would seem possible that the idea of a sharp-edged insert like that fitted to the drill steel bit shown in Figure 4.1 could well be applied to disc cutters and that this would be worth a try.

11.9 Roadheaders

While the developments on tunnelling machines that have just been described were taking place, an entirely different type of tunnelling machine was being developed independently in the coal-mining industry. During the underground extension of the workings of a colliery, permanent development tunnels are driven to provide access for the miners to different parts of the mine, to take in machinery and materials and to take out the coal. Before 1964 in Great Britain these development tunnels were driven by drill-and-blast methods and lined with steel arches and timber lagging. In 1964 the first machine for driving development tunnels was introduced into British coal mines: it was called a 'roadheader' because the development tunnels are called roadways or roads. By 1973 the National Coal Board had over 300 roadheaders.[43] A typical roadheader is shown in Figure 11.10. It consists of a chassis mounted on caterpillar tracks carrying at the front a massive boom on the end of which is mounted a cutting head fitted with picks. The cutting head rotates on the end of the boom and the boom can be moved both horizontally and vertically by hydraulic cylinders. The tunnel face is excavated by first making a horizontal cut across the bottom of the face – an operation referred to as 'sumping'. The rest of the face is then excavated by successively raising the boom vertically and making other horizontal cuts – referred to as 'slewing'. During sumping the cutter head is engaged in true rock cutting, but during slewing the rock often breaks out along existing joints and bedding planes so that less actual rock cutting by the head is done than when sumping. The spoil from the face falls to the floor of the tunnel where it is gathered on an apron at the front of the machine by a scraper-chain flight conveyor which also carries it to the rear of the roadheader and transfers it to a belt conveyor for removal down the roadway. The roadheader is powered electro-hydraulically and the whole operation of the machine is controlled by one man.[44] One great advantage of the roadheader over other kinds of tunnelling machines is that there is plenty of room around and above it to allow erection of the traditional colliery lining of steel arches and timber lagging (well shown in Figure 11.10).

The roadheader seems to have been used for the first time in 1932 in the coal mines of the Podmoskovnii basin in the Soviet Union where a

Figure 11.10 Dosco Mark 2A roadheader (1974)

machine designated the PK-1 was introduced. By the 1950s the development of roadheaders in the Soviet Union had led to the production of the PK-3 and, later, the PK-3R machines. By 1976 more than 1000 roadheaders of all types were in use in Russian coal mines.[45] In 1961 engineers from the National Coal Board went to the Soviet Union to look at the Russian roadheaders in action. As a result, a PK-3 roadheader was brought to England and given a trial at Ellington Colliery, Northumberland. In 1962–3 the Board built two roadheaders based on the design of the PK-3 and in 1963–4 roadheaders of similar type were in manufacture by a number of British engineering firms.

Having established themselves as a success in driving roadways in the coal-mining industry, roadheaders were soon taken up to drive tunnels by the tunnelling industry. One of the first tunnel schemes on which roadheaders were used was the Liverpool Loop and Link Tunnels. These tunnels were constructed between 1972 and 1977 to give travellers from Merseyside's main commuter areas fast rail access to Liverpool's business and shopping centres and to the inter-city railway station.[46] The 5 km long, 4.7 m internal diameter tunnels were originally intended to be driven through the medium strength sandstones that are present on this site by means of full-face tunnelling machines. In the event, the contractors chose to use roadheaders instead, the reason for this being that they considered that some of the lengths of drive were too short for a full-face machine, that the roadheader offered flexibility when it came to excavating the enlarged station sections of tunnel and that the roadheader allowed easy access to the face for erection of temporary support should it be required. The Dosco Mark 2A roadheader was chosen because of its good record in driving colliery roadways. The average rate of advance was 48 m per week with a best advance rate of 69 m per week.[47]

Roadheaders are now well established as an alternative to either drill-and-blast tunnelling or full-face tunnelling machines in the tunnelling industry. They are versatile and are at their best when excavating strata in which they can exploit existing bedding planes and joints, but they are not suitable for tunnelling in massive or very hard rocks. A type of tunnelling machine made by mounting a roadheader boom inside a shield will be described later, in Chapter 13. In the next Section comparison between a roadheader and a full-face tunnelling machine is made which highlights the advantages of each type.

11.10 The Selby Tunnels

Tunnel engineers have argued the relative merits of full-face tunnelling machines and roadheaders but while there have been instances of roughly similar drives, there has been no case until 1981 of drives by a full-face machine and a roadheader under identical conditions. The driving of two development tunnels at Selby has provided such a case. The Selby coalfield consists of an area of 110 square miles situated between Selby and York in North Yorkshire. The coalfield is a new one and the strategy that has been planned for developing it is to drive two main development tunnels, each 13.5 km long, from south-west to north-east through the heart of the coalfield. These will provide arterial feeder roadways at a depth of 70 m below the Barnsley Seam, which is the main coal seam to be worked. After being mined, the coal will be passed down a series of shafts to the roadways which will house belt conveyors that will carry the coal out of the coalfield and up to the surface via inclined adits at their ends.[48] The first step in opening up the coalfield was the driving of the development tunnels. Driving of the first, the North Tunnel, commenced in July 1980 using a Dosco Mark 3 roadheader which was replaced in August 1981 by a Titan 134 roadheader, and driving of the second, the South Tunnel, commenced in October 1981 using a Robbins full-face tunnelling machine. By November 1982 the roadheaders had driven the North Tunnel to chainage 5600 m, but the full-face machine, in spite of its much later start, had overtaken the roadheader and driven the South Tunnel to chainage 5800 m. Details of these two drives are given in Table 11.3. The tunnels are side by side, only 70 m apart and pass through an identical succession of Coal Measures strata, mainly mudstones, siltstones and sandstones. The size of the tunnels and the support are comparable, so that the only significant difference between the two is the method of excavation. Table 11.3 shows the superior performance of the full-face tunnelling machine – its rates of advance are about twice that of the roadheader. The conclusion is that for long lengths of drive in competent rock the full-face tunnelling machine will give a much faster rate of advance than the roadheader. This is not to say that it will always be the better choice. For short drives, of say less than 1 km, the roadheader will be the better economic choice because of its much lower capital cost and faster starting time. Also, where a non-circular or complex tunnel

Table 11.3. *Comparative performance of machines in the Selby Tunnels*

	Drive	
Details	North Tunnel	South Tunnel
Machine	Dosco Mark 3 roadheader replaced later by Titan 134 roadheader	Robbins full-face tunnelling machine
Shape and size of tunnels	Horseshoe 5.0 m wide, 3.7 m high	Circular 5.8 m diameter
Support	Wire mesh with steel rib arches at 1 m spacing	Wire mesh with steel ring beams at 1 m spacing
Best weekly advance	Dosco: 74 m Titan: 72 m	150 m
Average weekly advance	Dosco: 60 m Titan: 50 m	100 m

profile is required, as for station tunnels for example, obviously the roadheader will be the more suitable machine to use.

Another aspect of the Selby tunnels is worth commenting on. The North Tunnel was constructed with a horseshoe-shaped profile and the support was wire mesh and steel rib arches. The feet of the arches rested on wooden blocks placed on the flat rock surface which comprised the invert of the tunnel. The South Tunnel was, of necessity circular in profile because it was cut with the full-face machine, and the support was wire mesh and circular steel ring beams. The ring beams went all the way round the profile. Both tunnels are some 400 m below the ground surface and the stress on the openings at this depth is considerable. Already, the invert of the North tunnel has heaved in places, so much so that recutting of the invert has had to be carried out, but the South Tunnel with its circular profile and its all-round support has remained stable. It may well be, however, that if the invert of the North Tunnel had been fitted with a slightly curved steel beam, strutted across and bolted to the feet of the steel arch ribs, the floor heave would have been prevented or reduced. Be that as it may, the Selby Tunnels have demonstrated, under identical conditions, the superior stability of the fully supported circular cross section to the traditional horseshoe-shaped cross section with an unsupported floor.

The Robbins machine designed for use at Selby[49] benefited from the experience gained with the tunnelling machine used at Dawdon Colliery described in Section 11.8. In particular, the Robbins machine was designed so that the heading could be more adequately supported. This was done by a cutter head support which extended 3 m back from the face. Directly behind this was a canopy made from trailing head bars which was a further 3 m long. The wire mesh and steel rings were erected beneath the canopy close up behind the cutter head support. The rings were finally expanded tight against the tunnel wall after they had emerged from the canopy as the machine advanced. In this way the heading was supported at all times and the greatest distance of permanently tightened ring from the tunnel face was never more than 6.5 m. This system of support has contributed to the smooth stable tunnel profile and absence of overbreak in the length of tunnel so far completed.

In February 1983, however, the South Tunnel drive ran into an area of heavily faulted strata produced by the convergence of five major faults. So bad was the ground that the Robbins full-face tunnelling machine was able to advance only 25 m during the period from March to September 1983. The faults also extended into the North Tunnel, but were spaced wider apart and were easier to deal with on this account and also because the Titan roadheader being used in the North Tunnel allows better access to support the ground. As a consequence of the full-face machine being idle for so long, the roadheader overtook it in July 1983: the Titan roadheader having advanced to chainage 6500 m whilst the Robbins full-face machine remained stationery at chainage 6484 m.[50] This episode has demonstrated that in badly faulted ground the roadheader is still a better choice than a full-face machine in spite of the attempts made to provide more adequate support with the latter. After much difficulty the full-face machine was taken through the faulted zone. It made some further progress, advancing to chainage 7000 m, when it, along with the roadheader in the adjacent tunnel, was again brought to a halt, not by adverse ground conditions, but by the coal miners' strike of March 1984 to March 1985.

The Robbins full-face tunnelling machine used at Selby also featured an innovatory method of changing the cutters. With earlier full-face tunnelling machines the standard method of changing the cutters was for the cutter head to be withdrawn a short distance from the face so that men could enter the space so formed and remove the cutters from the front of the cutter head. Whilst they were doing this the face was, of course, unsupported and the men were exposed to the hazard of a possible face collapse. For the drive at Selby the Inspectorate of Mines and the National Coal Board decided this was an unacceptable risk and required Robbins to make other provision for changing cutters which would obviate the danger. Robbins' solution to the problem was to design a cutter head on which the individual cutter assemblies are replaced from the rear of the cutter head so that in the course of normal cutter changing, the need for men working on the face side of the machine has been eliminated altogether. The new method has proved both practicable and efficient on the Selby drive and is so convenient that it may well become the usual way of changing cutters on future machines. If so, we have here an example of an innovation, albeit a very minor one, being introduced originally to solve a safety problem but which turns out to have operational advantages as well.

11.11 Summary

The very first hard rock tunnelling machines were invented around 1850 to try to find a solution to the problem of slow and laborious progress by traditional tunnelling methods, so that they were prompted by the same need as were the compressed air rock drilling machines. However, the success of the latter coupled with the replacement of gunpowder by nitroglycerine gave the drill-and-blast tunnelling method a predominance that inhibited attempts to develop these or other tunnelling machines any further. The starting of the Channel Tunnel project in 1881 produced two new full-face tunnelling machines, one of which, the Beaumont–English, proved itself to be a practical success. Full-face tunnelling machines for driving through hard rock can be dated from the 1950s when successful models were used for driving tunnels for the Oahe Dam. From then on there has been steady development to the stage where the drill-and-blast method of tunnelling is no longer supreme. A remaining problem that has not yet been solved is to make full-face tunnelling machines work satisfactorily in broken or faulted ground. The whole of this development, from Maus' and Wilson's first machines at the Mont Cenis and Hoosac Tunnels, right up to the present day, has taken place within the tunnelling industry and is the result of the co-operative efforts of tunnel engineers, tunnelling machine designers and manufacturers and the users of tunnelling machines.

By contrast, the partial-face tunnelling machine, or roadheader, was an innovation that was developed entirely within the coal-mining industry. It was invented in the Soviet Union in the 1930s and brought to Great Britain for use in British coal mines in the 1960s. The idea was taken up by British manufacturers and roadheaders were soon introduced into the tunnelling industry where their versatility and the ease of erecting support around them right up to the face make them well suited to drives through broken or faulted rock. Like the idea of cast-iron tubbing a century before, the tunnelling industry has borrowed the roadheader from the coal-mining industry.

Notes and references for Chapter 11

1 Stephens, J.H. (1976) *The Guiness book of structures: bridges, towers, tunnels, dams.* Enfield (Guinness Superlatives Ltd) p. 129.

2 Full-face tunnelling machines are also sometimes referred to as 'tunnel boring machines' (often abbreviated to TBMs).

3 Often referred to as 'roadheaders' after the name used for them in the coal mining industry (see Section 11.9).

4 Stack, B. (1982) *Handbook of mining and tunnelling machinery.* Chichester (John Wiley and Sons) pp. 139–44.

5 Duluc, A. (1952) *Le Mont Cenis sa route, son tunnel.* Paris (Hermann et Cie) pp. 38–44.

6 Brierley, G.S. (1976) Construction of the Hoosac Tunnel 1855 to 1876. *Journal of the Boston Society of Civil Engineers Section, American Society of Civil Engineers*, **63**, (3), pp. 175–208.

7 Haupt, H. (1867) Tunnelling and mining by machinery. *Engineering*, **4**, 27 September, pp. 310–2.

8 Stack, B. (1982) *op cit*, Chapter 3.

9 There is a voluminous literature on the Channel Tunnel, for example see: Pugh, H.A. (1981) *Bibliography Series No 15: Channel Tunnel.* London (Departments of the Environment and Transport).

10 There are many books on the history of the Channel Tunnel but a good account of the history of the ill-fated project from its inception up till 1958 is given by:
Slater, H. & C. Barnett (1958) *The Channel Tunnel.* London (Alan Wingate).
An account of the scheme from the beginning of the nineteenth century up until 1967 is given by:
Travis, A.S. (1967) *Channel Tunnel 1802–1967.* London (Peter R. Travis).
Photographs of the 1880s Channel Tunnel workings are given by:
Harrington, J.L. (1949) *The Channel Tunnel and ferry.* South Godstone (The Oakwood Press).
Sketch maps showing the locations of the British and French 1880s Channel Tunnel works are given by:
Péquignot, C.A. (1965) *Chunnel.* London (CR Books Ltd) pp. 19–20.

11 In all, there were three British companies formed to build the British section

of the Channel Tunnel; they were:
(i) The Channel Tunnel Co, 1872.
(ii) The Anglo–French Submarine Railway Co, 1873.
(iii) The Submarine Continental Railway Co, 1881.

12 Association du Chemin de Fer Sous-marin entre la France et l'Angleterre, 1875.

13 This draft treaty is reproduced in:
Abel, D. (1961) *Channel underground.* London (The Pall Mall Press) Appendix 1, pp. 100–6.

14 Economic Advisory Council (1930) *Channel Tunnel Committee Report.* Cmnd 3513, London (HM Stationery Office) p. 11.

15 Anon (1882) The Channel Tunnel. *The Illustrated London News*, **80**, (2235), pp. 217–24.

16 Military opposition to the 1881 Channel Tunnel scheme was led by General Sir Garnet Joseph Wolseley (1883–1913), Adjutant-General at the War Office. Both at the time and since, writers have ridiculed Wolseley for this opposition, implying that he had no confidence in the British Army's ability to defend a front the width of the tunnel portal. However, Wolseley's fear was that a relatively small but determined French force of specially trained troops (commandos we would call them) would be landed one night from the sea to seize Dover and the area around it. The bridgehead, once established, could then be rapidly supplied with men and *matériel* via the Channel Tunnel without the Royal Navy being able to intervene, and the stage would be set for a successful invasion of southern England. Wolseley's memorandum on the danger of the Channel Tunnel is quoted by:
Haining, P. (1973) *Eurotunnel: an illustrated history of the Channel Tunnel scheme.* London (New English Library) pp. 73–86.

17 Muir Wood, A.M. (1980) A history of the Channel Tunnel. *Tunnels and Tunnelling*, **12**, (4), pp. 60–4.

18 Bruckshaw, J.M., J. Goguel, H.J.B. Harding & R. Malcor (1961) The work of the Channel Tunnel Study Group 1958–1960. *Proceedings of the Institution of Civil Engineers*, **18**, Paper No 6509, pp. 149–78.

19 Gould, H.B., G.O. Jackson & S.G. Tough (1975) The design of the Channel Tunnel. *The Structural Engineer*, **53**, (2), pp. 45–62.

20 Crosland, A. (1975) Parliamentary debates on the Channel Tunnel. *Weekly Hansard Commons*, **884**, (48), columns 1022 and 1102.

21 Morgan, J.M., D.A. Barratt & D.M. Tilly (1977) The tunnelling system for the British section of the Channel Tunnel Phase 2 works. *TRRL Report* LR734. Crowthorne (Transport and Road Research Laboratory).

22 Hayward, D. (1982) Channel link faces deadlock over finance guarantees. *New Civil Engineer*, 4 March, pp. 14–17.

23 Department of Transport (1982) *Fixed Channel Link: Report of UK/French Study Group.* Cmnd 8561, London (HM Stationery Office) pp. 22 and 46.

24 Ridley, N. (1984) Cross-Channel fixed link. *Weekly Hansard Commons*, **69**, (159), columns 380–1.

25 Brunton, J.D. & F. Trier (1881) On stone-dressing machinery. *Proceedings of the Institution of Mechanical Engineers*, **32**, pp. 133–44.

26 Anon (1869) Brunton's tunnelling machine. *Engineering*, **7**, 28 May, pp. 349 and 355.
Brunton's tunnelling machine was considered important enough at the time to figure in the popular work:

Small, J. (1877) *A hundred wonders of the world*. London (William P. Nimmo) p. 552.

27 The author has been able to find little about the inventor of this first Channel Tunnel machine, John Dickinson Brunton. He was active during the second half of the nineteenth century and seems to have been a prolific inventor. He is not the subject of, but may have been related to the contemporary John Brunton of:
Brunton, J. (1939) *John Brunton's book*. Cambridge (The University Press).

28 Temporal, J. & R.A. Snowdon (1982) The effect of hydraulic stiffness on tunnel machine performance. *Tunnels and Tunnellings*, **14**, (2), pp. 11–13.
Snowdon, R.A., M.D. Ryley, J. Temporal & G.I. Crabb (1983). The effect of hydraulic stiffness on tunnel boring machine performance. *International Journal of Rock Mechanics and Mining Science*, **20**, (5), pp. 203–14.

29 Beaumont, F.E.B. (1875) Tunnelling machinery. *British Patent* No 4166. London (Great Seal Patent Office).

30 English, T. (1880) Improvements in tunnelling machinery. *British Patent* No 4347. London (Great Seal Patent Office).

31 Stack, B. (1982) *op cit*, pp. 216–7.

32 Parkin, G.W. (1966) *The Mersey Railway*. South Godstone (The Oakwood Press) p. 11.

33 Anon (1881) The Channel Tunnel scheme. *The Engineer*, **51**, 24 June, p. 465.

34 Crampton, T.R. (1882) On an automatic hydraulic system for excavating the Channel Tunnel. *Proceedings of the Institution of Mechanical Engineers*, **33**, pp. 440–50.
Crampton, T.R. (1882) Crampton's system of excavating the Channel Tunnel. *The Engineer*, **54**, 6 October, pp. 255–6.

35 This little-known aspect of military history has been recorded by:
Barrie, A. (1962) *War underground*. London (Frederick Muller Ltd).

36 Anon (1923) The Whitaker tunnelling machine. *Engineering*, **65**, 26 January, pp. 102–3.

37 The use of tunnelling machines on the Oahe Dam project is described in:
Anon (1957) The mole comes through. *Engineering News Record*, **158**, 18 April, p. 28.
Anon (1959) Biggest tunneller goes to Oahe. *Engineering News Record*, **162**, 26 February, p. 26.

38 Harding, H.J.B. (1959) Notes on rotary excavators for tunnelling in shale. Oahe Dam, Pierre, South Dakota. (Unpublished report).

39 Stack, B. (1982) *op cit*, Chapter 5.

40 Rees, P.B., H.M. Hughes & J.D. Hay (1976) Full-face tunnelling machines in British coal mines. *Tunnelling 76*. London (The Institution of Mining and Metallurgy) pp. 413–22.

41 Snowdon, R.A. (1979) Performance of a full-face tunnelling machine in Coal Measures rocks at Dawdon Colliery. *MSc Thesis*, University of Newcastle-upon-Tyne.

42 Burgess, H. & J.G. Taylor (1980) Tunnel boring in a deep South African gold mine. *Transactions Section A of the Institution of Mining and Metallurgy*, **89**, pp. 84–98.

43 Scott, R. (1974) Drivages with Dosco Roadheaders. *Proceedings of Symposium on Mining Methods*, Paper No 4, pp. 1–14.

44 Further details of the roadheader are given in the brochure:
Dosco Overseas Engineering Ltd (1974) Dosco Roadheader Mk 2A tunnelling

machine. Aylesbury (Dosco Overseas Engineering Ltd).
45 Pokrovsky, N.M. (1980) *Underground structures and mines. Construction practices. Driving horizontal workings and tunnels.* Moscow (MIR Publishers) p. 234.
46 West, G. & A.F. Toombs (1978) Site investigation and construction of the Liverpool Loop and Link Tunnels. *TRRL Report* LR868. Crowthorne (Transport and Road Research Laboratory).
47 Johnson, A.D. & P.B. Parkes (1976) The use of roadheading machines in the construction of the Liverpool Link Line. *Proceedings of 2nd Australian Conference on Tunnelling.* Melbourne (Australian Tunnelling Assn) pp. 145–52.
48 The Selby Coalfield is described in the brochure:
National Coal Board (1981) Selby: a new world of mining. North Yorkshire Area (National Coal Board).
49 Tunnicliffe, J.F. (1983) Spine roads at Selby. *Tunnels and Tunnelling*, **15**, (3), pp. 57–9.
50 Wallis, S. (1983) Fullfacer vs roadheader: an investment in speed. *Tunnels and Tunnelling*, **15**, (6), pp. 35–8.
See also: *Tunnels and Tunnelling*, **15**, (9), World News p. 8 and Letters to the Editor p. 53.

Soft ground tunnelling machines

Machinery gifted with the wonderful power of shortening and fructifying human labour.

Full-face tunnelling machines designed for use in soft ground differ from those designed for use in hard rock in one important aspect – they must be able to support the ground as well as excavate it. For this reason soft ground tunnelling machines are basically tunnelling shields fitted with mechanical excavation and spoil handling systems. They have all the features of the tunnelling shield: the shield itself for supporting the ground, the tailskin for the safe erection of a prefabricated lining and jacks for shoving the shield forward off the lining. However, instead of the excavation of the face and the mucking out being done by manual labour these are done by machinery. There is another difference from hard rock machines which made the design of soft ground machines much simpler, and that is the less demanding nature of the excavation task. For hard rock machines the cutter head had to cut into the rock and break it out from the face in fragments of suitable size to be transported away by belt conveyor conveniently. As we have seen, this was a stringent requirement and in fact limited the successful application of tunnelling machines to the softer rocks until the 1950s. By contrast, the cutter head of a soft ground machine had a relatively straightforward task, excavation of clays, sands and gravels being fairly easy to achieve with simple steel tools shaped like picks or chisels. Once excavated, the material will usually crumble down to a size convenient for removal by belt conveyor. A soft ground tunnelling machine is, therefore, a much simpler design task than a hard rock machine and it is no surprise to find that successful examples appeared much earlier. It will be recalled that the hard rock tunnelling machines had to compete against the very successful drill-and-blast method; the soft ground tunnelling machine, by contrast, had only to

Figure 12.1 The Thomson excavator (1898)

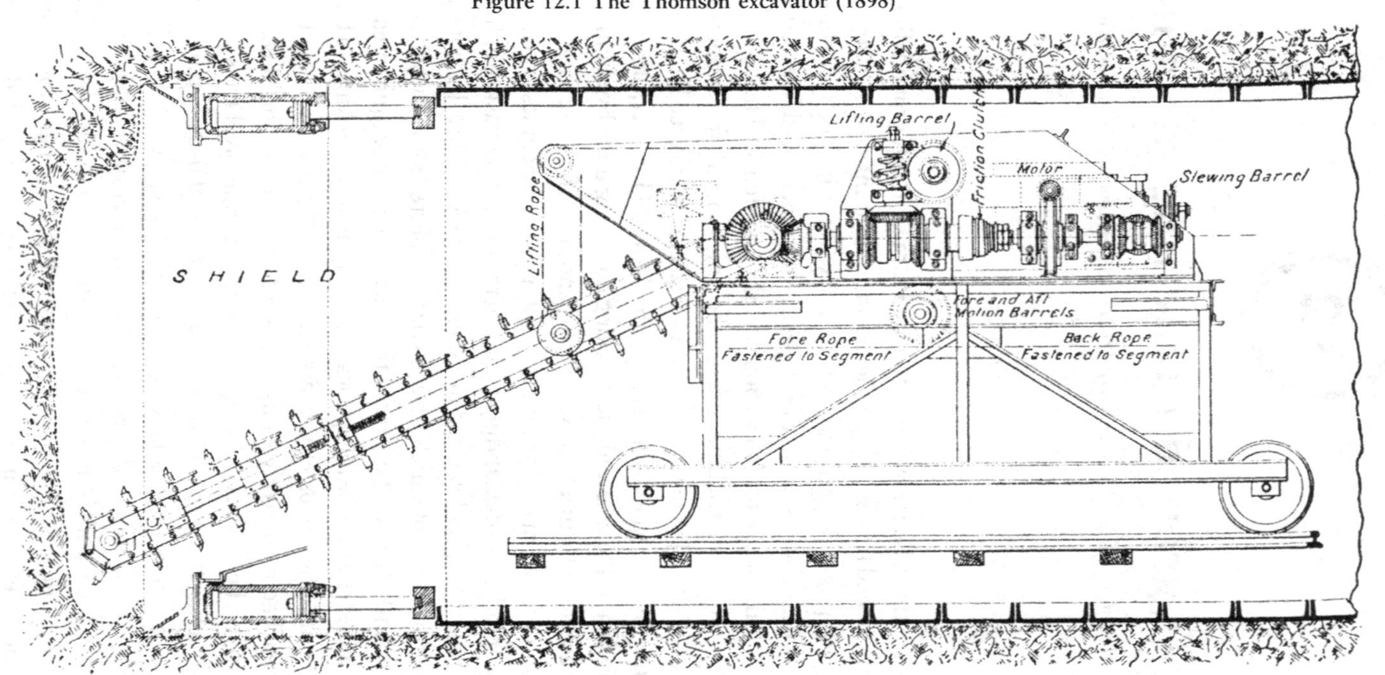

compete with hand excavation from within a shield. For small-diameter tunnels hand excavation was – and still is – a most effective process but for large-diameter tunnels manual operation becomes less effective because of the large number of miners required both to excavate the face and to remove the spoil. A tunnelling machine that could do both these jobs offered great benefit in saving labour and possibly also in speeding tunnel construction. For the first attempts at soft ground tunnelling machines we must return to London's Underground. By a curious coincidence the first two soft ground tunnelling machines to be designed and constructed were both used on different contracts of the same job – the Central London Railway.[1]

12.1 The Thomson excavator

The first of the soft ground tunnelling machines used on the Central London Railway was a ladder excavator mounted within a tunnelling shield. The excavator was designed by Thomas Thomson[2] who was the contractor's agent, and the machine was constructed by Scott and Mountain of Newcastle. Before the machine was designed, the following conditions were specified for its performance.[3] It would have to (i) excavate firm clay (the Central London Railway tunnels were to be sited mainly in London Clay), (ii) be able to cope with the occasional boulder-sized septarian nodules that were expected in the London Clay, (iii) have provision for discharging the spoil into wagons, (iv) not be an impediment to the other tunnelling operations (e.g. erecting the lining and grouting), (v) be readily removable in the event of breakdown so that hand working could be substituted, and (vi) not prevent the line and level of the shield from being checked in the usual manner during the drive. In the examination of the machine which follows we can note how the design allowed all these conditions to be met.

The machine is shown in Figure 12.1 and consists essentially of a dredger ladder, the working end of which can be moved vertically, horizontally and longitudinally. The lower part of the machine consisted of a large open travelling gantry (referred to as a 'Goliath') running on a 6 ft 3 in gauge rail track. The gantry was sufficiently wide and high to allow the tunnelling contractor's standard 2 ft gauge rail-borne muck wagons to run right underneath it and through the front. On top of the gantry was a carriage which revolved on a massive king post attached to the gantry.

Attached to the front of the carriage was a 17 ft long dredging bucket ladder, slung from a small crane which projected forward of the carriage. The carriage also housed a 100 A, 200 V electric motor which was mounted at the rear. The motor drove a longitudinal main shaft from which were driven by bevel gears the mechanisms for slewing the carriage from side to side, for raising and lowering the dredger ladder and for working the chain of buckets. Forward and backward motion of the gantry was effected by winding in lines fastened to segments of the tunnel lining. A fuse installed in the electric circuit prevented damage to the machine if the buckets hit an obstruction and stalled. The ladder was also fitted with a reversing switch so that it could be eased out of the face if it got stuck. The electricity was supplied by a 20 hp engine driving a dynamo at the surface. The buckets were more in the nature of scrapers, having a bottom and back only. On the backs of the buckets were fitted four or five wrought-iron chisel-shaped teeth. At first the buckets were made from cast-steel but these were replaced with gun-metal buckets after the cast-steel ones broke in use. All the levers and handwheels for controlling the machine were positioned together within easy reach of the driver who stood on a small platform at the front of the machine on the left hand side. The Thomson excavator was operated inside a standard 12 ft 8 in outside diameter tunnelling shield, the only modification being required was the replacement of the normal square door in the diaphragm with a 10 ft diameter circular opening.

In operation the machine was run up through the shield to the face and the front end of the dredger ladder was lowered to the invert. The bucket chain was started and after excavating into the face a short distance the ladder was slowly raised until it reached the top of its travel. The ladder was then lowered, the carriage was slewed a little to one side and another cut was taken from the face. The face was excavated in this manner, in a succession of vertical cuts, until a distance of 1½–2 ft in front of the shield had been excavated. Because of this method of cutting the face, the excavated profile was not circular but in the form of a broad cross. When the excavation was completed, the whole machine was run back down the heading a distance of about 10–12 ft and the shield was jacked forward in the usual manner. It was found that the cutting edge of the shield easily brought down the remaining pieces of clay outside the broad cross and gave the required circular finished profile to the tunnel. The cast-iron segmental lining was then erected and the Thomson excavator brought forward again for another cycle of operations. The time taken to excavate the $1\frac{1}{2}$–2 ft distance was about $1\frac{1}{2}$–$1\frac{3}{4}$ hours and pushing the shield forward and erecting a ring of lining segments took about the same time again. It was reported that eight rings had been built in two 10 hour shifts but the average was three rings per shift.[4] One ring was 1 ft 8 in wide so this gives an average advance rate of 10 ft per day assuming two shifts were worked. With the Thomson excavator the number of miners in the work force was eight, but sometimes six, compared to fourteen required for hand excavation. This shows the benefit of mechanised excavation and an incentive to introduce tunnelling machines in place of manual methods. Apart from the breakage of the original buckets referred to and some trouble with the electric motor and its connections, the excavator was reported to have operated satisfactorily. We will now see how well the machine met the original specification.

The dredger ladder was clearly able to excavate the firm clay satisfactorily, which was the condition of item (i), the only problem being the breakage of some of the original buckets which was cured by changing the metal from which they were made. The machine coped with septarian nodules, item (ii), by being protected with a fuse and having a reversing switch so that if the ladder became stuck against a boulder it could be eased out of the face. The gantry could then be run back a little way and the boulder removed by hand. Item (iii), the ability to discharge spoil into wagons was provided by mounting the machine's carriage on a gantry beneath which the wagons could pass to receive spoil from the rear end of the dredger ladder. Items (iv), (v) and (vi) were all provided for by arranging for the gantry to run on rails so that the whole machine could be withdrawn down the heading to allow lining and grouting, hand working if need be, and checking of the shield alignment to be carried out unobstructed.

After being used on one of the Central London Railway tunnels, the Thomson excavator seems never to have been used again. Indeed this kind of tunnelling machine could almost be said to have disappeared completely were it not for the fact that it surprisingly reappeared, again on the London Underground, as late as the 1970s. For on one of the contracts of the Jubilee Line a Westfalia Lünen boom cutter loader operating from within a shield was used to excavate the tunnel[5] and this machine was of a similar type to the Thomson excavator. The best performance of the Westfalia machine was an advance rate of 33 m per week but the average advance rate was only 14 m per week. It is interesting to note that, assuming a five-day week for comparison purposes, the Thomson excavator would have advanced 15 m in a week going at its average rate, showing that its performance was equal to that of the Westfalia machine built some 70 years later. In fact, as we shall see, the advance rate of the Westfalia Lünen boom cutter loader is considerably less than the advance rate of contemporary full-face soft ground tunnelling machines (see Section 12.4) and this explains why this type of machine has not found further application in soft ground tunnelling. In fairness, it should be said that the Westfalia machine was not intended for excavating London Clay, this type of boom cutter loader being designed for use in weak to medium strength rocks.

12.2 The Price rotary tunnelling machine

The Price rotary tunnelling machine was invented in 1896 for, and first used on, the Central London Railway tunnels by John Price, one of the contractors on the scheme. Price was born on 12 June 1846. In his earlier professional life he was involved in numerous railway construction projects, including tunnels, and was also contractor's engineer for sections of the Manchester Ship Canal. In 1893, as a contractor in partnership with T.J. Reeves and C.J. Wills, he carried out contracts on the Central London, Hampstead and Piccadilly tube railways. He later worked on the Rotherhithe Tunnel. He died at Wimbledon on the 25 June 1913. No account of the construction of the Central London Railway was ever published in the *Minutes of Proceedings of the Institution of Civil Engineers*, and the technical journalists from *Engineering* and *The Engineer* who visited the Central London Railway during construction were taken only to the contracts where the Thomson excavator was being used. Fortunately, however, the Price machine was used to drive the pilot tunnel for the Rotherhithe Tunnel in 1906, and a description of it is given in a paper on that tunnel.[6] This has been used as a basis for the following examination, supplemented in places by a description given by Copperthwaite,[7] although comments will be made based on an unpublished description of the machine for the Central London Railway written by Price himself.

Figure 12.2 is a photograph of a 12 ft 8 in diameter Price machine manufactured by Markham and Co Ltd. The tunnelling machine was constructed inside a conventional shield which was provided with ten hydraulic rams for jacking forward off the tunnel lining in the usual way. Across the shield were fixed two strong horizontal girders carrying bearings in which ran a steel shaft centred on the longitudinal axis of the shield. The shaft carried at its front end a heavy circular hub casting to which were bolted six radial arms made from steel channel. These

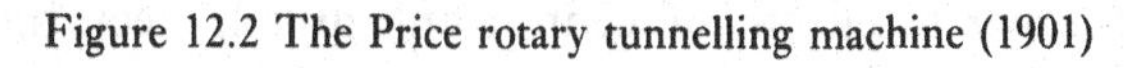
Figure 12.2 The Price rotary tunnelling machine (1901)

projected about 1 ft in front of the cutting edge of the shield and were held together by circumferential plates. A 10 ft diameter circular rack fitted with internal teeth was fixed to the back of the arms and was turned via reduction gearing by a 52 hp electric motor housed in the shield. The driving pinion that engaged the rack can just be seen inside the upper right quadrant of the shield in Figure 12.2. The motor ran at about 500 revolutions per minute and the speed of the excavator was about $1\frac{1}{2}$ revolutions per minute. An apron protected the teeth of the rack from becoming clogged by spoil. Later versions of the machine had a toothed V-shaped scraper fitted on the central hub to help keep the excavator steady on course. The radial arms carried renewable chisel-shaped steel cutters which projected about 6 in and, when the head rotated, excavated the ground by cutting concentric grooves in the tunnel face. In his 1896 proposal Price described the intended action of the cutting tools as follows:

> The T-steel arms carry a series of steel cutters having their cutting edges at right angles to the centre line of the shaft and so arranged, that starting from the centre, each following cutter is its own width further from the centre than the preceding cutter. By this arrangement each cutter has two free sides, which allows the material to fall away easily.[8]

This is important because it shows that Price had grasped the concept of 'cutting to relief' whereby instead of each cutter acting independently, every cutter was able to displace the clay cut off the face into the space produced by the passage of the preceding cutter. This point will be taken up again later. Each arm also carried a bucket which scooped up the spoil from the invert of the tunnel and discharged it into a chute in the upper part of the shield. A travelling belt conveyor delivered the spoil from the chute into wagons about 25 ft behind the shield. The belt conveyor was separately powered by its own 10 hp electric motor. A small compressor for the shield's hydraulic rams and the starting switches for the excavator and the conveyor were all mounted on the frame of the rail-borne conveyor carriage. The operator was seated on the carriage as well, with rheostats for controlling the excavator and conveyor motors and valves for controlling the hydraulic rams all to hand.

The method of operation was as follows. A ring of lining segments having been erected, the travelling conveyor was run up into position under the chute and started. The excavator motor was then started and any loose material removed by the buckets. Hydraulic pressure was then applied to the rams which advanced the shield together with the excavator into the face. The rate of travel of the rams was regulated by monitoring the electrical power required to turn the excavator, the operator of the hydraulic rams having an ammeter constantly under observation. The time taken to excavate the 1 ft 8 in length of a ring varied between 20 minutes for clay and 45 minutes for harder material. When this length had been achieved the excavator was stopped, the conveyor run back and another ring erected. It is interesting to note the similar appearance of the cutter heads of the Whitaker hard rock tunnelling machine (Section 11.6) and the Price soft ground tunnelling machine; Whitaker may have been influenced by Price's successful design.

The 1906 Price rotary tunnelling machine differed in a number of important ways from the original idea put forward by Price in his 1896 proposal. Firstly, in the original machine the excavator head was to be rotated by a motor turning the central steel shaft. Secondly, the prime mover was to be a twin-cylinder compressed air engine supplied with compressed air from a main in the tunnel. Thirdly, during excavation the shield was to remain stationary and the excavator head alone was to be advanced using a screw-thread mechanism acting on the rear end of the central steel shaft. When the head had advanced a few inches the shield was to be jacked forward and so on until the total advance of 1 ft 8 in had been made. Fourthly, rail-borne muck wagons were to run under the engine and the central shaft to right behind the excavator head where the chute was to deliver the spoil directly into them. The first machine that was made had the drive on the central shaft and had the excavator and shield as separate units as in Price's original proposal. This was not very successful and the machine design was changed to have peripheral drive and to be a combined unit as in the 1906 version and as shown in Figure 12.2.

The redesign was an immediate success and Price rotary tunnelling machines were used to drive long lengths of London's underground railway tunnels. On the Charing Cross, Euston and Hampstead Railway (now part of the Northern Line) and on the Great Northern, Piccadilly and Brompton Railway (now part of the Piccadilly Line) $32\frac{1}{2}$ miles of tunnel were constructed of which $11\frac{1}{2}$ miles were excavated with Price's rotary tunnelling machine. When the machines were first used, advance rates were 60 ft per week but by the time the tunnels were finished advance rates were up to over 180 ft per week. At first the machines had a tendency to run crookedly because of lack of attention to steering but by the end of the job 102 rings were completed in one week without a single ring being off line or level by more than $\frac{1}{2}$ in.[9] Ten miners were required when the Price machine was being used compared with fourteen for an ordinary tunnelling shield. During the period between the World Wars London's Underground was widely extended and for this work over forty Price rotary tunnelling machines were manufactured by Markham,[10] a photograph dating from 1923 in one of their trade brochures[11] showing ten under construction all at the same time.

The Price rotary tunnelling machine is important not only in its own right because so many were made and so much tunnel was driven with them, but because its basic design incorporates all the elements of later successful soft ground tunnelling machines so that it can be regarded as the forerunner of subsequent types. It is worth considering briefly why it was so successful compared with the contemporary Thomson excavator which, as we have seen, disappeared after its first trial on the Central London Railway. The Thomson excavator had an average advance rate of 10 ft per day, which in a 6 day week gives a weekly advance rate of 60 ft per week. This compares exactly with the advance rate of 60 ft per week for the Price rotary tunnelling machine when first used, although this rose to three times this figure later. So to begin with, there would seem little to choose between them. There is, however, one crucial difference. The Thomson excavator was not an automatic machine. The bucket ladder had to be under the continuous control of the operator at all times who had to direct it at different parts of the tunnel face in turn by a fairly complex sequence of operations. By contrast, the Price rotary tunnelling machine excavated the whole face area at once and was almost fully automatic – all the operator had to do was see that the rate of advance of the hydraulic jacks was not too much to overload the rotary head's electric motor. The advantages are obvious and it is no surprise that the rotary type of machine became the basic type that developed into the modern soft ground tunnelling machines of today.

12.3 The Kinnear Moodie drum digger

The Price type of rotary tunnelling machine became the standard method of driving soft ground tunnels and subsequent developments of it were little more than improvements in details, the basic elements as established by Price remaining the same. This was the case right up until 1955 when a further development took place. In January of that year construction of the Thames–Lee Valley water tunnel began: this was to be 19 miles in length and of 8 ft 6 in internal diameter, and was designed to be driven mainly through the London Clay.[12] It will be recalled that this tunnel, built for the Metropolitan Water Board, was the one on which the Don-Seg and wedge-block linings described in Section 8.7 were first used. Because of the long length of the tunnel, the use of tunnelling machines was considered desirable, but in the event the tunnel was constructed with fourteen hand shields and only two tunnelling machines. However, these two tunnelling machines were of a new type which will now be discussed. For a tunnel of this fairly small diameter, a Price rotary tunnelling machine was deemed to be out of the question because its central shaft, motor and reduction gearing would occupy so much space in the centre of the shield that there would be little room left for getting the spoil out from the cutter head. At this time, however, compact yet powerful hydraulic motors became available which allowed a new type of tunnelling machine to be designed and developed by one of the contractors for the Thomas–Lee Valley tunnel, Kinnear Moodie and Co Ltd; this machine was called a 'drum digger' for reasons that will soon become apparent.

The drum digger consisted of a conventional tunnelling shield inside which was a smaller cylinder, or drum, which was able to rotate on roller bearings attached to the inside of the shield. The outside of the drum was fitted with a peripheral rack which engaged with four pinions, each of which was directly driven by one hydraulic motor, the motors being small enough to fit into the space between the outside of the drum and the inside of the shield. These motors turned the drum at approximately 4 revolutions per minute. The inside of the drum formed an unobstructed opening from the front to the rear of the shield. On the front of the drum was fixed an annular plate which carried six radial cutting arms fitted with teeth, one pair of arms being extended to bridge across the central opening of the drum. As the drum rotated, the teeth excavated the clay and the

Table 12.1. *Comparison between drum diggers and hand shields on the Thames–Lee Valley tunnel*

	Machine	
Details	Kinnear Moodie drum diggers	Hand shields
Number used	2	14
Length of tunnel driven	5½ miles	13½ miles
Average advance rate	200 ft per week	150 ft per week
Best advance rate	360 ft per week	180 ft per week

debris fell to the invert where it was picked up by concave recesses in the sides of the cutting arms and directed into the central opening of the drum. When the spoil entered the drum it was directed by vanes fixed to the inner surface of the drum into the hopper of an inclined belt conveyor which discharged it into rail-borne muck skips running on 2 ft gauge track. Power for the drum digger was provided by a 120 hp electric motor which drove two hydraulic oil pumps which delivered oil to the four hydraulic motors in the head; the electric motor was mounted on a stage attached to the shield. The saving of space which enabled the whole of the inside of the drum to be available for spoil handling was made possible by dispensing with a centre shaft and by use of the small hydraulic motors which did not require the large train of reduction gears that on the Price machine occupied so much space in the centre of the shield.

Although there were only two of them, the drum diggers on the Thames–Lee Valley tunnel were an immediate success; compared with the hand shields the drum diggers gave impressive advance rates (see Table 12.1), the best rate of the drum diggers being twice that of the hand shields. This showed that, under favourable ground conditions, the soft ground tunnelling machine was superior in performance to the hand shield. However, the tunnelling machine will not always be the best choice; for short drives and for difficult ground a hand shield may be better than a machine.[13] The machines used on the Thames–Lee Valley tunnel also demonstrated the success of the drum digger design.

12.4 The Victoria Line

We have seen how the Kinnear Moodie drum digger was specifically designed as a tunnelling machine for small-diameter tunnels and how it

Table 12.2. *Comparison between tunnelling machines on the Victoria Line*

	Machine	
Details	Kinnear Moodie drum digger	McAlpine centre-shaft machine
Diameter of shield	13 ft 6 in	13 ft 6 in
Length of shield[a]	8 ft 8 in	10 ft 0 in
Cutter head drive	Four 50 hp motors	Four 60 hp motors
Cutter head type	Six arms with teeth	Four arms with teeth
Average advance rate	1.9 ft per hour	3.6 ft per hour
Best advance rate	3.9 ft per hour	5.4 ft per hour

[a]Neither shield had a tailskin because unbolted expanded precast concrete lining was used and each ring of segments was erected directly against the clay.

was very successful when used for this application. This led to the expectation that a scaled-up drum digger might be equally successful. For the construction of the Victoria Line for London's Underground between 1962 and 1968, the London Transport Board purchased a large-diameter version of the Kinnear Moodie drum digger together with a modern version of the Price type of tunnelling machine having a central shaft and peripheral drive that had been developed by Sir Robert McAlpine and Son Ltd. On the tunnel contract length between Ferry Lane and Walthamstow the Kinnear Moodie drum digger and the McAlpine centre-shaft machine were operating under identical conditions which made a comparison between them possible.[14] Table 12.2 shows that although the two machines were of similar size and of comparable power, the McAlpine centre-shaft machine, which was an up-to-date version of the Price type of tunnelling machine, performed better than the drum digger, having almost double the average advance rate. It was concluded that although the drum digger provided a fine practical solution to the problem of making a small-diameter tunnelling machine, the idea was not so beneficial when applied to a large-diameter machine. However, for tunnels of diameter up to about 2.5 m the drum digger type of machine, with the great advantage of its unobstructed central area, remains the preferred design.

Later machines built by Robert L. Priestley Ltd for small-diameter tunnels in clay at Oxford[15] and at Weymouth have both been similar to the drum digger type with one difference that instead of the spoil being deposited inside the cylindrical drum and then being fed onto a conveyor, it is collected directly from the cutter head by means of an extendable conveyor which can be run up right through the inside of the drum to receive the spoil straight from the mucking arms. On the Oxford job, which was a 2.54 m internal diameter sewer tunnel, an average advance rate of 113 m per week was achieved, with a best advance of 30 m being recorded in one 10 hour shift. Figure 12.3 shows the interior of the Priestley tunnelling machine used for driving the Oxford sewer tunnel; the view is from inside the rear of the shield looking forward towards the back of the cutting chamber. The shield diameter is only 2.8 m and yet there is an air of spaciousness inside. This is because the cutter head design, being similar to that of the drum digger, allows the central area of the shield to be kept free of machinery. In the centre of the cutting chamber can be seen the large opening through the inside of the drum into which the conveyor has been inserted to collect the Oxford Clay spoil directly from the arms of the cutter via a small chute. Also shown in the Figure are the front ends of the shove rams which are used to force the shield forward off the wedge-block lining (see Section 8.7) which was used to line the Oxford sewer tunnel; the lining segments were expanded directly against the clay, there being no tailskin to the shield. The only

Figure 12.3 Priestley tunnelling machine (1980)

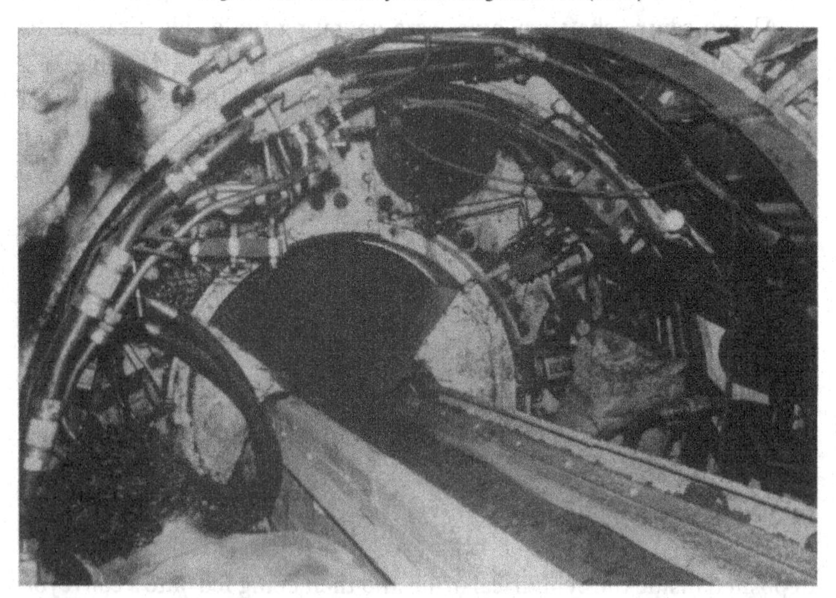

problem encountered with the Priestley machine on the Oxford drive was at the beginning of the job when the Oxford Clay squeezed in on the tunnel before the ring of lining segments could be placed and expanded.[16] This problem was eventually solved by fitting an oversize bead to the cutting edge of the shield and a modified trimmer to the rear edge of the shield. Until these had been fitted, the tunnel profile was trimmed by hand until the segments fitted – a somewhat laborious business. Research is in hand to try and determine the reason for the squeezing ground.[17]

It will be recalled that in the introduction to Chapter 11 the problems facing the designers of tunnelling machines were enumerated as (i) to devise a mechanical means of excavating the tunnel face, (ii) to devise a mechanical means of removing the spoil produced by the excavation, (iii) to devise a suitable power supply for operating the machine and (iv) to design machines robust enough for the job. We will now look at how these problems were solved on some of the soft ground tunnelling machines. All the soft ground tunnelling machines, except for the Thomson excavator, solved problem (i) by having a rotary cutter head fitted with arms carrying simple chisel-shaped teeth. This arrangement was hit upon early and has proved satisfactory up to the present for excavating soft ground, particularly when in clay. We have already drawn attention to the fact that Price in the design of the cutter head of his rotary tunnelling machine had arranged the layout of the cutters so that they were cutting to relief. The soundness of this concept has been confirmed by research in the 1970s with pilot-scale and full-scale tunnelling trials.[18] Again, all the machines except the Thomson excavator, solved problem (ii) by having some kind of spoil-collecting mechanism on the cutter head which delivered the spoil to a belt conveyor delivering into muck skips. This early arrangement has again served well to the present day. As regards problem (iii), that of power supply, the Thomson excavator and the Price rotary tunnelling machine were both driven by electric motor. In the case of the Price machine, which was to be the main type of soft ground tunnelling machine for the next 50 years, the electric motor drove the head via reduction gearing. When the drum digger was introduced the reduction gearing was eliminated and the head was rotated directly by four small hydraulic motors. However, the hydraulic pumps for these motors were driven by electric motor. The advantages of the electric motor for tunnelling machines were discussed in Section 11.6. Problem (iv), to make a machine robust enough for the job was fairly simple for a soft ground machine. The only difficulty seems to have been with the prototype of the Price machine where the electric motor drove the central shaft, 'driving power was applied in the wrong place', and where there was 'too much spring in the long axle'.[19] Both these faults were swiftly remedied by Price in the second model which, as we have seen, was an immediate and long lasting success. In contrast to the hard rock tunnelling machines, the design problems of soft ground machines were, understandably, solved very much faster because the task the machines had to perform was so much easier.

12.5 Later developments

A number of small improvements have since been made to the soft ground tunnelling machine. One of these was to provide a circular reaction ring between the shield and the lining. The reaction ring is expanded hydraulically until it is tight against the tunnel wall. It then supplies the reaction for the thrust jacks of the tunnelling machine shield. After the shield has advanced by the full stroke of the jacks, the reaction ring is contracted and moved forward, using the double-acting facility of the jacks, to its new position where it is again expanded ready for another cutting cycle to begin. The original reason for providing a reaction ring was to protect the tunnel lining from damage by the jacks when shoving forward and so enable thinner linings to be used than hitherto. Examples of machines with the reaction ring are the 5 m diameter tunnelling machine designed by Sir Robert McAlpine and Sons Ltd for the Chinnor Tunnelling Trials[20] in 1974, although in this case the purpose of the reaction ring was to enable only a temporary lining to be erected, and the 5.3 m diameter tunnelling machine built by Robert L. Priestley Ltd for the British 1975 Channel Tunnel works.[21] On the latter machine the purpose of the reaction ring was to allow machine advance and lining erection to be carried out independently of each other. The Priestley machine also had the unusual feature of built-in facilities for probing ahead of the face by means of two Diamec 250 rotary drill rigs.

A second development was the introduction of machines with oscillating cutter heads instead of the normal rotary ones. The idea behind the oscillating cutter head was that in loose sands and gravels or in soft silts the ground should be disturbed as little as was compatible with excavating the face. It was thought that an oscillating motion of the cutter arms over a limited distance would be better than the full sweep of the rotary head. The idea, it will be recalled, was used in the Universal soft ground tunnelling machine described in Section 9.3. The oscillating head type of tunnelling machine was first introduced by Calweld Inc in the United States in 1966, but later other manufacturers made machines of this kind; the oscillating head has also found application in pressurised face tunnelling machines. A third modification was a shield with a sloping front giving a sort of canopy over the cutter head. Here again, this was provided to give additional stability to the face in loose or soft ground and was often combined with the oscillator head. This idea, too, has found application in pressurised face tunnelling machines. It is interesting to note that the sloping front to a shield was used as early as 1899 on the River Spree Tunnel (see Section 9.2). Another form of soft ground tunnelling machine that has been developed is the digger shield but this will be examined later in Chapter 13.

It will be recalled that much of the early development of hard rock tunnelling machines took place in the Lower Chalk and it was remarked that in many ways this was the ideal rock type for the purpose. In the same way, it can be observed that all the early development of soft ground tunnelling machines took place in the London Clay – again ideal material for the purpose. Although all the machines that have been discussed in this Section were designed for excavation in clay, they could cope to some extent with other types of soft ground. For example, on the Rotherhithe Tunnel the Price rotary tunnelling machine coped with clay, sand and pebbly gravel, and even a bed of rock 3–5 ft thick in places. Because it was a subaqueous drive, compressed air at a pressure of $\frac{3}{4}$–$1\frac{1}{2}$ atm was used to keep the face dry. As an additional safety precaution, the shield of this particular machine was fitted with a water trap consisting of a pair of half diaphragms, the spoil being delivered through an aperture in the upper one. The aperture could be closed by a sliding door at a moment's notice should there be a sudden inflow of water at the face.[22] On other occasions these machines were unable to cope – as when a run-in of water-bearing gravel and sand put the Kinnear Moodie drum digger out of action on the Victoria Line;[23] however, without compressed air this would probably have happened with any tunnelling machine or hand shield except for the pressurised face types. Nevertheless, the tunnelling machines described in this Chapter were designed for, and always gave their best performance in, clays like the London Clay. The modern soft ground tunnelling machines either of the Price type or the drum digger type are almost at the limit of their development – they were perfected quickly and there is not much that can be done now to make them better. The further development of soft ground tunnelling machines was to lie in the direction of making machines that would excavate through gravels and sands below the water table, and the solution to this lay in inventing pressurised face types, the development of which has already been examined in Chapter 9.

12.6 The Mini Tunnel

Before leaving soft ground tunnelling machines, mention must be made of the Mini Tunnel. The Mini Tunnel is a system for driving small-diameter tunnels in soft ground and is essentially based on the tunnelling shield. The innovatory feature of the Mini Tunnel system does not lie in any particular machine or device, but in reproducing all the elements of shield tunnelling in a miniature form and so arranging the method of working that the whole operation could be carried out by only three men. The system was devised by Michael A. Richardson and brought into commercial practice by Mini Tunnels International Ltd of Old Woking, Surrey, in 1973.

Figure 12.4 shows the Mini Tunnel system as originally established[24] using hand excavation from within a tunnelling shield 1 as the method of driving the tunnel. The spoil was sent back down the tunnel in a skip 11 which was then hoisted up the shaft. The completed section of tunnel was lined with unbolted precast concrete segments 8, three to a ring. The space behind the ring was filled with pea gravel and this was then grouted with cement. The segments were brought up to the face on a rail-borne segment carriage 9. The whole operation can be carried out by only three men: one excavates the face from within a hand shield and erects the lining, the second drives the small locomotive 29 and handles the spoil and segments at the shaft bottom, and a third operates the crane and looks after loading and unloading at the surface. The complete tunnelling system was available in three internal diameters, 1, 1.2 and 1.3 m, standard segments of these three ring sizes being manufactured as stock items by special segment pressing plants.

In 1975 a small-diameter full-face soft ground tunnelling machine was produced for use with the 1.2 m diameter size of Mini Tunnel; this was called the dkr machine (Figure 12.5) and was designed for excavating stiff clay like London Clay and for one-man operation. The cutter head of this machine, which is driven by electro-hydraulic power, consists of three arms carrying chisel-shaped cutting teeth mounted on rocker bearings so that the direction of rotation is reversible. The cutter head is supported on a ring bearing and drive so that the central area is left unobstructed which allows a retractable belt conveyor to be run up to the head to collect the spoil; in this respect the machine is like the drum digger. The conveyor discharges the spoil into a rail-borne skip which is then dealt with as previously described.

The Mini Tunnel system was conceived, designed and produced to

Figure 12.4 Mini Tunnel system (1975)

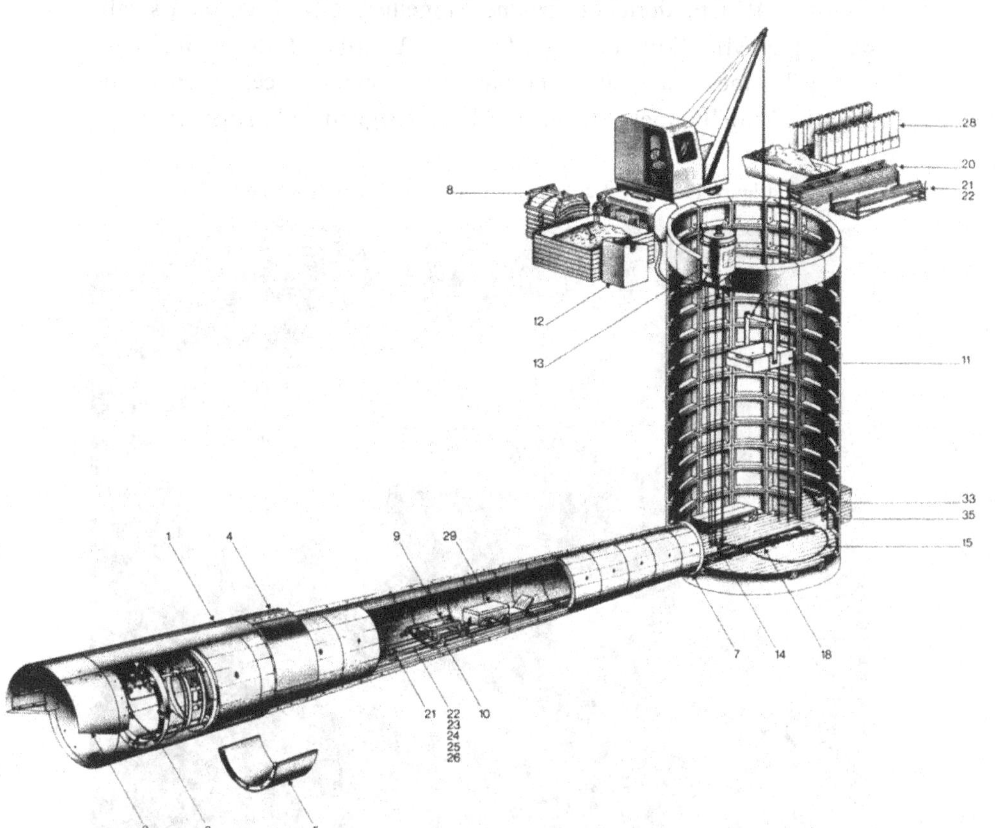

instal sewers, water mains, pipelines, cables and other services beneath urban areas. Hitherto all these had usually been installed by digging trenches from the surface and then installing the pipe or required service and backfilling the trench. This, as every town-dweller knows, is an immensely disruptive and inconvenient operation. The Mini Tunnel does away with all this disruption, all that is needed at the surface is a small working area around the top of the shaft which can usually be sited in some convenient out-of-the-way location. The Mini Tunnel, therefore, was produced to solve the same sort of problem that modern soft ground tunnelling itself was produced to solve, for it will be recalled from Chapter 6 how the previous very disruptive method of cut-and-cover construction was one of the factors that stimulated the search for effective methods of soft ground tunnelling. So the Mini Tunnel is not only a system of soft ground tunnelling in miniature, but was produced to solve one of the same problems that modern soft ground tunnelling was, albeit on a smaller scale. The Mini Tunnel won a Queen's Award to Industry in 1975.

The Japanese have taken the miniature tunnel concept even further than the Mini Tunnel in the Tele-Mole, designed and manufactured by

Figure 12.5 The 1.2 m diameter dkr tunnelling machine (1975)

Iseki Poly-Tech Inc of Tokyo. The Tele-Mole is a small-diameter soft ground full-face tunnelling machine with a pressurised face chamber and a spoil disposal system utilising bentonite slurry.[25] The whole machine is housed inside a shield of only 970 mm outside diameter. No miner is needed inside the shield because all the controls are brought to a control panel at the surface. A remotely controlled television camera is mounted in the shield so that operation of the hydraulic systems and the alignment and steering of the shield can be observed and corrected, if necessary, from a television monitor mounted on the control panel. The tunnel is lined using pipe jacked linings – a technique which will be described in the next Chapter – so that no miner is needed inside the tunnel to erect the lining. The market that the Tele-Mole is aimed at is the same one as the Mini Tunnel: the installation of sewers and water mains beneath urban areas. A total length of some 4.6 km of tunnel, on 19 different sites, has been driven with the Tele-Mole showing that the system is more than a mere novelty.

12.7 Summary

The soft ground tunnelling machine was a natural development of the tunnelling shield, being basically a tunnelling shield with mechanical excavation instead of hand excavation. The first ones were produced in 1896 for the Central London Railway and one of them, the Price machine, was an immediate success and became the forerunner of a long line of subsequent tunnelling machines. Compared with the hard rock tunnelling machines, the problem of excavation faced by the soft ground machine designers was very much easier and consequently soft ground machines were much earlier applied in the tunnelling industry. Their introduction coincided with the availability of heavy electric motors which are ideally suited to drive machinery in a tunnel. There has since been a steady development of soft ground tunnelling machines: originally they were designed to work mainly in clay, but the trend of development was to extend their applicability to other types of soft ground, particularly sand and gravel. For this purpose the pressurised face tunnelling machine was devised, as has already been discussed in Chapter 9. From the first machines of Thomson and Price, which were both invented by tunnelling contractors, to the present day, the development of soft ground tunnelling machines has taken place within the tunnelling industry, modern machines being the result of collaboration between consultants or contractors and tunnelling machine manufacturers. In this respect, the origin of hard rock and soft ground tunnelling machines is very similar, some tunnelling machine manufacturers making both types.

Notes and references for Chapter 12

1 The Central London Railway, construction of which commenced in 1896, ran from Shepherds Bush in the west to Liverpool Street Main Line Railway Station in the east; it now forms the middle section of the London Underground's Central Line. General descriptions of the Central London Railway scheme are given in the following two articles:
Anon (1898) The Central London Railway. *Engineering*, **65**, 18 February, pp. 214–6.
Anon (1898) The Central London Railway. *The Engineer*, **86**, 4 November, pp. 440–2.

2 Contemporary accounts spell his name as Thomson and Thompson; the former is chosen here as being the most often used form.

3 Anon (1898) The Central London Railway. *Engineering*, **65**, 22 April, pp. 485–6 and 500.

4 Anon (1898) The Central London Railway. *The Engineer*, **86**, 18 November, pp. 490–1.

5 Cuthbert, E.W., A.C. Lyons & B.L. Bubbers (1979). The Jubilee Line. *Proceedings of the Institution of Civil Engineers, Part 1*, **66**, Paper No 8250, pp. 359–406.

6 Tabor, E.H. (1909) The Rotherhithe Tunnel. *Minutes of Proceedings of the Institution of Civil Engineers*, **175**, Paper No 3743, pp. 190–208.

7 Copperthwaite, W.C. (1906) *Tunnel shields and the use of compressed air in subaqueous works*. London (Archibald Constable and Co Ltd) pp. 112–16.

8 Price, J. (1896) Central London Railway, Sections 1, 2 and 3. General description of the tunnelling machine proposed to be used on the above works, prefatory to the description enclosed. London (Unpublished manuscript).
Price, J. (1896) Central London Railway, Sections 1, 2 and 3. Description of tunnelling machine. London (Unpublished manuscript).

9 Dalrymple-Hay, H.H. (1909) Discussion on the Rotherhithe Tunnel. *Minutes of Proceedings of the Institution of Civil Engineers*, **175**, pp. 215–19.

10 Armstrong, A. (1968) Markham tunnelling shield equipment from 1887 to 1968. Reprint from *Broad Oaks Magazine*, Spring, pp. 1–8.

11 Markham and Co Ltd (1982) The history of Markham tunnelling shields. Chesterfield (Markham and Co Ltd).

12 Cuthbert, E.W. & F. Wood (1962) The Thames–Lee tunnel water main. *Proceedings of the Institution of Civil Engineers*, **21**, Paper No 6578, pp. 257–76. At the time it was constructed this was believed to be the longest soft ground tunnel driven in the world.

13 Bartlett, J.V. & J.R.J. King (1975) Soft ground tunnelling. *Proceedings of the Institution of Civil Engineers, Part 1*, **58**, Paper No 7826, pp. 615–28.

14 Clark, J.A.M. (1969) Aspects of mechanical shield tunnelling and comparison with hand shield tunnelling. *Proceedings of the Institution of Civil Engineers, Supplementary Volume*, Paper No 7270S (The Victoria Line), pp. 397–417.

In this paper a comparison was also made between tunnelling machines and hand shields used under identical conditions. Although tunnelling machines cost more to buy than hand shields they are faster, and it was concluded that they become economic for drives of upwards of one mile.
15 Watts, I.L., J.B. Northfield & L.F. Palfrey (1982) Wedge-block tunnel for Oxford's main sewer. *Tunnelling 82.* London (The Institution of Mining and Metallurgy) pp. 291–301.
16 *Ibid*, pp. 295 and 296–7.
17 Temporal, J. (1983) In Discussion on Tunnelling 82. *Transactions Section A of the Institution of Mining and Metallurgy*, **92**, pp. A56–7.
18 O'Reilly, M.P., F.F. Roxborough and H.J. Hignett (1976) Programme of laboratory, pilot and full-scale experiments in tunnel boring. *Tunnelling 76.* London (The Institution of Mining and Metallurgy) pp. 287–300.
19 Copperthwaite, W.C. (1906) *op cit*, pp. 112–3.
20 O'Reilly, M.P., S.G. Tough, N.D. Pirrie, H.J. Hignett & F.F. Roxborough (1979) Tunnelling trials in chalk. *Proceedings of the Institution of Civil Engineers, Part 2*, **67**, Paper No 8189, pp. 255–83.
21 The Channel Tunnel machine is described in the following manufacturer's brochure:
Robert L. Priestley Ltd (1976) Channel Tunnel project. Gravesend (Robert L. Priestley Ltd).
22 Tabor, E.H. (1909) *op cit*, p. 203.
23 Clark, J.A.M. (1969) *op cit*, p. 399.
24 This description is based in part on the manufacturer's brochures:
Mini Tunnels International Ltd (1975) *Design, Operation, Specification.* Old Woking (Mini Tunnels International Ltd).
Mini Tunnels International Ltd (1975) *The Mini Tunnel.* Old Woking (Mini Tunnels International Ltd).
25 Harding, P. (1981) Introducing the Tele-Mole, an innovation from Japan. *Tunnels and Tunnelling*, **13**, (2), pp. 26–30.

Figure 13.1 Pipe jacking (1980)

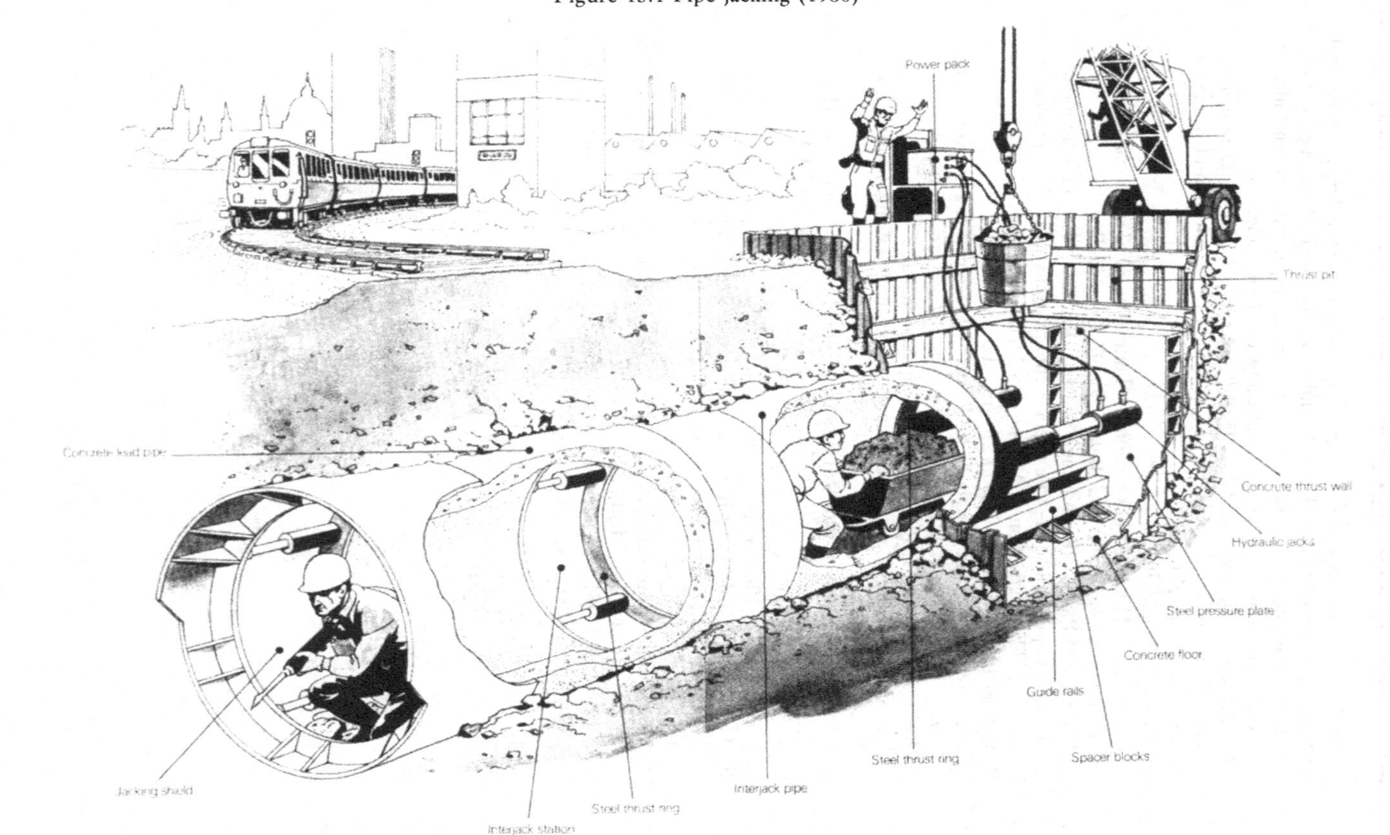

Other developments

Tomorrow to fresh woods, and pastures new.

The previous Chapter completes the examination of the main innovations that made the rise of the modern tunnelling industry possible. However, there have been a number of other developments in the industry that seem promising but about which it is too early to say whether or not they are major innovations. For completeness, it is proposed to deal with some of these in this Chapter, but the treatment will be shorter than for the previously discussed major innovations. Finally, an account will be given of what seems to be a most important development in the tunnelling industry – the rise of effective and innovative tunnelling in modern Japan. Again, in this Chapter, the accent will be on the innovations as solution to problems or as the meeting of needs.

13.1 Pipe jacking

In conventional soft ground tunnelling the lining for the tunnel is erected close behind the shield, near to the point where the ground is being currently excavated. By contrast, in pipe jacking although the ground is excavated from within a shield, the lining is formed by adding new sections at the rear end of the tunnel and then jacking the whole tunnel lining forward. Figure 13.1 shows the system in operation.[1] To do this a reaction frame capable of taking the reaction from the jacks which are thrusting the whole tunnel lining forward is necessary. This is a fairly simple matter to provide if the tunnel is being driven from the bottom of a shaft where the opposite wall of the shaft can provide the reaction, but if the tunnel portal is at ground level a substantial structure is required to provide the reaction. The great advantage of pipe jacking is that, being formed from complete sections of pipe, the completed tunnel has no longitudinal joints; it is therefore far easier to make watertight than a segmental lining and is also structurally stronger. The disadvantages of pipe jacking are that only straight or reasonably straight lengths of drive are possible and that there is a limitation on the length of drive which is determined when the frictional resistance of the ground on the pipe exceeds the thrust force available from the jacks. This limitation on length has been partially overcome by installing intermediate jacking stations along the tunnel so that the whole length of pipe need not be driven at once, and by lubricating the outside of pipe sections with bentonite slurry to reduce the ground friction.[2] As the name suggests, pipe jacking was originally used for installing small pipes for water, sewerage and other services but the method was soon found applicable to large-diameter pipes and for tunnels. It has been especially used for placing short tunnels, subways, culverts and services beneath existing trunk roads and railways, particularly when these are on embankments.

The earliest records of pipe jacking date back to the late nineteenth century in the United States but the method was not introduced into Great Britain until the 1950s, mainly for small-diameter pipes. During the 1960s rapid technical advances were made culminating in a method which was competitive with conventional tunnelling for large-diameter pipes and by 1967 the technique had been extended to the jacking of large rectangular box units.[3] Although pipe jacking for the larger sizes is clearly a method of tunnelling, a small pipe jacking industry has developed which is, to some extent, independent of the tunnelling industry. This is reflected by the fact that in 1974 the Pipe Jacking Association was formed to represent and promote the special interests of pipe jacking contractors. The Association has gathered together and published details of numerous pipe jacking jobs in Great Britain.[4] A sufficient number of pipe jacked tunnels has now been constructed for the subject to have attracted a state-of-the-art review.[5]

We will now look at the reasons why pipe jacking developed and at some of the problems to which it was the answer. Pipe jacking arose as a practical solution to some immediate tunnelling problems of the times. The factors which led to its growth during the 1950s and 1960s were as follows.[6] Firstly, there was a demand for pipelines for water mains, sewers and the new gas grid. Secondly, highway and railway authorities were becoming averse to disruption or disturbance to their increasingly heavily trafficked roads and railways with the result that trenching across these communication lines was no longer acceptable. Thirdly, driving small timbered headings – the traditional way of installing short lengths of pipe beneath barriers – was fast becoming a vanished skill. In addition, there was always the risk of collapse or settlement with the timbered heading and the progress was often slow. Fourthly, the lengths of drive for many of these jobs was small, as was the diameter of the pipes, making them unsuitable for shield tunnelling and segmental lining. All these factors combined together to provide the circumstances where pipe jacking filled a pressing need. The original need was for short drives and small diameters, but as we have seen, once the method proved itself, it soon developed to longer drives[7] and larger diameters where it became a direct competitor with traditional shield tunnelling.

A long-established technique in civil engineering is caisson sinking, in which a shaft is sunk by excavating the floor inside an open cylindrical or rectangular wall while at the same time the wall itself is allowed to penetrate slowly into the ground under its own weight or under a superimposed weight known as kentledge. The wall, called the caisson, can be made of cast-iron, concrete, masonry, brick or timber. As the caisson sinks, from time to time fresh sections are added at the top, any kentledge used being temporarily removed in order to do this. It can be seen that pipe jacking is very similar to caisson sinking except that a horizontal tunnel is formed instead of a vertical shaft. In pipe jacking, because no assistance is available from gravity, jacks have to be used to provide the thrust instead of the self-weight or kentledge used in caisson sinking.

13.2 The Unitunnel

The Unitunnel is a tunnelling system which combines the advantages of pipe jacking with those of shield tunnelling whilst eliminating some of the disadvantages of pipe jacking. The Unitunnel is shown in Figure 13.2. At the front end of the tunnel there is a tunnelling shield which is the same as a conventional shield except that it has no tailskin. Also, instead of jacks to shove it forward, the shield has an inflatable rubber bladder shaped like a bicycle inner tube. The tunnel is lined, as in pipe jacking, with short sections of pipe which are lowered down the shaft and added on to the rear end of the tunnel. Between each section of pipe is placed an inflatable

Figure 13.2 The Unitunnel (1981)

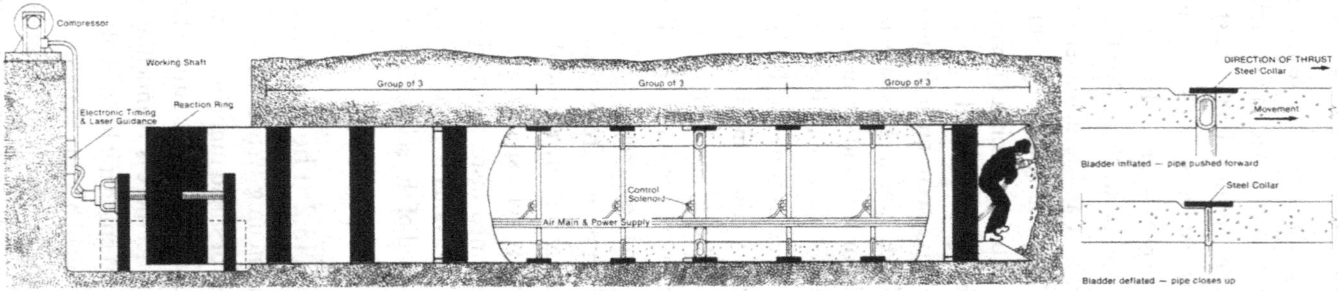

bladder of the same type as the one behind the shield. It can be seen that if an initially deflated bladder is inflated, the shield or the section of pipe adjacent to it will be pushed forward as shown in the diagram on the right of Figure 13.2. All these bladders, including the one behind the shield, are connected to a compressed air supply via individual solenoid control valves. Movement forward of the whole assembly is accomplished by simultaneously inflating every third bladder along the line (see main diagram of the Figure). This is because in any group of three pipes, application of a force at the joint between any two will force the single one away from the other two, owing to the fact that the frictional resistance of the ground on the pair of pipes is twice that on the single pipe. It should be noted that the reaction ring at the rear end of the tunnel has only to provide a reaction for the last group of three pipes and not for the whole length of pipes. By means of an electronic timing mechanism all the first, second and third pipes in each group of three are moved forward in turn, the sequence being repeated .automatically, so that the Unitunnel proceeds with a shuffling kind of movement through the ground. Three full cycles are completed every minute giving an average forward movement of 22 mm per minute. On completion of the drive, starting at one end of the tunnel the first bladder is inflated and then removed and sealant is placed into the joint. The second bladder is inflated to thrust the second pipe along to close the gap and make it watertight. The second bladder is then removed and the process is repeated. This procedure is continued down the line to give the smooth finished tunnel.[8]

The Unitunnel overcomes the problem of the limitation on length of drive which applies to pipe jacking, because with the Unitunnel the maximum frictional resistance offered by the ground is that over only one in every three sections of pipe and not over a long length as in pipe jacking. The shield of the Unitunnel is made exactly the same diameter as the pipe sections because there is no tailskin and, therefore, there is no annulus to grout as there is in conventional shield tunnelling, and no settlement at the surface arising from the difference in diameter between shield and lining. Also, with the Unitunnel the shield is steerable to some extent.

The first time the Unitunnel was used was for a short drive under the A38 trunk road at Slimbridge, Gloucestershire, which demonstrated that the system worked and enabled teething troubles to be rectified.[9] The first serious application of the system was for a 1.3 m internal diameter flood-relief sewer tunnel at Acton in West London in 1981.[10] The ground consisted of sandy clay overlying clayey gravel, the tunnel axis level being at the interface between the two strata; excavation was by a single miner operating a pneumatic spade. The overall progress rate was 18 m per week, but only one shift per day was being worked and tests were also being made on the system. With normal operation, 12 m per day advance was expected.

The inventor of the Unitunnel is Richardson, a man who had worked for many years in the tunnelling industry and had earlier devised the system of shield tunnelling for small-diameter tunnels known as the Mini Tunnel that was described in Section 12.6. Through his company, Marcon International, Richardson has sold the rights of the Unitunnel to the firm of contractors, Mowlem (Civil Engineering) Ltd, and it was Mowlem who operated the Unitunnel system on the Acton drive. The idea of the Unitunnel came to Richardson as a result of thinking about the action of the earthworm when moving along a burrow in soil.[11] The idea of moving pipe sections by inflatable bladders had occurred to Richardson some years before, but it was the idea of simultaneously moving one pipe section in every three by thrusting off the other two that was suggested to him by the action of the earthworm. The earthworm, *Lumbricus terrestris*, is long in relation to its diameter and if it had to move all at once, would probably be unable to overcome the frictional resistance of the ground all along its body. The earthworm, however, utilises the frictional resistance between the bristle-like *setae* on certain segments of its body and the surrounding ground as a means of obtaining the reaction necessary to thrust other parts forward, and by alternately lengthening and contracting its body the whole earthworm eventually moves along.[12] As in the case of the shipworm and the tunnel shield (see Section 6.5), so with the earthworm and the Unitunnel, we have a clear instance of a tunnelling innovation being suggested by a tunnelling animal.

In Chapter 1 reference was made to the Civil Engineering Innovation Competition, jointly sponsored by the magazine *New Civil Engineer* and the National Research Development Corporation. In fact, the Unitunnel was the innovation that won the first prize in this competition,[13] an outcome that was gratifying to the tunnelling industry since the competition was for innovation over the whole field of civil engineering and not just from one sector. Some observations can be made about this result and the judges' comments on it.[14] Firstly, the winning entry was from an individual who had already shown innovative ability; secondly, the inventor had had years of experience in the tunnelling industry and knew of the needs of the industry. Taking the first point, we have seen before in this account of innovation in the tunnelling industry that a particular person is sometimes responsible for more than one innovation, for example Greathead. Taking the second point, we see again how potent an externally suggested idea can be when it occurs to a person with a mind aware of the needs of the time and anxious to find a new solution to a problem. Lastly, it is important to note that although the Unitunnel won the Civil Engineering Innovation Competition, it was not invented to win the competition. What the competition did was to reward an innovation that was there already – the Unitunnel did not result from the stimulus of the competition.

In the case of the Unitunnel we actually know what motivated the inventor to develop it[15] and this is summarised as follows. It will be recalled that Richardson had earlier invented a small-diameter shield tunnelling system known as the Mini Tunnel. Now although Richardson was the inventor of this, the firm that had put up the money to develop it to the stage of a working proposition held all the rights to it, and although Richardson was a director, the company, not he, would reap the financial rewards. He therefore left the company and began developing a small-diameter air-driven tunnelling machine. However, he was then baulked by not being able to use with the machine, the three-segment lining that was the exclusive property of the firm that held the rights to the Mini Tunnel. This led him to experiment with complete sections of pipe with pieces of inflatable tubing placed between each pipe, and eventually to the idea of the Unitunnel which has been described. It is probable that if Richardson had not felt that he had obtained insufficient financial reward for his invention of the Mini Tunnel, he would not have invented the Unitunnel. Although the judges of the Innovation Competition considered that the Unitunnel filled an 'urgent need', this need could in fact, be filled by either the Mini Tunnel or by pipe jacking. It may be, of course, that the Unitunnel will turn out to fill the need for small-diameter tunnels better, or more economically, than either of these other methods – time will tell. What is clear from the inventor's own explanation is that the Unitunnel was developed in an effort to obtain the financial reward that he felt he had missed when inventing the Mini Tunnel, and not because of any serious problems that the other methods had encountered. The Unitunnel is, of course, aimed at a potentially large market as the nation's stock of nineteenth-century sewer tunnels comes up for renewal in the future so that there may be room enough for several methods.

13.3 The laser

As mentioned in Chapter 1, the subject of tunnel surveying is not covered in this book because, apart from the laser which will now be dealt with, tunnel surveying in general was based upon traditional civil engineering surveying methods and itself produced no innovations.[16] One of the principal operations in tunnel surveying is to determine the correct line and level of the tunnel while it is being driven. Traditionally, the tunnel line was carried along the tunnel by the surveyor using a theodolite sighting on to plumb bobs (or plumb lights) suspended from plugs set into the tunnel roof; the tunnel level was carried along by establishing bench marks periodically along the tunnel wall using a level and staff.[17] Information on the line and level of the tunnel had then to be passed to the leading miner at the face so that any corrections required could be made before the next advance. The whole process was fraught with difficulty. The tunnel environment is inimicable to careful instrumental work, and the survey plugs and bench marks are prone to being dislodged or demolished by the tunnel construction operations. All these problems were virtually eliminated at a stroke by the introduction of the laser as an underground surveying tool.

A laser is a device which amplifies or generates coherent light waves by stimulated emission of radiation.[18] Unlike the waves emitted by an ordinary lamp, the laser produces a light beam that does not diffuse; also, the laser emits waves travelling in the same direction at the same frequency – a property known as 'coherence'. The laser thus produces a highly parallel, intense beam of coherent radiation occupying a very narrow band of frequencies from the infra-red or visible portions of the spectrum. The first working laser was made by T.H. Maiman in 1960 following a proposal put forward in 1958 by A.L. Schawlow and C.H. Townes. Maiman's first laser used a crystal of synthetic ruby as the active material but both gases and semi-conductors are now used as the coherent source. The laser commonly used in tunnelling applications employs a helium-neon source.[19] The application for which the laser was developed was for communication purposes but other uses have since been found for it including welding, surgery and, of course, the use discussed here – tunnel surveying.

What the laser provides is a clearly visible bright beam of light which defines both the line and level of the tunnel, and which terminates in a small bright spot of light which can be clearly seen on the tunnel face or on the surface of a target fixed to the shield or to the tunnelling machine. The laser beam is set up and aligned using a theodolite in the conventional manner, but once this is done, and provided the laser itself is fixed in a protected position in the tunnel, then the beam itself is immune to disturbance from tunnelling plant and equipment and is present all the time. Lasers are usually fitted towards the upper part of the tunnel so that they are not obscured by plant. The miners have a clear indication of the correct line and level at all times and do not need to wait, as before, for an engineer to carry the survey forward. The only limitation is that when driving on a curve there comes a time when the laser must be repositioned. Also, on a very long straight tunnel the dispersion, due to dust, smoke and fumes in the tunnel atmosphere causes some broadening of the beam, so that the laser must occasionally be brought forward and repositioned. The laser, therefore, provides an accurate, constant, clearly visible and undisturbable survey line that projects itself forward as the tunnel advances with no further need for attention once it has been set up. It is an example of an innovation almost perfectly filling a need. In addition to its use as a means of setting out the line and level of a tunnel, the laser beam can be used to check the alignment and shape of rings of tunnel lining segments[20] and to control the steering and roll of tunnelling machines.[21] A laser which is proof against causing explosions has been developed for tunnels in which mine gases are present.[22]

There can be no doubt as to the great convenience that the laser has brought to tunnel surveying, lining alignment and direction control for tunnelling machines. All these hitherto time-consuming and awkward tasks have been made relatively quick and straightforward. However, boon though it is, the laser is a minor rather than a major tunnelling innovation.

13.4 Extruded linings

Chapter 8 discussed how prefabricated tunnel linings were developed in order to take full advantage of the tunnelling shield. It will be remembered that the lining had to be capable of taking the full thrust of the shield and the full overburden pressure of the ground as soon as it had been erected. However, erection of a prefabricated lining is a somewhat laborious process and tunnel engineers had long dreamed of a tunnel lining that could be extruded continuously from behind the shield as it advanced, leaving a smooth tube rather like the hard calcium carbonate lining of the shipworm's burrow (see Section 6.5). There were three technical problems that had to be solved before this idea could be realised and they all hinged upon the fact that concrete – which was the only practicable material for an extruded lining – needed a period of about 24 hours to set. The first problem was that by itself the freshly extruded lining would not be capable of providing any reaction for the hydraulic rams when the shield was thrust forward; the second was that it would not be able to support itself and the third was that it would not be able to support the ground. In fact, as we shall see, the three problems were all solved with one piece of plant.

The first use of extruded linings seems to have been in the Soviet Union some time before 1975.[23] For in that year, two Russian-manufactured type TSCB-3 mechanised tunnelling shields which incorporated equipment for constructing an extruded concrete lining were obtained for the construction of the 5.1 m internal diameter tunnels for the Prague Metro in Czechoslovakia.[24] To construct the extruded concrete lining a special piece of equipment was installed just behind the shield. It consisted of a large steel shuttering made up of ten detachable rings, a carriage mechanism for transporting one of these rings at a time, and pipework for transporting and placing the concrete. The front of the shuttering extended just inside the tailskin of the shield and the concrete was pumped in behind the shuttering at this position. The concrete was compacted by the action of a pressing ring which was actuated when the shield jacked forward. As the shield moved forward the concrete moved out from the tailskin and filled the whole of the space between the shuttering and the ground. A fresh ring of shuttering was detached from the rear, brought forward and added at the front position and the process was repeated. The whole operation depended upon the whole length of shuttering being sufficiently long so that when a ring of shutter had moved from the first position at the front to the tenth position at the rear, the concrete between it and the ground had set hard. It can be seen that this piece of shuttering solved all three problems. It effectively confined the fresh concrete so that although still in a fluid condition it could react against the jacking pressure of the shield applied to it via the pressing ring, it supported the concrete until it had set hard enough to support itself, and it also supported the ground at the same time. On the Prague Metro it was claimed that the extruded concrete lining was cheaper than a prefabricated lining and that less labour was needed but that the cost of operating the machine was somewhat higher.[25]

During the years 1974–7 a similar development was taking place independently in Japan. The solution was almost identical to the Russian system which has just been described, the concrete being placed behind detachable shuttering and compacted by the shield jacks acting through a pressing ring.[26] The Japanese system could incorporate reinforcing steel, if desired, into the lining and thereby produce extruded reinforced concrete lining. They also observed that because the fresh concrete flowed out behind the tailskin and filled the whole of the void between the shutter and the ground straight away, there was less settlement at the surface over the tunnel than with conventional shield tunnelling using segmental lining.

A different system of extruded concrete lining was used on the construction of the 6.9 m diameter tunnels for the Frankfurt Metro in the Federal Republic of Germany in 1982.[27] Here a 12-ring detachable shuttering was used in the same way as in the previously described jobs, but instead of using the shield jacks to compact the concrete via a pressing ring, the concrete was pumped into the space behind the shuttering and compressed by hydraulic pressure, a ring-shaped rubber seal preventing the pressurised concrete from leaking back into the shield. The shield used was a Westfalia Lünen blade-shield which is one in which the jacking forward of the shield does not rely on obtaining reaction from the lining. It works as follows: the shield consists of a large number of independently movable blades arranged around the periphery rather like the staves of a barrel. The shield moves forward by advancing single blades one at a time, utilising as a reaction the friction of all the remaining blades on the surrounding ground. Because of this the shield's forward movement is entirely independent of any reaction from the tunnel lining, which made it ideally suited to be combined with an extruded lining system.

More recently, in 1983, the firm Hochtief AG of Essen, Federal Republic of Germany, have produced an extruded lining system utilising the pressure ring principle and producing a continuous extruded lining of concrete reinforced by steel fibres (Figure 13.3). The fresh concrete is pumped into the annular space between the shuttering and the tunnelling wall via a supply pipe, the flow of concrete being indicated in the Figure by arrows. The shield is shoved forward off the pressure ring using one set of hydraulic jacks (see upper part of Figure) whilst a sliding front-form which supports the forward end of the extruded lining is controlled by another set of hydraulic jacks (see lower part of Figure). It is instructive to compare the smooth continuous concrete lining of Figure 13.3 with the smooth continuous calcium carbonate lining of the shipworm's burrow shown in Figure 6.4. In passing it can be noted that the Hochtief extruded lining system is built into a Hydroshield, making a link with Section 9.5.

The success of the tunnel schemes on which extruded concrete linings have been used has led the proponents of the method to state that in the

Figure 13.3 Extruded lining (1983)

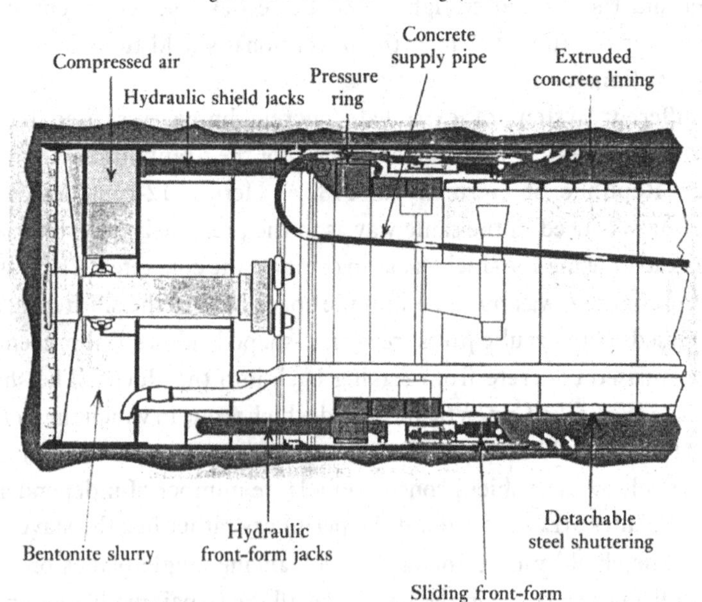

future most tunnels will be lined with continuously extruded concrete. It is too early to say whether or not this will be so, but it is fairly certain that the method would not be used for short tunnels, say less than 2 km, because of the cost of assembling and dismantling the shuttering and concreting plant in the tunnel.

13.5 New primary lining methods

In the construction of soft ground tunnels the erection of the permanent lining takes place as soon as the tunnel has been excavated, usually by the erection of a prefabricated segmental lining in the tailskin of the shield as described in Chapter 6. In the construction of hard rock tunnels, whether by drill-and-blast or by tunnelling machine, the construction of the permanent lining is often left until the whole tunnel has been excavated. After the tunnel has been driven, a lining train passes through the tunnel, consisting of a travelling shutter and concreting plant, and the permanent concrete lining is placed. However, if the rock will not stand unsupported, because of being broken up by closely spaced joints for example, then some temporary support, often called primary lining, is installed to keep the rock in place until the permanent lining is placed. If a primary lining has been used, then the permanent lining is sometimes called the secondary lining. Traditionally the primary lining was steel arches, with or without timber lagging, depending upon how much support was needed.

Since the 1950s, however, new methods of primary lining have been introduced and are gaining increasing application in hard rock tunnelling. These are rock bolts, wire mesh and sprayed concrete. The idea behind rock bolts, which were used in a slate quarry in North Wales as early as 1872, is quite simple: a loose, or potentially loose, block of rock in the tunnel roof or wall is fastened via the rock bolt to the main mass of intact rock. If a fan-shaped array of rock bolts is installed in the tunnel roof say, then a whole zone of loosened rock can be made stable. Rock bolts are made of steel rod, typically 20 mm in diameter. They range in length from 2–5 m and have some method of securely attaching the furthermost end into the intact rock mass; mechanical expansion or resin grouting are the commonest methods of attachment. A metal plate is put on the threaded end that projects into the tunnel and this is bolted up tight.[28]

Wire mesh is also a very simple idea. Sheets of wire mesh are placed around the tunnel wall and held in place either with steel arches or rock bolts. Wire mesh provides little structural support but it does prevent small pieces of rock falling out from the space between the rock bolts or arches and can assist in preventing progressive spalling away of the rock, termed 'ravelling', from occurring on the tunnel walls. Sheets of wire mesh are typically made from wire of about 4 mm diameter welded into a mesh of spacing about 100 mm.

Sprayed concrete has been used in civil engineering since 1909, and was used underground in Pittsburgh, USA, in 1914. It consists of spraying a mixture of aggregate, sand, cement and water under pressure from a gun onto the surface to which it is desired to apply a layer of concrete.[29] Because of its method of application, sprayed concrete is often referred to as 'shotcrete' or 'gunite'. As used in tunnels, sprayed concrete is applied to the tunnel walls as soon as possible after excavation. It is often used in conjunction with arches or rock bolts and wire mesh. Typically, shotcrete is sprayed on until a thickness of 80–300 mm has been built up. When it sets it provides some structural support, but more important, if adhesion to the rock surface is good it prevents ravelling of the rock completely and forestalls the initiation of progressive failure. It can be noted that for some tunnels where the rock is in sound condition and a good tunnel profile has been formed during excavation, the primary lining of sprayed concrete also forms the final lining, there being no secondary lining placed afterwards. If sprayed concrete is applied through a layer of wire mesh, then the mesh contributes to the strength of the sprayed concrete by acting as reinforcement. Sometimes sprayed concrete is applied in two layers with wire mesh between them, put there specifically as reinforcement. Sprayed concrete is said to be derived from the use of sprayed plaster on wire frames to provide a base for mounting animal skins, a technique developed by Carl Akeley of the Smithsonian Institution in Washington in the early 1900s. This was adapted to commercial use in the form of the Allentown Cement Gun by that firm when it bought Akeley's patent in 1915. Subsequent developments in the 1930s made better pumps available for pumping concrete to the nozzle of the gun.

These new measures of primary support came into use gradually in different parts of the world during the 1950–60 period, but in Austria they became incorporated into a design philosophy which became known as the New Austrian Tunnelling Method (often abbreviated to NATM), so called to distinguish it from the old Austrian System of tunnelling[30] which was one of the nineteenth-century methods of tunnelling with timber supports. The design philosophy of the New Austrian Tunnelling Method has been critically reviewed by E.T. Brown,[31] who also provides an extensive bibliography of the method. The essential features of it are (i) the early installation of a semi-flexible primary support system utilising steel arches or rock bolts and wire mesh, in combination with sprayed concrete, and (ii) the monitoring of the deformation of the primary lining so that the lining measures can be increased if required. As has been pointed out, the individual elements in item (i) are not new, and item (ii) is no more than the 'observational method' which has for many years been used in civil engineering.[32] In assessing whether these new primary support methods constitute a major innovation in tunnelling, we can conclude that in themselves they do not, but that as used in the New Austrian Tunnelling Method in which they are all brought together into a design concept, then they probably do.

13.6 Composite tunnelling machines

It will be recalled from Section 11.9 that the roadheader, introduced into the tunnelling industry from the coal-mining industry, offered a versatile ready-made tunnelling machine. However, a problem remained in that the roadheader itself had no means of supporting the ground around the heading if this was required, and the usual method of doing this was to install steel arches as a separate operation. It was not long, therefore, before the idea of mounting a roadheader boom inside a shield occurred, and by the 1970s this type of machine was being offered by a number of manufacturers. Figure 13.4 shows an example of one of these machines made by Dosco Overseas Engineering Ltd. It is the Dosco TM 1800 machine which is made up of the boom of a Dosco Mark 2A roadheader mounted in a shield.[33] These machines are sometimes called boomheader shields. The advantages of a machine of this kind are obvious. It combines the cutting and excavating power of the roadheader boom with the ability of supporting the ground of the conventional tunnelling shield. Standard segmental lining can be erected within the security of the tailskin of the shield if required, as shown in the Figure. Furthermore, the roadheader boom can be mounted much more rigidly inside a shield than when on a normal roadheader machine chassis, and this results in a higher degree of stiffness of the overall system which, we have already commented, is a desirable property for a tunnelling machine to have. A good example of the use of a boomheader shield was on the Southern Foul Water Interceptor tunnels in the city of Bristol.[34] Here a Dosco TM 1800

Figure 13.4 Dosco TM 1800 boomheader shield (1974)

machine with a 3 m diameter shield was used to drive the tunnels which were lined with a bolted concrete segmental lining erected within the tailskin of the shield. Advance rates of 60 m per week were achieved on this job. The ground consisted of Keuper Marl which at this site included thick bands of hard sandstone; the roadheader boom was able to excavate the hard sandstone bands with ease while the shield was essential for supporting the ground as a whole.

Another form of composite tunnelling machine is one which consists of a shield in which an articulated arm is mounted, on the end of which is fitted a ripper, a bucket armed with ripping teeth. These machines are sometimes called digger shields. Figure 13.5 shows a digger shield manufactured by the American firm Zokor International Ltd.[35] The machine is built inside a tunnelling shield which has all the usual facilities for the erection of a segmental lining in the tailskin and for jacking itself forward off the erected lining. The excavator consists of a powerful articulated arm on the end of which is mounted a bucket fitted with ripper teeth. The base of the arm rotates on a turntable which can turn through a

Figure 13.5 Zokor digger shield (1979)

complete circle and thus direct the bucket to any part of the face. All the arm and bucket movements are controlled hydraulically. The excavator bucket is capable of dealing with clays, sands and gravels including the capacity to handle cobbles and some boulders, and certain rocks such as shales provided they are already well broken up by joints or bedding plane partings. The spoil falls to the invert where, as the shield is jacked forward, it rides up an apron on the floor of the shield and is collected by a belt conveyor which takes it well to the rear of the shield and tips it into rail-borne mine cars. Zokor digger shields are made in diameters ranging from 7.5 to 12 m; the Figure shows a 9 m diameter machine.[36] The Zokor digger shield also incorporates some other interesting features. First, around the crown of the shield are provided eleven poling plates. These can be pushed forward a distance of 1.5 m in front of the shield independently, and while the shield is stationary, to support the roof of the excavation while the face is being dug out with the bucket. One poling plate is shown slightly advanced at the top of the shield in Figure 13.5. Second, the shield is provided with seven breast plates. These can be used to support the upper part of the tunnel face from collapse while the bucket is being used to excavate the lower half of the face. Both the poling plates and breast plates are automatically retracted when the shield is jacked forward. Third, the digger shield is provided with two probe drills, located in the upper portion of the shield and capable of drilling up to 8.5 m forward of the shield edge. This is a facility not often provided as a purpose-made item on a tunnelling machine and shows foresight on the part of the machine designer to the needs of the tunnelling engineer. With most tunnelling machines, if probe drilling is required it has to be done as an afterthought by fitting a drill rig into the head of the machine as best one can.

Although the boomheader shield and the digger shield are both tunnelling machines of fairly recent origin so that it is difficult to tell how long lived these types will be, it would seem that they may well find a permanent place because they fill an important gap, being designed to excavate tunnels in ground that is unsuitable either for full-face hard rock tunnelling machines or for soft ground tunnelling machines. The boomheader shield and the digger shield are ideal for the soft rocks or hard ground conditions that often prove troublesome for the other machines. The tunnel engineer now has a choice of machine – a full-face tunnelling machine, roadheader, boomheader shield, digger shield, soft ground tunnelling machine – and can make his choice to suit the particular ground conditions at the site. Composite machines, therefore, seem likely to become innovations of some importance.

13.7 Innovation in tunnelling in modern Japan

Before the mid-nineteenth century, Japan was a medieval feudal state in which civilisation and culture were essentially derived from China; indeed the Japanese written language is based on characters borrowed entirely from Chinese. The opening of Japan to influences from the West began in 1853 and 1854 when naval missions led by the American Commodore Matthew Calbraith Perry (1794–1858) opened the country to foreign trade. In 1868 the Meiji Restoration took place in which the hitherto vestigial position of the Emperor was restored to power at the head of a central system of government that set about to reproduce in Japan most of the elements of Western civilisation. The navy was modelled on British lines, the army, the educational system and the political institutions were based on German examples, and from France was borrowed the legal system. By the end of the century the process of modernisation was virtually complete. Heavy industry for iron and steel production, ship building, manufacture of railway locomotives and rolling stock, electrical equipment, cement, paper, china, glass and textiles had been well established, as well as a host of smaller factories and workshops for light industry engaged in the production of a whole range of consumer goods. A railway network had been extended to most parts of the Japanese islands and by 1906 some 5000 route miles were in use.

During the first part of the twentieth century Japan continued with the process of borrowing ideas from the West, so much so that by the 1920s Japan was to be considered a World Power, especially in regard to its naval predominance in the Western Pacific. At the same time a militarist faction was gaining influence in political circles which sought to gain by conquest an overseas empire which would provide Japan with the fuel and raw materials that the rapid pace of industrialisation in the homeland islands was increasingly demanding. Even this ambition could be seen as imitation of the Western nations, many of whom had colonial empires. The Japanese leaders considered that Japan's 'hour of destiny' had struck in December 1941. They saw the Western powers who had colonies in the

Far East and Pacific, France, the Netherlands and Great Britain, either defeated by Germany or apparently on the verge of defeat, and the United States isolated in uneasy neutrality. With audacity and ruthlessness they tried to seize for Japan an Empire to match that of any possessed by the West. After an initial series of brilliant conquests Japan was eventually defeated by an alliance in which the United States played the major part, the final surrender in August 1945 being precipitated by atomic bombs dropped on the cities of Hiroshima and Nagasaki.

The year 1945 was not only the year that saw the end of Japan's short-lived Imperial ambitions but was also a year that was a turning point in the whole history of the country.[37] As well as opening the way for social and political changes it also opened the way for profound changes in Japanese technology. Before 1945, as we have seen, the Japanese had concentrated on borrowing or copying technological ideas from the West. So pervasive was this that Japanese manufactured goods were commonly regarded as being wholly imitative. After 1945 all this changed. The destruction of plant by bombing during the war ensured that Japanese industry, when it arose again, was equipped with up-to-date methods and machinery. By the 1960s, the West began to take notice of the high quality of Japanese products. Motor cycles, cameras and transistor radios were just some of the many mechanical, optical and electrical goods that began to gain for Japanese goods a reputation for quality, good design and reliability. And this had to be obtained in the face of the prejudice against prewar Japanese goods. Significantly too, the Japanese began taking out patents and complaining about others copying their designs! This birth of technological innovation also extended to the tunnelling industry.

The postwar expansion of tunnelling in Japan[38] was produced by factors that were very similar to those that caused the rise of the tunnelling industry in Europe and the United States some 100 years before. As the larger cities in Japan expanded and the population became more concentrated in them, the need for soft ground tunnels for water supply and sewerage arose, particularly as the cut-and-cover methods that were first tried proved more and more difficult due to disruption of urban life. This paralleled the nineteenth-century experience of London. Also the expansion of Japan's railway network called for many hard rock tunnels because the Japanese islands are characterised by mountainous terrain calling for frequent tunnels. The same was true for the road tunnels required for the postwar highway construction programme. Construction of a Metro began in 1967 in Tokyo of which 62 km had been completed by 1987. The Japanese have also driven the longest undersea tunnel in the world, excavation of which was completed in March 1985. This is the Seikan Undersea Tunnel[39] between two of the main islands of Japan: Honshu and Hokkaido (Figure 13.6). The 54 km long tunnel will carry the Japan National Railways line below the 23 km wide Tsugaru Strait to replace the present ferry crossing which is often suspended in stormy weather. The first trains are expected to run in early 1988.

To create this underground infrastructure the Japanese have fostered a vigorous tunnelling industry. In the traditional way, they have borrowed ideas from the West, but in addition they have made advances in tunnelling technology themselves. In Section 9.7 the Japanese pressurised face tunnelling shields were described which demonstrate their innovation in the field of soft ground tunnelling. But the Japanese have been innovative in hard rock tunnelling as well, having developed advanced drilling jumbos having hydraulic rock drills and featuring

Figure 13.6 Location of the Seikan Tunnel

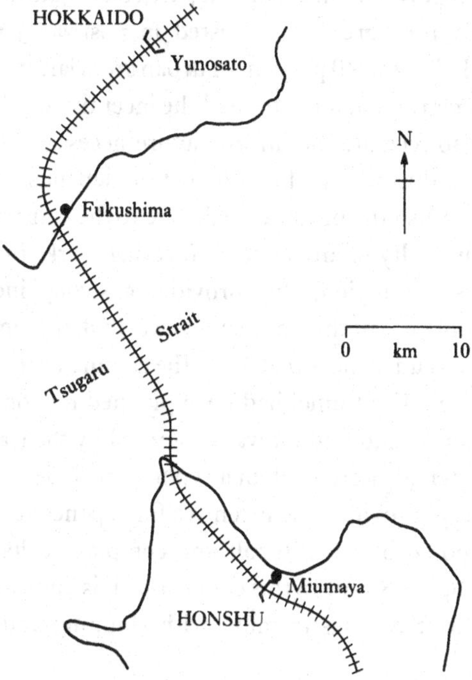

automated and computerised control systems.[40] In the development of both soft ground and hard rock tunnelling equipment the Japanese have one eye on the home market and the other on the export market for, as remarked in Section 9.7, Japanese engineers are now able to win tunnelling contracts abroad. As a further example of pioneering, on the Seikan Undersea Tunnel during the 1970s, Japanese engineers practised long-hole exploratory horizontal drilling up to 1000 m ahead of the face including geophysical borehole logging.[41] To show how important Japan now is as a tunnelling nation, Table 13.1 compares sixteen of the world's longest tunnels: of these eight are Japanese, seven having been constructed since 1945. For comparison purposes the proposed Channel Tunnel is included.

Having seen that the Japanese have become so innovative in tunnelling, we can now enquire why this should be. There have been a number of rather superficial explanations given which will not be considered here – instead attention will be given to the results of systematic and thoughtful enquiry into Japanese innovation in the transportation construction industry. This has been carried out by Boyd C. Paulson who devotes four chapters of his report[42] to tunnelling in particular. Paulson first reviewed the incentives for progress that existed in postwar Japan. With two million people killed and 40 per cent of urban areas laid waste by bombing the Japanese construction industry had the incentive of a huge rebuilding job ahead of it to recreate the infrastructure necessary for the Japanese manufacturing industry, on which the nation depends, to get on its feet again. There was also the physical challenge to the construction industry posed by the difficulty of any civil engineering works in a mountainous and volcanic island region; this provides a strong incentive to find solutions to problems. In this context it is worth digressing for a moment to consider the circumstances that led to the commencement of the Seikan Undersea Tunnel. The tunnel had been planned for some time, but the decision to go ahead and build it was prompted by the tragic loss of 1414 lives when five ferries were sunk in a typhoon in 1954,[43] an event which also seems to have provided the incentive for Japanese engineers to meet the challenge posed by the difficult and complex geological conditions beneath the Tsugaru Strait. In this connection it is impossible not to think of our own Dover Strait swarming with shipping and cross-Channel ferries.

Table 13.1. *World's longest tunnels*

Name	Country	Type	Date completed	Length km
Land Tunnels				
Daishimizu	Japan	Railway	1982	22.2
Simplon	Italy and Switzerland	Railway	1906	19.8
Apennines	Italy	Railway	1931	18.6
St Gotthard	Switzerland	Road	1980	16.3
Rokko	Japan	Railway	1971	16.2
Haruna	Japan	Railway	1982	15.4
St Gotthard	Switzerland	Railway	1882	15.0
Nakayama	Japan	Railway	1982	14.8
Lötschberg	Switzerland	Railway	1913	14.5
Subaqueous Tunnels				
Seikan	Japan	Railway	1985	53.9
Channel	Great Britain and France	Railway	—	51.8
Shin Kanmon	Japan	Railway	1973	18.7
Severn	Great Britain	Railway	1886	7.0
Mersey	Great Britain	Railway	1886	4.9
Mersey	Great Britain	Road	1934	4.2
Kanmon	Japan	Railway	1942	3.6
Kanmon	Japan	Road	1958	3.5

Turning to the reasons for the advancement of the Japanese transportation construction industry, Paulson concluded[44] that this is without doubt because of the establishment of large, well-staffed and well-funded research laboratories throughout the companies comprising the industry. Since many of these companies are heavily involved in tunnelling, this must also account for the innovation in tunnelling technology that has been so marked in Japan. In fact the level of research activity in Japanese tunnelling firms is remarkable. For example, in the firm of Ohbayashi-Gumi Ltd alone, a main contractor for underground works, there were 110 research engineers in 1975, and in the 26 top Japanese transportation construction companies there were no less than 1770 research engineers in all – and this total did not include development engineers, or field engineers engaged in trouble-shooting. Paulson observed that these figures far exceed the numbers of research engineers in similar fields in the United States, and the same is true of Great Britain. All the large Japanese construction firms have research laboratories, the result of which is that innovation and technological development become an integral part of each firm's operations, and by a transferring process the benefits are passed down to the smaller firms which do not have research laboratories or research staff of their own; the overall result is that the transportation construction industry in Japan has become the most technologically advanced in the world. To summarise, two factors are seen to have been important in the rise of innovation in the Japanese tunnelling industry. First, a very pressing need for tunnels of all kinds following the destruction during the war, coupled with the challenge of difficult tunnelling conditions due to the geography and geology of the country. Second, the wherewithal to produce the technological innovations that would meet the need and the challenge: this was found by the deliberate practice of establishing research laboratories, not only in universities or in central government institutes, but in the construction firms where the problems arose and where the solutions could be applied. These research laboratories, originally set up to help initiate postwar reconstruction, have been the means of advancing Japanese tunnelling technology to the front rank.

There has also been another factor which must be considered when discussing Japanese tunnelling. That is the willingness of the clients, usually public bodies, to allow novel machines or new methods to be used on short tunnel drives. In Great Britain it is usually considered that it is uneconomic to use a tunnelling machine on a drive of less than 1 km in length; a hand shield would usually be used instead. However, in Japan tunnelling machines are frequently used for drives of shorter than 1 km; for example, many of the drives discussed in Sections 9.6 and 9.7, and this allows Japanese tunnel machine designers and contractors to try out new ideas and to build up experience with novel methods where the financial consequences of failure would be small. This in turn gives them the confidence to tackle large jobs with innovative methods in the knowledge that they have at least been tried out on the smaller jobs where any major practical problems should have come to light.

Notes and references for Chapter 13

1 Most of the details of the operation of pipe jacking are given in the two following guides:
Pipe Jacking Association (1980) *Jacking concrete pipes*. London (Pipe Jacking Assn).
Pipe Jacking Association (1981) *A guide to pipe jacking design*. London (Pipe Jacking Assn).
There is also a description of pipe jacking and allied techniques in:

Thomson, J.C. (1967) Horizontal earth boring. *Proceedings of the Institution of Civil Engineers*, **36**, Paper No 6820, pp. 819–35.
2 Another disadvantage of pipe jacking is the possibility that the horizontal passage of the tunnel lining may cause horizontal movement of the ground. Such movements have been measured and reported in the following paper: Toombs, A.F. & G. West (1980) Ground movement as a result of thrust boring through a railway embankment. *Tunnels and Tunnelling*, **12**, (8), pp. 11–16.
3 Clarkson, T.E. & J.W. T. Ropkins (1977) Pipe jacking applied to large structures. *Proceedings of the Institution of Civil Engineers, Part 1*, **62**, Paper No 8058, pp. 539–61.
4 Pipe Jacking Association (1979) *Case histories in pipe jacking*. London (Pipe Jacking Assn).
5 Craig, R.N. (1983) Pipe-jacking: a state-of-the-art review. *CIRIA Technical Note* 112. London (Construction Industry Research and Information Assn).
6 Thomson, J.C. (1982) Whatever happened to British pipe jacking? *Tunnels and Tunnelling*, **14**, (3), pp. 42–3.
7 What was thought at the time to be the longest pipe jacked tunnel in the United Kingdom was a 460 m long surface-water sewer tunnel constructed at Feltham, West London, in 1982. This long length was achieved with the aid of intermediate jacking stations together with bentonite slurry to lubricate the passage of the pipes through the ground. Details are given in:
Byles, R. (1982) Longest pipejack makes a breakthrough. *New Civil Engineer*, 1 July, pp. 14–5.
In 1983 this record was broken. The current longest United Kingdom pipe jacked tunnel is 520 m long: it was a 1.8 m internal diameter sewer tunnel at Poole, Dorset. A 560 m long pipe jacked tunnel is planned to follow on from it at the same site. These are described in:
Wallis, S. (1983) UK pipe jacking pushed to greater lengths. *Tunnels and Tunnelling*, **15**, (10), pp. 59–60.
8 Most of the details of the operation of the Unitunnel are given in the brochure:
Mowlem (Civil Engineering) Ltd (1981) Unitunnel. Bracknell (Mowlem (Civil Engineering) Ltd).
9 Richardson, M. & J. Scruby (1981) The Unitunnel in action. *Tunnels and Tunnelling*, **13**, (4), pp. 29–33.
10 Ferguson, H. (1981) Tunnelling breakthrough for pneumatic shuffler? *New Civil Engineer*, 13/20 August, pp. 24–5.
11 Richardson, M. & J. Scruby (1981) 'Earthworm' system will threaten conventional pipe jacking. *Tunnels and Tunnelling*, **13**, (3), pp. 29–32.
12 Hatfield, E.J. (1939) *An introduction to biology*. Oxford (The Clarendon Press) pp. 154–5.
13 Ferguson, H. (1981) Unitunnel. *New Civil Engineer*, 15 October, pp. 38–9.
14 The Judges' comments were:

> We considered this to be an outstanding example of the kind of innovation we were looking for. The idea is brilliant in its simplicity, yet clearly rooted in years of tunnelling experience. It is particularly suitable for the smaller size tunnel and it comes at a time when there is an urgent need, in Britain and overseas, to develop safe and economic techniques for renewing worn-out or overloaded urban sewers and other underground services.

They were reported in the following article:

Anon (1981) A healthy spirit of innovation. *New Civil Engineer*, 15 October, p. 22.

15 Ferguson, H. (1981) *op cit*, pp. 38–9.

16 Tunnel surveying, including setting out, is a very important part of tunnelling and requires great skill and accuracy when the tunnel is long, is in mountainous terrain, or is undersea. The history of tunnel surveying is dealt with in:
Sandström, G. (1963) *The history of tunnelling*. London (Barrie and Rockcliff) Chapter 11.

17 These techniques are described in:
Richardson, H.W. & R.S. Mayo (1941) *Practical tunnel driving*. New York (McGraw-Hill Book Co Inc) Chapter 3.

18 The term 'laser' is an acronym for *l*ight *a*mplification by *s*timulated *e*mission of *r*adiation.

19 The helium-neon laser produces a thin bright beam of red light that is clearly visible in tunnels in which it is in use; the beam is also sometimes visible in colour photographs taken inside the tunnels.

20 Irwin, K. (1979) Tunnelling applications for pentaprisms. *Tunnels and Tunnelling*, **11**, (7), pp. 76–8.

21 Renshaw, A. (1981) Laser line-up beams tunnels on course. *Tunnels and Tunnelling*, **13**, (9), pp. 60–2.

22 Van Suilichem, P. (1979) Explosion proof laser aligns Austrian tunnel. *Tunnels and Tunnelling*, **11**, (10), p. 61.

23 Although there is a fairly long tradition of tunnelling in the Soviet Union – the Moscow and Leningrad Metros are world famous – information on Russian tunnelling is scanty in the West. The Soviet Union is not a member of the International Tunnelling Association and information on Russian tunnelling methods and machinery usually comes to light as a result of their being used in other Eastern bloc countries.

24 Gran, J. (1981) The mechanical shield and press-crete lining. *Proceedings of Conference on La Recherche d'Economies dans les Travaux Souterrains, Nice.* Paris (Association Français des Travaux en Souterrains) pp. 245–50.

25 Further details on the Russian method of forming extruded linings are given in the following English language translation of a Russian textbook:
Pokrovsky, N.M. (1980) *Underground structures and mines. Construction practices. Driving horizontal workings and tunnels.* Moscow (MIR Publishers) pp. 395–8 and 404–8.

26 Matsushita, Y., Y. Kanbe, S. Kurosawa & S. Matsuo (1981) Cast-in place lining method. *Proceedings of Conference on La Recherche d'Economies dans les Travaux Souterrains, Nice.* Paris (Association Français des Travaux en Souterrains) pp. 251–7.

27 Martin, D. & W.M. Braun (1982) Blade shield tunnelling machine extrudes its own lining. *Tunnels and Tunnelling*, **14**, (2), pp. 54–6. Note: in this article the wrong machine was illustrated; the correct machine is shown in a *corrigendum* given in *Tunnels and Tunnelling*, **14**, (5), p. 55.

28 Schach, R., K. Garshol & A.M. Heltzen (1979) *Rock bolting: a practical handbook.* Oxford (Pergamon Press) *passim.*
Hoek, E. (1982) Geotechnical considerations in tunnel design and contract preparation. *Transactions Section A of the Institution of Mining and Metallurgy*, **91**, pp. A101–9.

29 Hoek, E. & E.T. Brown (1980) *Underground excavations in rock.* London (The

Institution of Mining and Metallurgy) pp. 353–64.
Bickel, J.O. & T.R. Kuesel, Editors (1982) *Tunnel engineering handbook*. London (Van Nostrand Reinhold Co) Chapter 12.
30 Sandström, G.E. (1963) *op cit*, pp. 119–24.
31 Brown, E.T. (1981) Putting the NATM into perspective. *Tunnels and Tunnelling*, **13**, (10), pp. 13–17.
32 Terzaghi, K. & R.B. Peck (1948) *Soil mechanics in engineering practice*. London (Chapman and Hall Ltd) p. xviii.
33 Described in the following brochure:
Dosco Overseas Engineering Ltd (1974) Dosco TM 1800 tunnelling machine. Aylesbury (Dosco Overseas Engineering Ltd).
34 This job is described in the following two papers:
Martin, D. (1982) Sewerage scheme will make River Avon 'shipshape and Bristol fashion'. *Tunnels and Tunnelling*, **14**, (3), pp. 26–8.
Anon (1982) Tunnelling record at Bristol, UK. *Tunnels and Tunnelling*, **14**, (9), pp. 29–30.
35 Described in the following manufacturer's specification:
Zokor International Ltd (1979) Excavator tunnelling system. Aurora, Illinois (Zokor International Ltd).
36 It will be noticed that the American, Iranian and French flags are in front of the machine shown in Figure 13.5. This is because this particular digger shield was one to be supplied to excavate a proposed underground railway in Teheran for which the consulting engineers were French. The design had been done and the project was due to start in the spring of 1979 but the Iranian revolution supervened and swept away all plans for a Teheran Metro.
37 Referring to the year 1945 in the history of Japan, the historian W.G. Beasley has written:

> So sharp was the break with what had gone before that one is tempted to regard it as the end, not of a chapter, but of a story, to treat all that followed as something new. Indeed in many ways it was. For defeat seems to have been a catharsis, exhausting the emotions which Japanese had hitherto brought to their relations with the outside world, as well as opening the way for experiments in social and political institutions.

This passage is from:
Beasley, W.G. (1971) *The modern history of Japan*. London (Weidenfeld and Nicolson) p. 280.
38 Japan Tunnelling Association Secretariat (1978) Tunnelling in Japan. *Tunnels and Tunnelling*, **10**, (5), pp. 19–22.
39 Anon (1978) Seikan Tunnel: hazardous undersea bore is world's longest. *Construction Industry International*, **4**, (5), pp. 18–24.
Fujita, M., T. Nishimura & A. Kitamura (1982) Seikan tunnel, Japan: present position. *Tunnelling 82*. London (Institution of Mining and Metallurgy) pp. 93–9.
Anon (1983) Seikan pilot tunnel opens the way for Japan's 23 km undersea rail link. *Tunnels and Tunnelling*, **15**, (7), pp. 24–5.
40 Harding, P. (1981) Japan: a new sun rises upon the underground horizon. *Tunnels and Tunnelling*, **13**, (2), pp. 20–3.
41 Seikan Tunnel Technical Investigation Committee (1972) *Data of Boring Speciality Committee*. Hakodate, Japan (Japan Railway Construction Public Corporation, Seikan Construction Bureau) (in Japanese).
42 Paulson, B.C. (1980) Transportation construction in Japan. *Report* No

DOT/RSPA/DPB-50/80/9. Washington DC (US Department of Transportation, Research and Special Programs Administration) Chapters 4–7.

43 Megaw, T.M. & J.V. Bartlett (1981) *Tunnels: planning, design, construction.* Volume 1. Chichester (Ellis Horwood Ltd) p. 270.

44 Paulson, B.C. (1980) *op cit*, Chapter 3.

Conclusion

His [Aristotle's] greatest contribution was the idea of classification.

In this final Chapter some general observations on technical innovation arising from the study will be made, and the classification of technical innovation that emerged as a result of the work will be presented and discussed.

14.1 General observations on technical innovation

It will be recalled that in the Introduction it was remarked that technical innovation and the ability of the nation to generate it is frequently the cause of concern,[1] and that attempts are often made to stimulate innovation. To this end a constant stream of papers and articles appears in the technical or general press examining the process of innovation and exhorting various branches of British industry to be more innovative. And various committees are appointed to achieve the same object. Amongst the many remedies prescribed for this state of affairs are: better management, Government financial support,[2] more research and development, better, longer or different training for engineers,[3] transfer of engineers to management, and competitions, prizes and awards. Although all these measures may be admirable in themselves, the conclusions of this work indicate that they may not be successful in stimulating innovation, for we have found that nearly all the innovations have been in response to problems arising in the course of trying to extend the scope of tunnelling or to make it more effective. It would seem, therefore, that efforts to stimulate innovation artificially are misguided, because innovation arises from the need to solve problems or overcome difficulties standing in the way of progress in an industry. If the right conditions are provided for industry to flourish then engineers will be naturally innovative when problems arise that need solving. These conditions existed in nineteenth-century Britain in the field of tunnelling and the natural response of British engineers was to be inventive. Similar circumstances arose in Japan in the period following the Second World War and the natural response of Japanese engineers was to be inventive as well.

At the time of writing the British tunnelling industry – along with the whole of the British civil engineering industry – is in the doldrums. When the present economic recession passes, conditions will be right for a renewed programme of tunnelling in Great Britain. Not only is there the urban infrastructure of sewers, water and gas mains needing replacement – many are now over a century old – but there will be demands for highway tunnels[4] to take arterial roads into city centres and there will be further needs for underground railways since any city with a population of over one million can justify the construction of an underground railway.[5] In Great Britain, London, Glasgow, Newcastle and Liverpool have underground railways at present, but by this criterion Birmingham and Manchester could support underground railways as well. Furthermore, London Underground's proposed Jubilee Line extension through east London to Woolwich, which has been put in abeyance,[6] will need to be built sooner or later. And if the proposed National Water Grid is started there will be a need for water transfer tunnels across watersheds. The storage or disposal of radioactive waste will call for the excavation and construction of specially engineered facilities deep underground.[7] When all this work is put in hand the British tunnelling industry will have to expand again to cope with it. Some of the problems will be similar to the old ones but there will be new problems as well – mainly concerned with the siting of new tunnels in ground already occupied by existing tunnels and underground services. In particular there will be a need for devices to probe ahead of the tunnel face and locate existing tunnels and underground services because experience has shown that many of these are unmapped or incorrectly mapped. But the new problems will call for new innovations by a future generation of tunnel engineers. If the hypothesis advanced here is correct, these innovations will arise naturally in response to the problems: the difficulties will stimulate inventive solutions in the minds of some of the engineers grappling with the problems. Therefore, the best way to stimulate innovation in an industry is to give that industry plenty of work to do and not to worry about trying to stimulate innovation artificially by other methods. In the final analysis, the users of nearly all tunnels are the general public and the promoters are, in one way or another, public bodies of some kind. For the tunnelling industry, then, the work will have to come mainly from the public sector which in turn means investment of public funds. A warning has been given by J.V. Bartlett[8] about the present neglect of the national infrastructure, particularly highlighting the case of London, and measures to remedy the situation have been proposed including the setting up of a statutory 'Construction Authority' for the nation. This body would have the necessary powers to raise funds and construct major civil engineering works for the public good, which when complete would be handed over to an operating authority.[9] This measure, if adopted, would enable the tunnels that the nation needs to be constructed and would also provide the right conditions for innovative tunnelling to flourish by removing some of the impediments to the introduction of innovations that will be discussed later in this Section.

The Institution of Civil Engineers' Infrastructure Planning Group[10] has reviewed the whole of the United Kingdom's infrastructure and has concluded that although its past development has been satisfactory, there is concern about its adequacy to meet present and future needs. There are some signs that attention is now being given to the Capital's infrastructure. The Docklands Light Railway, although a much smaller scheme than the Jubilee Line extension referred to above, is now under construction and a 53 km long 2.5 m diameter ring main for London's water supply, all in tunnels some 30–50 m deep, is shortly to be constructed. The scheme is planned to be completed by 1994.

With increasing concentration of population in cities the scope for future tunnelling on a world-wide basis is, of course, enormous. There are no less than 63 cities in which a Metro system is either under construction or planned, and of these 40 are estimated to be completed during the five year period from 1983.[11] Looking to the future, one of the countries with the greatest potential for tunnelling is China. With a population estimated to be one billion or more, much of it concentrated in cities, the scope for tunnelling to provide the infrastructure for services and transport is immense. China is, in this respect, now in a similar position to Great Britain in the mid-nineteenth century and Japan in the mid-twentieth century. In 1979 China joined the International Tunnelling Association, and since then information on tunnelling in the People's Republic has increasingly become available.[12] For the present, Chinese engineers are in the main adapting the well-tried tunnelling techniques established in the West, in particular having made good use of tunnelling shields in soft ground and of the New Austrian Tunnelling Method in hard rock, but there must be every expectation that as their experience and confidence grows they will themselves develop innovatory techniques to suit the problems that they encounter in their own particular geological and site conditions.

It is not only in Great Britain that the innovatory capability of industry has been under scrutiny and unfavourable comparison; in the United States the tunnelling industry has been the subject of much heart searching and enquiry for its lack of innovation. We have seen in this book how in the past the Americans were particularly successful in pioneering many important innovations in the tunnelling industry, for example: compressed air rock drills and nitroglycerine at the Hoosac Tunnel, immersed tube tunnelling at the Detroit River and hard rock full-face tunnelling machines at the Oahe Dam. This has not escaped the Americans themselves, of course, and the more recent lack of innovation in the American tunnelling industry has been compared unfavourably with the past record. The reasons for the decline in innovatory tunnelling in the United States are complex, but there seems to be a consensus of opinion that the main cause is the 'adversarial approach' that has crept into relations between owner and contractor and which has resulted in a reluctance by engineers to take the risk inevitably associated with introducing an innovation.[13] In a review made in 1981 of innovatory tunnelling techniques directed at mainly American readers, it was noted that with the exceptions of some support techniques and hard rock tunnelling machines, most of the innovations came from Japan and Europe.[14] It is likely that if this non-technical problem could be put right, and public funding were available for construction, American tunnelling would again become innovatory because there is great scope for tunnelling in the United States and the problems encountered would produce the innovations.

The study has clearly identified one thing that inhibits innovation and this is the existence already of a very effective method of doing something. Examples are drill-and-blast tunnelling and shield tunnelling with compressed air. We have seen that the very effective system of drill-and-blast tunnelling was so successful that it prevented the development of practical hard rock tunnelling machines until after the Second World War. Also, shield tunnelling under compressed air was so effective a method of subaqueous tunnelling in the hands of British contractors that immersed tube tunnels were precluded from use in Great Britain until 1970 although there were a number of occasions when the method could have been used, particularly for some of the later Thames crossings.

The study has also shown that there is sometimes a critical period in the life of an innovation when success or failure hangs in the balance, and this is during the development period when engineers are trying to apply the new idea for the first time. It is clear that these first trials must be made where ground and site conditions are such that the new method has a reasonable chance of being successful. An example of where this was not the case is at Warrington where the bentonite tunnelling machine was being used on its first commercial contract (see Section 9.4). It is almost certain that any tunnelling method would have run into difficulties on this drive, the site and ground conditions being so unsuitable for tunnelling. As it was, the British bentonite tunnelling machine took the blame for the difficulties – so much so that it was never used again. It was left to the Germans and Japanese to develop the bentonite type of machine successfully, mainly because they took care not to expose it to unsuitable conditions before the teething troubles had been put right.

It is also at the development stage of an innovation that financial help from central and local government and other public bodies is most useful. Nearly all tunnels, whatever their purpose, are built for the benefit of the man in the street and it is right that if he is to benefit from more effective tunnel construction arising from some technical innovation in tunnelling, then he should be prepared to contribute something via the public purse towards the cost of producing the innovation. There is some evidence that the Japanese may do this by financing the use of new tunnelling machines or novel tunnelling techniques on short lengths of tunnel for which the use of these new methods would not be justified on strict economic grounds. However, by doing this they give Japanese engineers the chance to try out new methods and perfect them so that when a suitable job comes along, most of the teething troubles have already been solved. Furthermore, and most importantly, it enables Japanese tunnelling contractors and tunnelling machine manufacturers to put forward impressive portfolios of case histories of successfully completed tunnel drives when tendering for contracts abroad. The Japanese are now able to offer a wide range of types of tunnelling shield, together with the experience of their use.[15]

In 1844, describing the method of installing his proposed compressed air hammer, C. Brunton wrote:

> By some such means I am of opinion, that the apparent difficulties would nearly vanish in practice under the hands of a miner, possessing a mechanical turn of mind, and some resources; for I believe that it will never be brought into practical operation but by such a person, encouraged of course by the parties by whom he is employed.[16]

This highlights an important point regarding the introduction of an innovation and that is that the prototype must be first put to work under the control of people who are strongly motivated to make it a success and who understand the mechanical principles involved and how to make adjustments or modifications on site. It will be recalled that a major factor in the lack of success of the Universal soft ground tunnelling machine (Section 9.3) was because it passed out of the control of the contractor for whom it was made and of the machine designers and manufacturers. Indeed, this is not a new problem: P. Mathias has described how during the eighteenth and early nineteenth centuries many of the new machines, devices and processes sent from England to the Continent remained unused because of the scarcity of skilled fitters from British workshops to set them up, maintain and operate them.[17] If competent mechanics are not available the innovation may appear to be a failure when all it needs is knowledgeable attention coupled with a commitment to make it work.

On a number of occasions innovations have been thought of long before they were actually put into practice. For example, Greathead's patent specification for a pressurised face tunnelling shield dating from 1884 (see Section 9.1), was not put to practical use at the time and it was not until 1967 that this concept was used in a working machine. Similarly, Crampton's idea for the hydraulic transport of spoil thought of in 1882 (see Section 11.5), remained only an idea until it was eventually used in the bentonite tunnelling machine in 1967. Sometimes an innovation was put into practice at the time of its invention but then fell into disuse, only to be rediscovered many years later. Brandt invented a very successful hydraulic rotary drill which was used in drill-and-blast tunnelling during the years 1880–1906 (see Section 5.1). It then disappeared from the tunnelling scene, but the principle reappeared again in the 1970s in the form of the Diamec 250 drill – which, however, was not used in tunnelling but in exploration drilling for mining. Immersed tube tunnelling was successfully demonstrated by Hawkins in 1811 with his brick cylinders in the Thames (see Section 10.2), but the first immersed tube tunnel was not constructed until 1906 when the Detroit River Tunnel was built using this method. In 1981 proposals for a buoyant tube tunnel are the same as those proposed in 1855 (see Section 10.8), and the use of hydrostatic pressure to close immersed tube sections tightly was thought of as long ago as 1861 (see Section 10.7). Instances of this kind show that the engineer today who is confronted by a seemingly intractable problem may well find a solution by looking through the nineteenth-century tunnelling literature. This is true of concepts as well as devices: for example, in 1881 Brunton and Trier enunciated the principle of the rolling disc cutter and stated the need for stiffness in the machine (Section 11.3); both these ideas were rediscovered in the present century. This approach in itself could be regarded as one reason for the study of the history of engineering and technology – as witness the admittedly rather special interpretation of the motto of the Newcomen Society: 'That the future may learn from the past'.

Sometimes the solution to one problem threw up another. For example, in Section 4.4 we saw how the successful introduction of the sintered tungsten carbide drill bit so prolonged the working life of drill steels that they began to suffer from fatigue failure. This is an instance of a phenomenon often observed in technological development – improvement in one element of a system produces inadequacies in another. When the same problem of fatigue stress in drill steels cropped up again following the introduction of the hydraulic rock drilling machine (see Section 5.3) it was solved by incorporating a shock absorber inside the drilling machine itself.

Several times the same idea has occurred simultaneously in different places. For example, the compressed air rock drilling machine appears to have been thought of independently in Europe and the United States (Chapter 2), and the idea of using compressed air to exclude water from the works in subaqueous tunnels was independently thought of by Colladon, Cochrane and perhaps Brunel (Chapter 7). Similarly, a number of firms all began to consider introducing hydraulic rock drilling machines at about the same time (Chapter 5). All this supports the belief that it is the problem to be solved which is the principal stimulus for the innovation to solve it: when the same problem cropped up in different places, or was considered by different men, the same or very similar solutions were forthcoming. Having said this, the human factor cannot be entirely dismissed for sometimes a particular individual seems to have been innovative: Greathead is a good example from the nineteenth century (Chapters 6, 8 and 9) with his contributions to shield tunnelling, and Richardson for the twentieth century (Sections 12.6 and 13.2) with the Mini Tunnel and the Unitunnel.

We can note the importance of a suitable tunnelling medium on the successful establishment of innovative tunnelling. There can be no doubt that the existence of the London Clay beneath the city of London was extremely conducive to the development first of the tunnelling shield and later of the soft ground full-face tunnelling machine. In much the same way, the Lower Chalk at the Channel Tunnel site provided an ideal material for the development of the early rock full-face tunnelling machines. The reason for this is that an ideal tunnelling medium provides the innovatory technique with the best chance of success and an opportunity to overcome teething troubles; once the technique has proved itself it is then easier to apply it to less tractable ground conditions. The firm of Markham, the old-established British firm of tunnelling machine manufacturers, got their start in this business by the manufacture of Price-type rotary tunnelling machines for work in the London Clay during the early days of the construction of London's Underground.[18] Other aspects of the effect of geology on innovation have been discussed in detail where relevant, but it perhaps can be most clearly discerned when viewed in general terms. The pre-eminence of British innovations in soft ground tunnelling during the nineteenth century was undoubtedly due to the concentrated tunnelling activities in London – particularly in the London Clay, but also in the overlying alluvial deposits. Similarly the lead taken by engineers in Europe and the United States in hard rock tunnelling reflected their early experience in tunnelling through the Alps and the mountains of the east coast of the

United States.[19] The development and application of immersed tube tunnelling in the Netherlands since the Second World War is a clear consequence of the geography, if not the geology, of that country and its suitability for this method of tunnel construction.

The study has clearly shown that although a pressing problem or a deeply-felt need has often been the stimulus for producing an innovation, in many cases the successful production of the innovation itself could only be achieved after a whole suite of ancillary problems had been solved. This is well shown in the mechanical-type innovations such as the compressed air rock drilling machine and the hard rock and soft ground tunnelling machines where, once the idea had been thought of, there was still a large number of mechanical problems that had to be solved before the particular machine could be made to work successfully. This was sometimes also true of innovatory systems or methods of working: for instance, the development of immersed tube tunnels required the solving of a number of operational problems, one of which, the hazard to navigation, still remains a difficulty. The pressurised face tunnelling shield was another innovation that required considerable development until it could be made to work, the problems of dealing with boulders and cobbles in particular having been discussed. Indeed the problem of boulders has yet to be solved – if it ever can be – satisfactorily. These instances show that it is not sufficient just to have a good idea. For an innovation to be introduced successfully into engineering practice, the will and ability to overcome the subsidiary or consequential problems that follow in the train of the main idea are also required.

After the Thames Tunnel was flooded in January 1828 (see Section 6.4 and introduction to Chapter 7) Brunel received many unhelpful suggestions as to how he should proceed, of which he remarked 'In every case they made the ground to suit the plan and not the plan to suit the ground.' This makes the important point that for an innovation to be successful it must be a solution to the problem *as it actually exists* and not to a problem that is tailored to suit the solution, nor to a form of the problem that exists only in the mind of the inventor.

One innovation, compressed air tunnelling (Chapter 7), whilst in itself being the solution to the problem of subaqueous tunnelling, produced other problems of health to the workforce which have proved very difficult to overcome satisfactorily. The solution lay in a further innovation – the pressurised face tunnelling shield (Chapter 9). Compressed air tunnelling is still allowed, and although the health hazards are lessened if the air pressure is kept below 1 atm, the practice is being superseded by the use of pressurised face shields.[20]

Another innovation, the precast concrete tunnel lining, has led to the production of a large number of types, of both the bolted and unbolted forms. This multiplicity of solutions seems to be an unnecessary complication and attempts were made in 1979 in Great Britain by the industry itself to standardise these linings, at least in dimensions if not in details of assembly, in order to reduce the number to a manageable size. As well as the obvious advantage of economy in the production of fewer types, one further advantage that would follow from standardised tunnel linings is that tunnelling shields and machines, which of course have their tailskin and segment handling and erecting systems tailor-made to deal with particular linings, could be used on a number of jobs, and not just on one or two where a particular lining was specified.

The expanded form of the precast concrete tunnel lining – the wedge-block lining – has proved very successful in suitable ground conditions, namely in clays, and provides a good example of how an analysis of the behaviour of the lining in the ground led to a better design (see Sections 8.6 and 8.7). Similarly, applied research carried out by drilling machine manufacturers led to the successful development of the hydraulic rock drill (Chapter 5). Both these are examples of innovations brought about by analysis or research applied to the problem within the industry itself. This may well be a method by which innovation will come about increasingly in the tunnelling industry in the future. Indeed, in the same way that the applied study of soil mechanics has influenced the development of prefabricated tunnel linings for soft ground tunnelling, the more recent applied study of rock mechanics[21] can be expected to have an influence on the design of new methods (both temporary and permanent) for the support and lining of tunnels in hard rock, and the first signs of this are already apparent.[22]

The results of the Civil Engineering Innovation Competition have already been discussed in Section 13.2 where it was remarked that although Richardson won the first prize with his Unitunnel, the competition played no part in stimulating the invention that won it, but only rewarded one that had already been made. A measure of caution is necessary in drawing a general conclusion from this particular circumstance. We can note that this competition was for any innovation drawn from the whole field of civil engineering and not for a specific purpose or objective. There is some evidence from technology more generally that competitions or prizes offered for an innovation for some *specific purpose* are sometimes more successful in stimulating innovation. Probably the most famous example is that of the marine chronometer as the solution for solving the problem of determining longitude at sea,[23] but there is another example from a field closer to the subject of this book. That is the prize offered by the Société d'Encouragement pour l'Industrie Nationale in France in 1826 to anyone who could apply satisfactorily on a large scale, water turbines with curved blades. The prize was won in 1833 by the hydraulic engineer Benoit Fourneyron (1802–67), and there is some evidence that the offering of the prize stimulated some competition amongst engineers to find a solution.[24]

14.2 Origins of the innovations

During the course of the work described in this book, a simple classification of technical innovation has emerged based on the perceived origin of the innovation. This will now be described, first for the major innovations that form the main part of the study and then for the other developments outlined in Chapter 13.

The compressed air rock drilling machine was the first major innovation to be considered, and in Section 2.7 its origins were described. Both the American and European versions of this innovation derived directly from the steam engine, the evidence for this being not only documentary but also in the nature of the drill mechanisms themselves. We can, therefore, see this as an example of a category in the classification of technical innovation we can call *borrowed technology*.

By contrast nitroglycerine explosive, the origin of which is described in Section 3.5, was the result of independent scientific research in a chemical laboratory, there being no thought at the time of seeking an explosive. It was only after its properties had been discovered that this use was made of it. Nitroglycerine explosive is, therefore, an example of a category in the classification system that we can call *science-based innovation*. If we now look at the discovery that nitroglycerine could be safely handled when frozen, this provides an instance of another category which can be called *accidental innovation.* Giving Nobel the benefit of the doubt and allowing that dynamite was the product of his research laboratory, this discovery is then another science-based innovation.

The use of sintered tungsten carbide to form a very hard tip for drill steels was the third major innovation in drill-and-blast tunnelling. In Section 4.6 it is pointed out that sintered tungsten carbide was a material wholly developed by the mechanical engineering industry and which was later borrowed by the tunnelling industry. From the standpoint of the tunnelling industry, therefore, sintered tungsten carbide is an example of borrowed technology. Having said this, it can be noted that from the point of view of the mechanical engineering industry tungsten carbide can be regarded as an example of science-based innovation, because the development of the sintered material could not take place until the scientific discovery of the carbide had been made.

The hydraulic rock drilling machine was the last major innovation in drill-and-blast tunnelling, and its origin was discussed in Section 5.5. It was shown that the development of this machine took place entirely within the rock drilling machine industry, which we can regard as part of the tunnelling industry, and was based on advances made in technology itself. It therefore provides us with a further category in the classification system – one that can be called *technology-based innovation.*

The first, and in many ways the most important, of the major innovations in soft ground tunnelling was the tunnelling shield. We have seen that the first true tunnelling shield is Brunel's Thames Tunnel shield and that the origin of this was Brunel's encounter with the shipworm and the observations that he made of its mode of action (Section 6.11). Here then we have another category in the classification system, one which can be called *innovation based on an external idea*, the point being that the innovation was suggested by something from quite outside the industry or technology in which the innovation was applied.

Compressed air tunnelling was the second major innovation in soft ground tunnelling; it permitted tunnels to be safely driven in ground beneath the water table or beneath rivers, especially when used in conjunction with the shield. The origin of this innovation is discussed in Section 7.8 where it was argued that Cochrane, the inventor, probably derived the ideas for it from the diving bell and the canal lock. If this is the case, then compressed air tunnelling is an example of borrowed technology, both the diving bell and the canal lock being from technologies outside the tunnelling industry.

The prefabricated tunnel lining was an innovation that was a necessary concomitant of the tunnelling shield with which the traditional brick lining could not be used. The first type introduced was the bolted cast-iron segmental lining which was clearly derived from the bolted cast-iron segmental colliery shaft tubbing of Buddle the elder (Section 8.8). It is, therefore, an example of borrowed technology, coming from the mining industry. The unbolted precast concrete segmental lining is, however, an example of technology-based innovation, stemming as it does from developments wholly within the tunnelling industry following an enquiry into the design principle of the bolted lining.

The next major innovation in soft ground tunnelling was the pressurised face tunnelling shield which was introduced to overcome the disadvantages of compressed air tunnelling. In Section 9.8 we have seen that both the concept and the various practical machines were all developed within the tunnelling industry itself. The pressurised face tunnelling shield is, therefore, a technology-based innovation from inception to realisation. However, in one particular version of the pressurised face shield, the bentonite tunnelling machine, the use of bentonite as the working fluid was suggested by its previous use in the civil engineering and petroleum industries – this use is, therefore, borrowed technology.

The immersed tube tunnel (Section 10.9) was an innovation that provided an alternative to the use of the shield combined with compressed air for river or estuary crossings. The idea dated back to the beginning of the nineteenth century but its realisation had to wait until the beginning of the twentieth century. Nearly the whole of the development of immersed tube tunnelling took place entirely within the tunnelling industry so that it can be seen to be a technology-based innovation. The exception to this is that the steel tube type of immersed tube tunnel was constructed in shipyards using ship fabrication methods and techniques. This branch of the innovation can, therefore, be thought of as borrowed technology.

The origin of the hard rock tunnelling machine is described in Section 11.11 where it was noted that the first machines were intended to replace the old laborious hand methods of tunnelling. However, the success of the innovations in the drill-and-blast system prevented tunnelling machines from being introduced for a long time. When it did take place, the development of the full-face hard rock tunnelling machine was made entirely within the tunnelling industry and this is, therefore, another example of technology-based innovation. On the other hand, the development of the roadheader took place within the coalmining industry and these machines were simply taken over by the tunnelling industry. The roadheader is, therefore, an example of borrowed technology from the point of view of the tunnelling industry.

The soft ground tunnelling machine is the last of the major innovations that are examined in this book, and its origin is discussed in Section 12.7. Soft ground tunnelling machines enabled tunnels to be driven in suitable ground at a faster rate and with fewer men than the hand shields from which they were a natural development. They were a success from the start, partly because the design problems were fairly simple matters to solve compared with hard rock tunnelling machines. From the first machines of Thomson and Price to the present, the development of soft ground tunnelling machines took place within the tunnelling industry, providing another instance of technology-based innovation. We can note that the success of the first machines was partly due to the coincidental availability of the industrial electric motor, which was so well suited to driving underground machinery. This, of course, came from the electrical engineering industry so can be considered a small item of borrowed technology.

We can examine the origins of the other developments in the same way as we did those of the major innovations. Firstly, there is pipe jacking. In Section 13.1 it was seen that pipe jacking is a form of horizontal caisson sinking so that it is a technique clearly taken from civil engineering; hence it can be classified as borrowed technology. In the Unitunnel (Section 13.2) we have an innovation, which like Brunel's tunnelling shield, was inspired by a tunnelling organism. Here then is an innovation whose origin was an external idea. Section 13.3 describes the laser, which was considered to be a minor, rather than a major innovation in spite of its undoubted usefulness in tunnel surveying. The laser is clearly a scientific discovery which has been adapted to technological uses, including that of a survey instrument; it is, therefore, a science-based innovation from the point of view of the tunnelling industry. The extruded lining, described in

Section 13.4, was the only challenge to the prefabricated tunnel lining for soft ground tunnelling, and was developed wholly within and by the tunnelling industry – an example, therefore, of a technology-based innovation. The new primary lining methods – rock bolts, wire mesh and sprayed concrete – (Section 13.5) were not considered to be major innovations in themselves, but when used together in the design concept known as the New Austrian Tunnelling Method, they probably can be considered to be so. This then provides an example of a type of innovation we have not seen before: one which is an innovation by reason of bringing together a number of separate components and making of them something that is greater than the sum of the parts. This can be called *composite innovation* and it constitutes a new category in the classification system. We can note in passing that sprayed concrete, a technique derived from museum practice, is an example of borrowed technology. Composite tunnelling machines, the last of the individual technical advances considered in the other developments (Section 13.6), were of two kinds, the boomheader shield and the digger shield. Each of these was formed by bringing together the tunnelling shield and, in one case, the roadheader boom, and in the other case the digger arm. They are, therefore, further examples of composite innovation.

14.3 The classification system and its uses

The categories in this simple classification system are then (i) science-based innovation, (ii) technology-based innovation, (iii) borrowed technology, (iv) accidental innovation, (v) innovation based on an external idea and (vi) composite innovation. We can now give formal definitions of these categories in line with the principles demonstrated in Section 14.2:

(i) *Science-based innovation* is innovation that is essentially dependent on an advance or new discovery in some branch of science or on the new application of some existing scientific knowledge.

(ii) *Technology-based innovation* is innovation that essentially arises from within a technology itself and owes little or nothing to science directly.

(iii) *Borrowed technology* occurs when an innovation made in one particular technology is found to be useful in another and is therefore borrowed. Sometimes it is little changed and sometimes adaptation is required.

Table 14.1. *Tunnelling innovations classified according to origin*

Innovation	Classification category					
	Science-based	Technology-based	Borrowed technology	Accidental	External idea	Composite
Major innovations						
Compressed air rock drilling machines			+			
Nitroglycerine explosive	+					
Nitroglycerine safely handled when frozen				+		
Dynamite	+					
Sintered tungsten carbide drill steel bits			+			
Tungsten carbide (for mechanical engineering industry)	+					
Hydraulic rock drilling machines		+				
Tunnelling shields					+	
Compressed air tunnelling			+			
Prefabricated tunnel linings						
Bolted cast-iron segmental lining			+			
Unbolted precast concrete segmental lining		+				
Pressurised face tunnelling shields		+				
Use of bentonite in bentonite tunnelling machine			+			
Immersed tube tunnels		+				
Steel tube type of immersed tube tunnel			+			
Hard rock tunnelling machines						
Full-face tunnelling machines		+				
Roadheaders			+			
Soft ground tunnelling machines		+				
Electric motors for driving tunnelling machines			+			
Other developments						
Pipe jacking			+			
Unitunnel					+	
Laser	+					
Extruded linings		+				
New Austrian Tunnelling Method						+
Sprayed concrete			+			
Composite tunnelling machines						+
Totals 26	4	7	10	1	2	2

(iv) *Accidental innovation* occurs when an innovation is made as the result of an accident or chance occurrence, usually by showing the effect of something that would not have been considered had the accident not occurred.

(v) *Innovation based on an external idea* occurs when an innovation is due to something from quite outside the particular technology providing a new approach or solution to a problem. (This is close to (iv) above and needs careful distinction from it.)

(vi) *Composite innovation* is innovation that results from bringing together two or more pre-existing elements to form a combination that is not only different but much more effective than any one of the elements by itself.

An amplifying comment can be made regarding categories (iv) and (v); it is that an accident or an external idea cannot by itself produce an innovation. Louis Pasteur's dictum that 'chance favours only the prepared mind' is applicable to these situations. Unless someone who is actively seeking the solution to a problem or who has the insight to recognise the significance of a chance occurrence is present to observe the external idea or the accident, then the innovation will not be made.

We will now suggest some uses that might be made of the classification system, starting with the tunnelling industry. Table 14.1 shows the innovations in the modern tunnelling industry and the classification of them by their origin using the categories listed above. The Table allows a comparison to be made of the relative importance to the tunnelling industry of the different categories of innovation. Thus it can be seen that borrowed technology with ten entries has been the most frequent type of innovation closely followed by technology-based innovation with seven entries. Science-based innovation accounts for four entries whilst external idea and composite innovation have two each and accidental innovation has one. This shows that in the modern tunnelling industry, innovation has come about predominantly by borrowing ideas from other closely related industries such as civil and mechanical engineering and mining, or by producing the ideas within the tunnelling industry itself. Accidents and external ideas, although much beloved by popular writers, play a very small part in the overall scheme. This outcome is perhaps not unexpected given the make-up of the tunnelling industry. Tunnel engineers and tunnelling machine designers would naturally look to the engineering industries for ideas that they could adopt, and to the mining industry which has many similarities technologically to the tunnelling industry.[25] The other great source of innovation has been from within the industry itself. Here again we should not be surprised; in an engineering-type industry it must be mainly up to the industry itself to find the answers to the problems it faces, and the tunnelling industry has shown that it is well able to respond by being innovative when faced with technical difficulties. This kind of analysis could be applied to technical innovation in any industry, but of course, with different results for different industries; for example, an examination of a chemical industry such as pharmaceuticals would be expected to provide many more instances of science-based innovations than the tunnelling industry has done. In fact, the classification could be used to see to what extent science has influenced the development of a particular industry since this is a question that is often discussed, both among historians of science and technology and in wider circles.

In making an analysis of this kind it should be borne in mind that each of the innovations will not necessarily carry equal weight. For instance, the invention of the tunnelling shield was of more crucial importance to the tunnelling industry than, say, the invention of the hydraulic rock drilling machine. The fact that not all innovations are of equal significance must be carefully considered when applying the classification. A sophisticated form of analysis might attempt to give some kind of 'weighting' to the innovations reflecting their relative importance to the industry. Also, there will be geographical variation to take into account. For example, as events turned out immersed tube tunnelling was an innovation of little importance to the British tunnelling industry, but of considerable importance abroad.

Sometimes the assignment of a particular innovation to a category of the classification system will have to be made on the best evidence available, which may be subject to change. For instance, should new information come to light which proved that the apocryphal story given in Chapter 3, Note 21, concerning the accidental leakage of nitroglycerine into kieselguhr packing was true, then it would be appropriate to reclassify the invention of dynamite as an accidental innovation instead of a science-based innovation.

We can now look in broad terms at how the classification might be applied outside the tunnelling industry to technology more generally. It can be seen that categories (i) and (ii) can not only be applied to particular examples of innovation, but to whole industries. For instance, the electronics industry can be considered as science-based since the essential underlying basis of the industry lies in a fundamental understanding of the science of electricity. By contrast, in the early days the aviation industry could be considered as essentially technology-based since powered flight was invented, practised and developed on an empirical basis by practical engineers well before the science of aerodynamics to which it gave rise was available. The remaining categories are more properly restricted to particular technical inventions. The following examples are given to show how the classification can be applied to particular innovations from a wide range of technologies; they are intended to be illustrative only and therefore not supported by detailed evidence.

An example of a science-based innovation is a nuclear reactor which clearly could not be conceived of, let alone designed and constructed, without a fundamental understanding of the principles of nuclear physics. By contrast, an example of a technology-based innovation is the invention and early development of the steam engine; the origin of this owed little to science but much to the ingenuity of a mining engineer, Thomas Savery (1650–1715), an ironmonger, Thomas Newcomen (1663–1729) and an instrument maker, James Watt (1736–1819). On the other hand science owes much to the steam engine, study of the operation of which gave rise to the science of thermodynamics. The centrifugal governor is an instance of borrowed technology. This device had been long in use to regulate the speed of millstones in watermills and windmills; Watt adopted it to regulate the speed of the steam engine and it was later used to control the internal combustion engine. Accidental innovation can be illustrated by the fact that Charles Goodyear (1800–60) discovered how to vulcanise rubber by accidentally overheating a mixture of rubber, sulphur and white lead.[26] An example of an innovation based on an external idea is provided by the 'float glass' process, introduced by Pilkington Brothers Ltd in 1959, which consists of forming a sheet of glass on the surface of a pool of molten tin. This was said by one of the inventors to have been inspired by noticing the layer of fat which formed on the surface of some water in a domestic frying pan on cooling. Finally, a good instance of a composite innovation is the dreadnought battleship, named after *HMS Dreadnought* launched in 1906, in which the combination of a high speed (21 knots), armour plate and the very big gun produced a warship which was not just a bigger battleship, but one so different that it immediately rendered obsolete all others existing when it was introduced.

The classification system must be applied with common sense because the categories are not unique and mutually exclusive like those of the biologists' Linnaean classification for plants and animals; the system is different in that a particular technical innovation may have had a mixed origin and thus fall into more than one category. For example, although the steam engine has been cited as an example of a technology-based innovation, it must not be overlooked that science may have had a small part to play, because Watt knew of the work of Joseph Black (1728–99) on latent heat which may have influenced his idea of the separate condenser. In a similar way, to describe the nuclear reactor as a science-based innovation is not to deny that a high level of technology was also necessary to develop working nuclear power sources. These examples are given by way of illustration. It is recognised that such complex items will include many separate innovations which may have had different provenances.

Moreover, if the classification system were applied to other industries it may well be that further categories would emerge. This because it is unlikely that the modern tunnelling industry has representatives within it of every possible kind of major innovation. The fact that the technical innovations of the tunnelling industry provided the six categories given above does not mean that these are the only categories, although it is likely that these do include the ones of primary importance to all industries – categories (i), (ii) and (iii).

The classification of innovation might be useful in the following ways. Firstly, for the historian of technology such a classification would provide some insight into the relative importance of the different classes of innovation for any given industry, and also a means of comparing different industries or even different firms within an industry. Secondly, for the historian concerned with the relationship between science and technology the classification would provide a means of assessing the extent to which a particular industry had depended on science for its advancement. Thirdly, for those presently engaged in the technical development of an industry the classification may show how in the past the progress of their industry was influenced and suggest fruitful ways for future progress. Fourthly, in the teaching of history and in particular the history of technology the classification would provide a framework for both teachers and students to understand and analyse technological change.[27]

The foregoing has demonstrated how fruitful the concept of looking at technical innovations as solutions to problems has been for the tunnelling industry. What at first sight might seem to be a simplistic approach has in the event provided an extremely versatile method of examining a whole range of different developments in the technical progress of an industry. It has been possible to use it at various levels: to look at whole processes like, for example, pressurised face tunnelling shields, and at small details such as the parts of individual machines. Moreover, looking at technical innovations as solutions to problems has shed light on other, more general, issues such as the role of technical innovation in the rise of an industry such as happened in tunnelling both in nineteenth-century Britain and in twentieth-century Japan, and perhaps on a more fundamental question – the cause of technical innovation itself.

The proposed scheme for classifying technical innovation that emerged as a result of studying the tunnelling industry has also been illuminating. It has shown, amongst other things, how the industry has drawn for much of its technical innovation upon other technologies. Although the scheme is tentative, only having been worked out for one industry, it is a great advance on anything of the kind that has been available hitherto. However, a classification is not an end in itself, and will only be accepted and used if it provides a method of analysing or understanding what is classified – and this classification is no different in that respect. It is considered that the proposed classification will prove a useful tool in examining technical development in other industries.

The book, therefore, demonstrates two methods of approach which may be of general application in the history of technology, and if they are taken up by historians and found useful the author would be well pleased.

Notes and references for Chapter 14

1 Pavitt, K., Editor (1980) *Technical innovation and British economic performance.* London (The Macmillan Press Ltd) *passim.*

2 The British government's schemes for financially supporting technical

innovation are described in the brochure:
Anon (1983) Support for innovation. London (Department of Industry).
3 Corfield, Sir K. (1982) Attitudes to engineers must change. *Civil Engineering* (UK), November, pp. 11 and 16.
4 Muir Wood, A.M. (1972) Tunnels for roads and motorways. *Quarterly Journal of Engineering Geology*, 5, (1 and 2), pp. 111–26.
5 Girnau, G. (1982) Costs and benefits of underground railway construction. *Underground Space*, **6**, (6), pp. 323–30.
6 West, G. (1983) Comparisons between real and predicted geology in tunnels: examples from recent cases. *Quarterly Journal of Engineering Geology*, **16**, (2), pp. 113–26.
7 Hudson, J.A. & J.B. Boden (1982) Geotechnical and tunnelling aspects of radioactive waste disposal. *Tunnelling 82*. London (Institution of Mining and Metallurgy) pp. 271–81.
8 Bartlett, J.V. (1983) Presidential address 1982. *Proceedings of the Institution of Civil Engineers, Part 1*, **74**, February, pp. 3–13.
9 It has been suggested that the revenues from North Sea oil and gas should be used to fund the reconstruction and expansion of the nation's infrastructure; see the following article:
Anon (1983) Expanding the infrastructure message. *New Civil Engineer*, 10 November, editorial p. 5.
10 Institution of Civil Engineers (1986) *Second report of the Infrastructure Planning Group*. London (Institution of Civil Engineers).
11 Goldsack, P.J. (1983) Subways of the world. *Mass Transit*, **10**, (10), pp. 14–72.
12 Freeman, S.T., R. Hamburger, D.J. Lachel, R.S. Mayo, J.L. Merritt & B.L. Smith (1982) Tunnelling in the People's Republic of China. *Underground Space*, 7, (1), pp. 24–30.
Li Shi-Zhong (1983) Tunnelling in China. *Tunnels and Tunnelling*, **15**, (6), pp. 49–51.
13 Lane, K.S. (1983) What happened to innovation in US tunnelling? *Underground Space*, 7, (3), pp. 161–6.
14 Clough, G.W. (1981) Innovations in tunnel construction and support techniques. *Bulletin of the Association of Engineering Geologists*, **18**, (2), pp. 151–67.
15 Yoshikawa, T. (1985) Soft ground tunnel shields in Japan. *Tunnels and Tunnelling*, **17**, (7), pp. 43–7.
16 Brunton, C. (1844) Design of a wind hammer for boring rocks. *Mechanics Magazine*, **41**, 21 September, pp. 203–4.
17 Mathias, P. (1975) Skills and the diffusion of innovations from Britain in the eighteenth century. *Transactions of the Royal Historical Society*, **25**, pp. 93–113.
18 Armstrong, A. (1968) Markham tunnelling shield equipment from 1887 to 1968. Reprint from *Broad Oaks Magazine*, Spring, pp. 1–8.
19 The influence of geology on tunnelling is an important field of study in its own right, see for example:
Legget, R.F. (1962) *Geology and engineering*. London (McGraw-Hill Book Co Inc) 2nd edition, Chapter 9.
Legget, R.F. (1973) *Cities and geology*. London (McGraw-Hill Book Co) Chapter 6.
Legget, R.F. & P.F. Karrow (1983) *Handbook of geology in civil engineering*. London (McGraw-Hill Book Co) Chapter 20.

20 Jones, M. (1986) Working under pressure. *Tunnels and Tunnelling*, **18**, (4), pp. 75–9.
21 Brown, E.T. (1982) Holes in the ground: applications of engineering mechanics. *Royal School of Mines Journal*, (32), pp. 19–26.
22 Ward, W.H. (1978) Ground supports for tunnels in weak rocks. *Géotechnique*, **28**, (2), pp. 133–71.
23 Pledge, H.T. (1966) *Science since 1500*. London (HM Stationery Office) 2nd edition, p. 70.
24 Smith, N.A.F. (1977) The origins of the water turbine and the invention of its name. *History of Technology*, **2**, pp. 215–59.
25 Boden, B., G. West & C.J. Harrad (1975) Tunnels in mining and civil engineering – common ground. *Tunnels and Tunnelling*, **7**, (3), pp. 60–7.
26 Derry, T.K. & T.I. Williams (1973) *A short history of technology*. London (Oxford University Press) p. 527.
27 The role of innovation in the teaching of history and in the teaching of the history of technology has been discussed by the two following writers:
Lilley, S. (1965) *Men, machines and history*. London (Lawrence and Wishart).
Buchanan, R.A. (1978) History of technology in the teaching of history. *History of Technology*, **3**, pp. 13–27.

Glossary

Terms adequately defined in the text when first used have not been repeated here.

TUNNELLING TERMS

Adit See *drift*.

Bench A step left in the heading of a rock tunnel which is blasted afterwards by drilling vertical holes.

Blow Sudden, often catastrophic, loss of air from a compressed air tunnel or loss of fluid from a pressurised face shield.

Chainage Distance (in ft or, after 1969, in m) along tunnel as measured from the portal.

Competent rock Rock that can stand for a relatively long period with no support or little support when a tunnel is excavated through it.

Cut-and-cover Method of soft ground tunnelling in which the tunnel is formed by excavating a trench, putting a roof on and then backfilling.

Crown The top or roof of a tunnel.

Drift An inclined access tunnel driven to a main tunnel or to underground workings from the surface.

Driftway (nineteenth century) A small construction tunnel.

Drill-and-blast Method of hard rock tunnelling in which the rock is excavated by drilling holes in the face, placing explosive in them and then firing.

Drill steel Drill rod used for drilling shotholes in drill-and-blast tunnelling.

Face The vertical rock or ground surface exposed at the end of the heading during excavation.

Gallery Horizontal passage in a mine.

Grillage A foundation for a load, consisting of two layers of joists laid at right angles to each other.

Heading The working section at the end of a tunnel under construction where excavation and lining erection are in progress.

Hydrocyclone A cone-shaped device for removing sand from bentonite slurry by centrifugal separation.

Incompetent rock The opposite of competent rock, often found in fault zones or where the rock is weathered, decomposed or closely jointed.

Invert The bottom or floor of a tunnel.

Jumbo A carriage for mounting rock drilling machines used in drill-and-blast tunnelling.

Kentledge Loading, usually of stone, concrete or pig-iron to give weight to a caisson or shaft lining during sinking.

Lagging Wooden boards, wire mesh or corrugated steel sheet placed behind the steel arches in a tunnel to prevent pieces of rock falling out.

Miners The workmen who work in the heading carrying out the excavation and erecting the lining; use of this term has extended from the early days of tunnelling down to the present time.

Muck See *spoil.*

Overbreak The extent of rock or ground excavated beyond the intended cross-sectional profile of the tunnel.

Pilot tunnel A small-diameter tunnel driven in advance of the main tunnel to prove ground conditions and locate hazards.

Poling boards (nineteenth century) The wooden boards put up to support the tunnel face temporarily; today they would be called 'face boards' or 'breast boards'.

Poling boards (today) The boards driven forward into the face of tunnel at the crown to support the roof.

Portal The entrance to a tunnel.

Probing ahead Drilling boreholes from the tunnel face ahead of the drive to obtain information on ground and ground water conditions.

Rib See *steel arch.*

Ring One complete circle of prefabricated tunnel lining segments.

Roadway A main tunnel in a coal mine.

Screed A bed of gravel, sand, concrete or mortar laid to form a level base for some structure.

Shaft A vertical passage from the surface to a tunnel or to underground workings.

Shore A timber support set at a slant.

Shove Movement forward of the shield, usually by jacking off the lining that has just been installed.

Shuttering Panels erected to contain and support concrete during placing and hardening.

Skip A container for carting away spoil from the heading. Often mounted on a rail bogie.

Spoil The rock debris or loose material produced by excavation at the tunnel face.

Steel arch A steel joist formed into a horseshoe shape; used as support in a rock tunnel.

Tailskin Projection at rear of tunnelling shield to provide protection for miners erecting prefabricated lining.

Tremie pipe A pipe for placing fresh concrete under water so as to prevent segregation and dilution of the concrete mix.

Tunnel An horizontal underground passage, usually open at both ends, for the transport of people, materials or services.

GEOLOGICAL TERMS

Bath Stone Pale yellow limestone quarried near Bath and much used there as a building stone.

Bedding plane A surface in a sedimentary rock that is parallel to the surface of deposition.

Blue Lias A deposit which consists of alternating thin beds of limestone and clay; when fired produces a natural cement.

Boulders Rock particles over 200 mm in size.

Bunter Sandstone Thick stratum of red sandstone of moderate strength occurring extensively in the midlands of England.

Calcite Crystalline calcium carbonate, $CaCO_3$.

Chalk A fine-grained soft white limestone.

Clay Particles less than 0.002 mm in size.

Claystone A rock material consisting of clay with some cementing agent such as calcium carbonate or iron oxide.

Coal A rock consisting almost entirely of carbonaceous material derived from the remains of vegetation.

Coal Measures A division of strata containing the most important coal seams of Great Britain and Europe; for the most part the rocks are sandstones, siltstones and mudstones.

Cobbles Rock particles 60–200 mm in size.

Conglomerate A sedimentary rock consisting predominantly of rounded or subrounded fragments of pre-existing rocks.

Dolomite A limestone consisting of, or containing, the mineral dolomite, $CaMg\,(CO_3)_2$.

Dolerite A medium-grained basic igneous rock.

Feldspar The most important single group of rock-forming minerals; essentially silicates of aluminium.

Flint A hard brittle rock material consisting of silica; occurs as nodules in the Chalk.

Gneiss A coarsely crystalline banded metamorphic rock.

Granite A coarsely crystalline acid igneous rock consisting essentially of quartz, feldspar and mica.

Granitoid Having a texture in which the minerals do not possess the outlines characteristic of their crystals.

Gravel Rock particles 2–60 mm in size.

Gritstone Coarse sandstone.

Joint A naturally occurring break or crack in the rock mass.

Keuper Marl Thick stratum of soft red mudstone occurring extensively in the midlands of England.

Limestone A sedimentary rock consisting predominantly of calcium carbonate $CaCO_3$.

London Clay Thick stratum of firm grey clay occurring in south-east England, particularly in the London basin; a perfect medium for shield tunnelling.

Lower Chalk Thick stratum of slightly clayey chalk occurring extensively in south-east England and extending under the Channel to France; a perfect medium for machine tunnelling.

Mica A rock-forming mineral which has the property of splitting down into thin elastic plates.

Micaceous Containing the mineral mica.

Mohs' scale of hardness Rank order scale of hardness based on the ability of a harder mineral to scratch a softer one; runs from talc = 1 to diamond = 10.

Montmorillonite One of the clay minerals; a hydrated aluminium silicate and the raw material of bentonite.

Mudstone A sedimentary rock consisting predominantly of clay-size particles.

Oxford Clay Thick stratum of firm grey clay occurring in southern England.

Pea gravel Single-sized gravel consisting of particles about the size of peas.

Quartz A crystalline form of silica, SiO_2.

Quartzite A metamorphic or sedimentary rock consisting predominantly of quartz.

Sand Particles 0.06–2 mm in size.

Sandstone A sedimentary rock consisting predominantly of sand-sized particles.

Schist A banded metamorphic rock that can be split into flakes.

Septarian nodules Boulder-sized concretions made of claystone and often containing calcite-filled cracks. Found in the London Clay.

Shale A mudstone or siltstone having well-marked bedding plane partings.

Siliceous Containing silica, SiO_2.

Silt Particles 0.002–0.06 mm in size.

Siltstone A sedimentary rock consisting predominantly of silt-sized particles.

Slate A metamorphic rock formed by the action of pressure on clay or mudstone.

Stratum (pl. *strata*) Originally horizontally deposited layer or bed of geological material.

Syenite A coarse grained intermediate igneous rock characterised by the presence of alkali feldspars.

Water table The level of water in the ground.

Acknowledgment of sources of figures

The following sources of figures are gratefully acknowledged:

Atlas Copco (Great Britain) Ltd: 4.3, 4.4 and 5.4
de Vries, L. (1975) *Victorian inventions.* London (John Murray): 6.7 and 11.3
Dosco Overseas Engineering Ltd: 11.10 and 13.4
The Engineer: 12.1
Fontana New Naturalist (Collins): 6.4
Géotechnique: 7.1
Ground Engineering: 2.8, 2.9 and 2.10
Hochtief AG: 13.3
The Illustrated London News: 6.2 and 6.6 (lower)
Institution of Civil Engineers: 8.1, 8.2, 8.3, 8.5 and 10.2
Institution of Mechanical Engineers: 11.2
London Underground Ltd: 8.7
Marcon International Ltd and Mowlem (Civil Engineering) Ltd: 13.2
Markham and Co Ltd: 11.4 (redrawn), 11.6 and 12.2
Mini Tunnels International: 12.4 and 12.5
Ingenieursburo ir V.L. Molenaar bv, 4837 BM Breda-Holland: 10.3 and 10.4
Montabert SA: 5.3
Courtesy of *New Civil Engineer*: Frontispiece
Nobel's Explosives Co Ltd: 3.2
Edmund Nuttall Ltd: 9.2 and 11.7
Pipe Jacking Association: 13.1
The Robbins Co (UK) Ltd: 11.8 and 11.9
Trustees of the Science Museum (London): 8.4
Smithsonian Institution: 2.4, 2.6 and 7.3
Spie Batignolles: 11.5
Tekken Construction Co Ltd: 9.5
Transportation Research Board, National Research Council, Washington DC: 2.1
Tunnels and Tunnelling: 9.3 and 9.4
Zokor International (UK) Ltd: 13.5

Index

Printed in the United States
By Bookmasters